A·N·N·U·A·L E·D·

GEOGRAPHY
Fourteenth Edition

99/00

EDITOR

Gerald R. Pitzl
Macalester College

Gerald R. Pitzl, professor of geography at Macalester College, received his bachelor's degree in secondary social science education from the University of Minnesota in 1964 and his M.A. (1971) and Ph.D. (1974) in geography from the same institution. He teaches a wide array of geography courses and is the author of a number of articles on geography, the developing world, and the use of the Harvard case method.

Dushkin/McGraw-Hill
Sluice Dock, Guilford, Connecticut 06437

Visit us on the Internet
http://www.dushkin.com/annualeditions/

World Map

This map has been developed to give you a graphic picture of where the countries of the world are located, the relationship they have with their region and neighbors, and their positions relative to the superpowers and power blocs. We have focused on certain areas to more clearly illustrate these crowded regions.

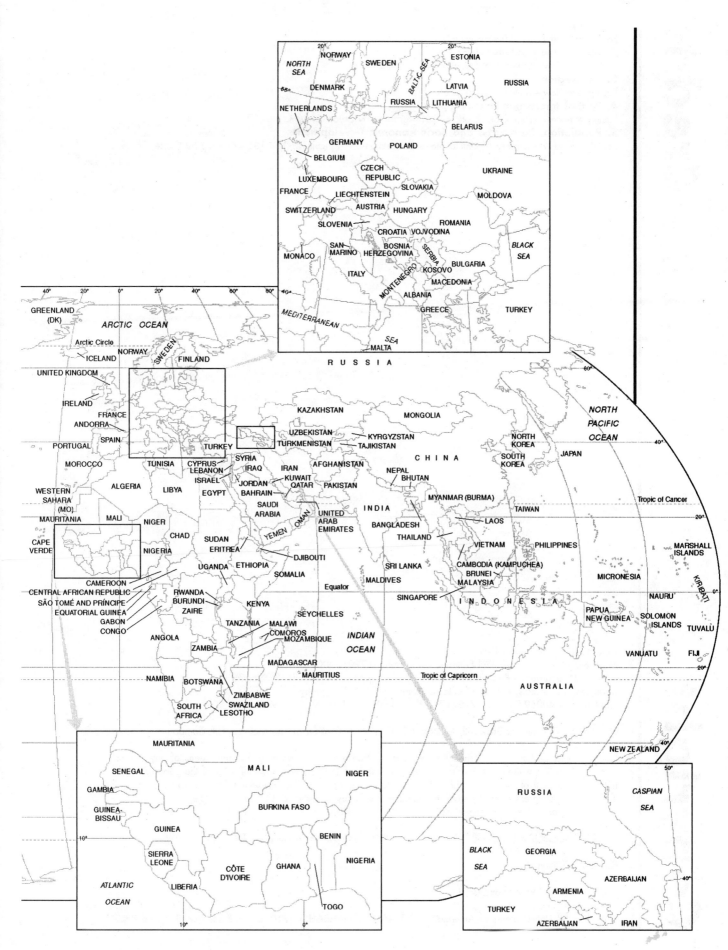

Credits

1. Geography in a Changing World
Facing overview—© 1998 by PhotoDisc, Inc.
2. Land-Human Relationships
Facing overview—United Nations photo.
3. The Region
Facing overview—United Nations photo.
4. Spatial Interaction and Mapping
Facing overview—Digital Stock photo. 158—graphic by Bryan Christie.
5. Population, Resources, and Socioeconomic Development
Facing overview—courtesy of United States Department of Agriculture. 184-185—graphic by Laurie Grace.

Copyright

Cataloging in Publication Data
Main entry under title: Annual Editions: Geography. 1999/2000.
    1. Geography—Periodicals. 2. Anthropo-geography—Periodicals. 3. Natural resources—Periodicals. I. Pitzl, Gerald R., comp. II. Title: Geography.
ISBN 0-07-040069-5   87-641715   ISSN 1091-9937   910'.5

© 1999 by Dushkin/McGraw-Hill, Guilford, CT 06437, A Division of The McGraw-Hill Companies.

Copyright law prohibits the reproduction, storage, or transmission in any form by any means of any portion of this publication without the express written permission of Dushkin/McGraw-Hill, and of the copyright holder (if different) of the part of the publication to be reproduced. The Guidelines for Classroom Copying endorsed by Congress explicitly state that unauthorized copying may not be used to create, to replace, or to substitute for anthologies, compilations, or collective works.

Annual Editions® is a Registered Trademark of Dushkin/McGraw-Hill, A Division of The McGraw-Hill Companies.

Fourteenth Edition

Cover image © 1999 PhotoDisc, Inc.

Printed in the United States of America   1234567890BAHBAH54321098   Printed on Recycled Paper

Members of the Advisory Board are instrumental in the final selection of articles for each edition of ANNUAL EDITIONS. Their review of articles for content, level, currency, and appropriateness provides critical direction to the editor and staff. We think that you will find their careful consideration well reflected in this volume.

EDITOR

Gerald R. Pitzl
Macalester College

ADVISORY BOARD

Paul S. Anderson
Illinois State University

Robert S. Bednarz
Texas A&M University

Sarah Witham Bednarz
Texas A&M University

Roger Crawford
San Francisco State University

James Fryman
University of Northern Iowa

Allison C. Gilmore
Mercer University

J. Michael Hammett
Azusa Pacific University

Vern Harnapp
University of Akron

Miriam Helen Hill
Auburn University–
Main

Lance F. Howard
Clemson University

Artimus Keiffer
Indiana University–
Purdue University

David J. Larson
California State University
Hayward

Mark Lowry
Western Kentucky University

Tom L. Martinson
Auburn University–
Main

John T. Metzger
Michigan State University

Peter O. Muller
University of Miami

Robert E. Nunley
University of Kansas

Ray Oldakowski
Jacksonville University

James D. Proctor
University of California
Santa Barbara

Donald M. Spano
University of Southern Colorado

Eileen Starr
Valley City State University

Wayne G. Strickland
Roanoke College

John A. Vargas
Pennsylvania State University
Du Bois

Daniel Weiner
West Virginia University

Randy W. Widdis
University of Regina

EDITORIAL STAFF

Ian A. Nielsen, Publisher
Roberta Monaco, Senior Developmental Editor
Dorothy Fink, Associate Developmental Editor
Addie Raucci, Senior Administrative Editor
Cheryl Greenleaf, Permissions Editor
Joseph Offredi, Permissions/Editorial Assistant
Diane Barker, Proofreader
Lisa Holmes-Doebrick, Program Coordinator

PRODUCTION STAFF

Brenda S. Filley, Production Manager
Charles Vitelli, Designer
Lara M. Johnson, Design/
Advertising Coordinator
Laura Levine, Graphics
Mike Campbell, Graphics
Tom Goddard, Graphics
Juliana Arbo, Typesetting Supervisor
Jane Jaegersen, Typesetter
Marie Lazauskas, Word Processor
Kathleen D'Amico, Word Processor
Larry Killian, Copier Coordinator

To the Reader

In publishing ANNUAL EDITIONS we recognize the enormous role played by the magazines, newspapers, and journals of the public press in providing current, first-rate educational information in a broad spectrum of interest areas. Many of these articles are appropriate for students, researchers, and professionals seeking accurate, current material to help bridge the gap between principles and theories and the real world. These articles, however, become more useful for study when those of lasting value are carefully collected, organized, indexed, and reproduced in a low-cost format, which provides easy and permanent access when the material is needed. That is the role played by ANNUAL EDITIONS.

New to ANNUAL EDITIONS is the inclusion of related World Wide Web sites. These sites have been selected by our editorial staff to represent some of the best resources found on the World Wide Web today. Through our carefully developed topic guide, we have linked these Web resources to the articles covered in this ANNUAL EDITIONS reader. We think that you will find this volume useful, and we hope that you will take a moment to visit us on the Web at **http://www.dushkin.com/** to tell us what you think.

The articles in this fourteenth edition of *Annual Editions: Geography* represent the wide range of topics associated with the discipline of geography. The major themes of spatial relationships, regional development, the population explosion, and socioeconomic inequalities exemplify the diversity of research areas within geography.

The book is organized into five units, each containing articles relating to geographical themes. Selections address the conceptual nature of geography and the global and regional problems in the world today. This latter theme reflects the geographer's concern with finding solutions to these serious issues. Regional problems, such as food shortages in the Sahel and the greenhouse effect, concern not only geographers but also researchers from other disciplines.

The association of geography with other fields is important, because expertise from related research will be necessary in finding solutions to some difficult problems. Input from the focus of geography is vital in our common search for solutions. This discipline has always been integrative. That is, geography uses evidence from many sources to answer the basic questions, "Where is it?" "Why is it there?" and "What is its relevance?" The first group of articles emphasizes the interconnectedness not only of places and regions in the world but of efforts toward solutions to problems as well. No single discipline can have all of the answers to the problems facing us today; the complexity of the issues is simply too great.

The writings in unit 1 discuss particular aspects of geography as a discipline and provide examples of the topics presented in the remaining four sections. Units 2, 3, and 4 represent major themes in geography. Unit 5 addresses important problems faced by geographers and others.

Annual Editions: Geography 99/00 will be useful to both teachers and students in their study of geography. The anthology is designed to provide detail and case study material to supplement the standard textbook treatment of geography. The goals of this anthology are to introduce students to the richness and diversity of topics relating to places and regions on Earth's surface, to pay heed to the serious problems facing humankind, and to stimulate the search for more information on topics of interest.

I would like to express my gratitude to Barbara Wells-Howe for her continued help in preparing this material for publication. Her typing, organization of materials, and many helpful suggestions are greatly appreciated. Without her diligence and professional efforts, this undertaking could not have been completed. Special thanks are also extended to Ian Nielsen for his continued encouragement during the preparation of this new edition and to Addie Raucci for her enthusiasm and helpfulness. A word of thanks must go as well to all those who recommended articles for inclusion in this volume and who commented on its overall organization. Robert S. Bednarz, Vern Harhapp, James Hathaway, Artimus Keiffer, John Metzger, Peter O. Muller, Wayne G. Strickland, and Randy W. Widdis were especially helpful in that regard.

In order to improve the next edition of *Annual Editions: Geography*, we need your help. Please share your opinions by filling out and returning to us the postage-paid *Article Rating Form* on the last page of this book. We will give serious consideration to all your comments.

Gerald R. Pitzl
Editor

Contents

World Map	ii
To the Reader	vi
Topic Guide	4
⊚ Selected World Wide Web Sites	6
Overview	8

1. **The Four Traditions of Geography,** William D. Pattison, *Journal of Geography,* September/October 1990. — 10
   This key article, originally published in 1964, was reprinted, with the author's later comments, in the 75-year retrospective of the *Journal of Geography.* It is a classic in the **history of geography.** William Pattison discusses the four main themes in geography that have been the focus of work in the discipline for centuries—the spatial concept, area studies, land-human relationships, and earth science.

2. **The American Geographies,** Barry Lopez, *Orion,* Autumn 1989. — 14
   The American **landscape** is nearly incomprehensible in depth and complexity, according to Barry Lopez. To truly understand American **geography,** one must seek out local experts who have an intimate knowledge of **place,** people who have a feel for their locale that no outsider could possibly develop.

3. **NASA Readies a 'Mission to Planet Earth,'** William K. Stevens, *New York Times,* February 17, 1998. — 19
   NASA's novel new project, the Earth Science Enterprise, will orbit an information-gathering satellite to study **climate change** and variability, atmospheric chemistry, **ozone** depletion, **natural hazards,** and changes in **land use.**

4. **Human Domination of Earth's Ecosystems,** Peter M. Vitousek, Harold A. Mooney, Jane Lubchenco, and Jerry M. Melillo, *Science,* July 25, 1997. — 22
   "Human alteration of the Earth," the authors contend, "is substantial and growing." Earth's **ecosystems** are being affected by human intervention through **land transformation** and biotic and chemical changes. Recommendations are made to reduce the rate of change, to better understand Earth's ecosystems, and to assume more responsibility for managing the planet.

5. **Counting the Cost: The Growing Role of Economics in Environmental Decisionmaking,** Paul R. Portney, *Environment,* March 1998. — 28
   Questions are now emerging about expenditures on **environmental** quality and whether these costs are justified. More and more, **economics** is becoming an important factor in "the mother of all environmental problems," **global climatic change.**

UNIT 1

Geography in a Changing World

Eight articles discuss the discipline of geography and the extremely varied and wide-ranging themes that define geography today.

The concepts in bold italics are developed in the article. For further expansion please refer to the Topic Guide and the Index.

6. **The Coming Climate,** Thomas R. Karl, Neville Nicholls, and Jonathan Gregory, *Scientific American,* May 1997. 33
Global warming is a widely accepted phenomenon. The consequent changes in ***climate*** are directly connected to ***anthropogenic*** (human-induced) factors. The major culprits are ***greenhouse gases,*** aerosols, and a number of other substances.

7. **The Emergence of Gasohol: A Renewable Fuel for American Roadways,** William D. Warren and Dennis Roberts, *Focus,* Spring 1998. 39
The article highlights, through ***choropleth maps,*** state use rates for gasohol, a renewable fuel for automobiles and trucks. The Midwest, with its abundance of corn production, is the clear ***regional*** leader in gasohol use.

8. **America's Rush to Suburbia,** Kenneth T. Jackson, *New York Times,* June 9, 1996. 46
While ***urban*** places globally are achieving higher ***population*** densities, many cities in the United States are losing population. Moves to ***suburban*** settings are exacerbating urban sprawl. By 1990, there were more people living in suburbs than in rural areas and center cities combined.

UNIT 2

Land-Human Relationships

Six articles examine the relationship between humans and the land on which we live. Topics include the destruction of the rain forests, desertification, pollution, and the effects of human society on the global environment.

Overview 48

9. **The Season of El Niño,** *The Economist,* May 9, 1998. 50
This article recounts the momentous damage in Latin America from the ***El Niño*** of 1997–98. Floods, ***drought,*** and extensive fires in the Amazon destroyed nearly 3,300 square kilometers of ***forest.*** In Peru and Argentina alone, 500,000 people were driven from their homes.

10. **The Great Climate Flip-Flop,** William H. Calvin, *The Atlantic Monthly,* January 1998. 53
A group of scientists are suggesting that ***global warming*** may well lead to a period of abrupt global cooling. Such a cooling could take place if the North Atlantic Current changes its course, something that has occurred in the past. This article goes into detail about the possible consequences of cooling and ways to prevent this shift.

11. **Temperature Rising,** Taras Grescoe, *Canadian Geographic,* November/December 1997. 62
Canada's Mackenzie Basin is one of three major ***regions*** with significantly rising average temperatures. These changes are a clear indication of ***global warming.*** Rising temperatures have melted permafrost and increased the incidence of forest fires.

The concepts in bold italics are developed in the article. For further expansion please refer to the Topic Guide and the Index.

12. **"Dammed If You Do . . . ,"** *World Press Review,* 67
    August 1997.
    The article summarizes the controversies raging over megadamming projects to enhance **economic development** and **energy** production. Economic gains are offset by **population** relocation and **environmental** degradation.

13. **Past and Present Land Use and Land Cover in the USA,** William B. Meyer, *Consequences,* Spring 1995. 73
    Cities, towns, and villages now cover slightly more than 4 out of every 100 acres of the land in the continental United States, and every day the percentage grows. This article looks at both the positive and negative **consequences of changes in land use.**

14. **China Shoulders the Cost of Environmental Change,** Vaclav Smil, *Environment,* July/August 1997. 82
    Economic development in China is occurring at staggeringly high rates. **Environmental** protection is taking a back seat to dynamic growth and modernization. **Ecosystem** degradation is taking a big bite out of China's growing gross domestic product.

Overview 88

15. **The Importance of Places, or, a Sense of Where You Are,** Paul F. Starrs, *Spectrum: The Journal of State Government,* Summer 1994. 90
    Paul Starrs writes eloquently of the importance of **place** as a geographical concept and the value of **regions** in our common thought. "Places matter," he states. "And that, most of all, is why regions are relevant."

16. **The Rise of the Region State,** Kenichi Ohmae, *Foreign Affairs,* Spring 1993. 97
    The **global** market has dictated a new, more significant set of **region states** that reflect the real flow of economic activity. The old idea of the **nation-state** is considered an anachronism, an unnatural and dysfunctional unit for organizing either human activity or emerging economic systems.

17. **Metropolis Unbound: The Sprawling American City and the Search for Alternatives,** Robert Geddes, *The American Prospect,* November/December 1997. 101
    Cities in the United States, it is argued, are becoming city-regions as **urban sprawl** increases. New York City, like many others, is at risk of expanding too much, leading to more traffic congestion, inefficient public **transportation,** water and air pollution, and social segregation.

UNIT 3

The Region

Nine selections review the importance of the region as a concept in geography and as an organizing framework for research. A number of world regional trends, as well as the patterns of area relationships, are examined.

18. **Greenville: From Back Country to Forefront,** 108
Eugene A. Kennedy, *Focus,* Spring 1998.
Eugene Kennedy traces the rise of Greenville, South Carolina, from a back country **region** in its early days, to "Textile Center of the World" in the 1920s, to a supremely successful, diversified **economic** stronghold in the 1990s. Greenville's location within a thriving area midway between Charlotte and Atlanta gives it tremendous growth potential.

19. **Does It Matter Where You Are?** *The Economist,* July 30, 1994. 114
As the computer affects society, it appears that it is possible to run a business from any location. The truth, however, as this article points out, is that people and businesses are most effective when they have a center or base of operations; where one is located does, indeed, matter greatly.

20. **Low Water in the American High Plains,** David E. Kromm, *The World & I,* February 1992. 115
The serious depletion of **groundwater** in the Ogallala aquifer highlights a key example of the question of **sustainable development** in midwestern North America. There has been widespread concern that this important source of water is drying up. If excessive use of this water supply is not curtailed, **drought** conditions could result in this key region of **agriculture.**

21. **Water Resource Conflicts in the Middle East,** Christine Drake, *Journal of Geography,* January/February 1997. 120
Wars in the future may well be fought over **accessibility** to fresh water. The Middle East, an **arid land,** is blessed with an abundance of petroleum but very little fresh water. The seeds of potential conflict are ever-present.

22. **Boomtown Baku,** Richard C. Longworth, *The Bulletin of the Atomic Scientists,* May/June 1998. 130
There are 100–200 billion barrels of petroleum beneath the Caspian Sea. With the demise of the USSR, a number of countries are vying for rights to extract and ship this valuable form of **energy.** The scramble for position has unleashed a major series of **geopolitical** skirmishes.

23. **Demographic Clouds on China's Horizon,** Nicholas Eberstadt, *The American Enterprise,* July/August 1998. 135
China's current **population** policy limiting the number of children per family will produce an increase in average age in the next century. With this change will come a decrease in **economic** growth due to declining manpower. Finally, there will be a severe imbalance in male-female ratios resulting from the current preference for boys in one-child families.

The concepts in bold italics are developed in the article. For further expansion please refer to the Topic Guide and the Index.

Overview   **138**

24. **Transportation and Urban Growth: The Shaping of the American Metropolis,** Peter O. Muller, *Focus,* Summer 1986.   **140**

    Peter Muller reviews the importance of **transportation** in the growth of American **urban places.** The highly compact urban form of the middle nineteenth century was transformed in successive eras by electric streetcars, highways, and expressways. The city of the future may rely more on **communication** than on the actual movement of people.

25. **Bridge to the Past,** Scott Elias, *Earth,* April 1997.   **148**

    Accurate dating of the Bering Sea Land Bridge, the route of **accessibility** for the first New World inhabitants, has been established at 12,000 years B.P. Possible scenarios for that time include animal **migrations** by grazers from the Siberian steppes to Alaska and the Yukon.

26. **County Buying Power, 1987–97,** Brad Edmondson, *American Demographics,* August 1998.   **154**

    The map and narrative in this article indicate the winners and losers in county buying power in the United States. A **choropleth map** of the country points out that the hot spots are **suburbs** and small cities in the Sunbelt **region.**

27. **Puerto Rico, U.S.A,** Brad Edmondson, *American Demographics,* July 1998.   **156**

    Puerto Rico is a $12.2 billion **economic** market for U.S. goods, and it may become the 51st state. The **choropleth map** illustrates changes in population between 1990 and 1996.

28. **Do We Still Need Skyscrapers?** William J. Mitchell, *Scientific American,* December 1997.   **158**

    The need for skyscrapers, the signature structure in America's great **urban** places, is called into question. With rapid **communication** through e-mail, the Internet, phone, and fax, is it necessary to bring people face-to-face? The article answers with a resounding "yes" and speculates that even taller buildings are coming.

UNIT 4

Spatial Interaction and Mapping

Seven articles discuss the key theme in geographical analysis: place-to-place spatial interaction. Human diffusion, transportation systems, urban growth, and cartography are some of the themes examined.

The concepts in bold italics are developed in the article. For further expansion please refer to the Topic Guide and the Index.

29. **Indian Gaming in the U.S.: Distribution, Significance and Trends,** Dick G. Winchell, John F. Lounsbury, and Lawrence W. Sommers, *Focus,* Winter 1997. 161
Indian reservations have become centers for gambling in all U.S. *regions.* The article presents historical backgrounds to the rapid rise of gaming as an *economic* activity and uses several *choropleth maps* to illustrate the widespread distribution of this activity.

30. **For Poorest Indians, Casinos Aren't Enough,** Peter T. Kilborn, *New York Times,* June 11, 1997. 172
Casinos may be bringing large amounts of money to Indian reservations, but the newfound wealth does not extend to every *location.* The Pine Ridge Reservation in South Dakota suffers because it is remote from *population* centers and with limited *accessibility* to them. *Poverty* still haunts this place.

UNIT 5

Population, Resources, and Socioeconomic Development

Nine articles examine the effects of population growth on natural resources and the resulting socioeconomic level of development.

Overview 174

31. **Before the Next Doubling,** Jennifer D. Mitchell, *World Watch,* January/February 1998. 176
The article warns of the possibility of another doubling of world *population* in the next century. Food shortages, inadequate supplies of fresh water, and greatly strained *economic* systems will result. Family planning efforts need to be increased and a change in the desire for large families should be addressed.

32. **The End of Cheap Oil,** Colin J. Campbell and Jean H. Laherrère, *Scientific American,* March 1998. 183
Petroleum, a prominent source of *energy,* will become a scare *resource* in coming years. Some experts, however, suggest that undiscovered reserves will eventually be found, that advanced technologies will make extraction of oil more efficient, and that oil shales and "heavy oil" can someday be exploited. A decline in oil production would have serious *economic* and *geopolitical* repercussions.

33. **Reseeding the Green Revolution,** Charles Mann, *Science,* August 22, 1997. 188
If sufficient food is to be available for the rapidly growing world *population,* a new Green Revolution is needed. *Agricultural* systems today are hard-pressed to produce sufficient amounts of food. The problem will worsen in coming years without significant breakthroughs in technology.

34. **How Much Food Will We Need in the 21st Century?** William H. Bender, *Environment,* March 1997. 194
Expanding human *population* will require greater food supplies. Efforts toward *sustainable development* in the *agricultural sector* are essential to ensure adequate food supplies in the next century.

The concepts in bold italics are developed in the article. For further expansion please refer to the Topic Guide and the Index.

35. **The Changing Geography of U.S. Hispanics, 1850–1990,** Terrence Haverluk, *Journal of Geography,* May/June 1997. 198
    This article chronicles the rise in population of the **Hispanic** culture group in the United States, from 1 million in 1930 to over 32 million in 1997. Hispanics represent the fastest-growing ethnic group in the United States.

36. **'Hispanics' Don't Exist,** Linda Robinson, *U.S. News & World Report,* May 11, 1998. 211
    The growth of the **Hispanic population** in the United States represents a dramatic demographic shift: The number of Hispanics is increasing at four times the rate of the rest of the **population.** The article aims at clearing up the ambiguities in the term "Hispanic."

37. **Russia's Population Sink,** Toni Nelson, *World Watch,* January/February 1996. 215
    The **demographic transition** in Russia has taken another deadly turn. Since 1986, death rates in Russia have slowly risen, and, beginning in 1991, birthrates have slipped below death rates. The result for Russia is a decline in population from 1991 to the present. Breakdowns in the social fabric are to blame, along with serious **environmental** problems.

38. **Vanishing Languages,** David Crystal, *Civilization,* February/March 1997. 217
    The language **map** of the world is changing steadily. When **populations** are decimated by new diseases, their languages die as well. In other instances, the draw of **economic** gains in another region will lower the numbers of people using a language.

39. **Risky Business: Who Will Pay for the Growing Costs of Global Change?** *Options,* Spring 1998. 222
    The combination of **natural disasters** and human-induced **global** change in the 1990s is unprecedented in its **economic** impact. The costs associated with natural and human-induced catastrophes are expected to increase dramatically.

Index   225
Article Review Form   228
Article Rating Form   229

Topic Guide

This topic guide suggests how the selections and World Wide Web sites found in the next section of this book relate to topics of traditional concern to geography students and professionals. It is useful for locating interrelated articles and Web sites for reading and research. The guide is arranged alphabetically according to topic.

The relevant Web sites, which are numbered and annotated on pages 4 and 5, are easily identified by the Web icon (◉) under the topic articles. By linking the articles and the Web sites by topic, this ANNUAL EDITIONS reader becomes a powerful learning and research tool.

TOPIC AREA	TREATED IN	TOPIC AREA	TREATED IN
Accessibility	21. Water Resource Conflicts 25. Bridge to the Past 30. For Poorest Indians ◉ **14, 20, 31**		23. Demographic Clouds on China's Horizon 27. Puerto Rico, U.S.A. 29. Indian Gaming in the U.S. 31. Before the Next Doubling 32. End of Cheap Oil 38. Vanishing Languages 39. Risky Business ◉ **1, 5, 8, 11, 12, 17, 20, 30**
Agriculture	20. Low Water in the American High Plains 33. Reseeding the Green Revolution 34. How Much Food? ◉ **16, 21, 27, 30**		
		Ecosystem	4. Human Domination 14. China Shoulders the Cost ◉ **2, 3, 5, 6, 7, 8, 9**
Anthropogenic	6. Coming Climate ◉ **7, 8**	El Niño	9. Season of El Niño ◉ **2, 3, 5, 7, 9**
Arid Lands	21. Water Resource Conflicts ◉ **11, 12, 13**	Energy	7. Emergence of Gasohol 12. "Dammed If You Do . . ." 22. Boomtown Baku 32. End of Cheap Oil ◉ **1, 2, 3, 5, 8**
Cartography	See Maps		
Climate Change	3. NASA Readies a 'Mission to Planet Earth' 5. Counting the Cost 6. Coming Climate 10. Great Climate Flip-Flop ◉ **7, 8**	Environment	5. Counting the Cost 12. "Dammed If You Do . . ." 14. China Shoulders the Cost 37. Russia's Population Sink ◉ **2, 3, 5, 8, 9, 12, 16, 17, 18, 31**
Commun-ication	24. Transportation and Urban Growth 28. Do We Still Need Skyscrapers? ◉ **19, 20**	Forests	9. Season of El Niño ◉ **3, 5, 7, 31**
		Geography	2. American Geographies 19. Does It Matter Where You Are? ◉ **1, 2, 20**
Demographic Transition	23. Demographic Clouds on China's Horizon 37. Russia's Population Sink ◉ **27, 30, 31**	Geopolitical	22. Boomtown Baku 32. End of Cheap Oil ◉ **16, 20, 21**
Development	12. "Dammed If You Do . . ." 14. China Shoulders the Cost ◉ **11, 12, 13, 27**		
Drought	9. Season of El Niño 20. Low Water in the American High Plains ◉ **2, 5, 13, 20**	Global Issues	5. Counting the Cost 6. Coming Climate 10. Great Climate Flip-Flop 11. Temperature Rising 16. Rise of the Region State 19. Does It Matter Where You Are? 39. Risky Business ◉ **16, 20, 30**
Economic	5. Counting the Cost 12. "Dammed If You Do . . ." 14. China Shoulders the Cost 18. Greenville 19. Does It Matter Where You Are?		

TOPIC AREA	TREATED IN	TOPIC AREA	TREATED IN
Greenhouse	6. Coming Climate ◦ 5, 8	**Population**	8. America's Rush to Suburbia 12. "Dammed If You Do..." 23. Demographic Clouds on China's Horizon 30. For Poorest Indians 31. Before the Next Doubling 33. Reseeding the Green Revolution 34. How Much Food? 36. 'Hispanics' Don't Exist 37. Russia's Population Sink 38. Vanishing Languages ◦ 22, 23, 24, 25, 26, 27, 31
Groundwater	20. Low Water in the American High Plains ◦ 14, 16, 21
Hispanics	35. Changing Geography of U.S. Hispanics 36. 'Hispanics' Don't Exist ◦ 2, 3, 6, 11, 14, 25, 28
History of Geography	1. Four Traditions of Geography ◦ 2, 3
Poverty	30. For Poorest Indians ◦ 25, 27, 28
Indian Reservations	29. Indian Gaming in the U.S. 30. For Poorest Indians ◦ 3, 4, 9, 25	**Region**	7. Emergence of Gasohol 11. Temperature Rising 18. Greenville 19. Does It Matter Where You Are? 26. County Buying Power 29. Indian Gaming in the U.S. 38. Vanishing Languages ◦ 9, 14, 20, 25, 29, 31
Land Transformation	4. Human Domination ◦ 9
Landscape	2. American Geographies 3. NASA Readies a 'Mission to Planet Earth' 13. Past and Present Land Use ◦ 2, 3, 7
Region State	16. Rise of the Region State ◦ 16, 20
Location	19. Does It Matter Where You Are? 30. For Poorest Indians ◦ 2, 9, 20	**Resources**	32. End of Cheap Oil ◦ 2, 27, 30
Suburban	8. America's Rush to Suburbia 26. County Buying Power ◦ 14, 25
Maps	7. Emergence of Gasohol 26. County Buying Power 27. Puerto Rico, U.S.A. 29. Indian Gaming in the U.S. 38. Vanishing Languages ◦ 22, 23, 24, 25, 26	**Sustainable Development**	20. Low Water in the American High Plains 34. How Much Food? ◦ 16, 20, 21
Transportation Systems	17. Metropolis Unbound 24. Transportation and Urban Growth ◦ 19, 20
Migration	25. Bridge to the Past ◦ 22, 23, 24, 25, 26
Nation-State	16. Rise of the Region State ◦ 20	**Urban**	8. America's Rush to Suburbia 17. Metropolis Unbound 24. Transportation and Urban Growth 28. Do We Still Need Skyscrapers? ◦ 17, 19, 20, 21
Natural Hazards	3. NASA Readies a 'Mission to Planet Earth' 39. Risky Business ◦ 7, 27, 28
Ozone Depletion	3. NASA Readies a 'Mission to Planet Earth' ◦ 7	**Weather**	9. Season of El Niño ◦ 2, 3, 5, 7, 9, 11, 12
Place	2. American Geographies 15. Importance of Places ◦ 2, 3, 9

Annual Editions: Geography

The following World Wide Web sites have been carefully researched and selected to support the articles found in this reader. If you are interested in learning more about specific topics found in this book, these Web sites are a good place to start. The sites are cross-referenced by number and appear in the topic guide on the previous two pages. Also, you can link to these Web sites through our DUSHKIN ONLINE support site at http://www.dushkin.com/online/.

The following sites were available at the time of publication. Visit our Web site—we update DUSHKIN ONLINE regularly to reflect any changes.

General Sources

1. The Association of American Geographers
http://www.aag.org/
Surf this site of the Association of American Geographers to learn about AAG projects and publications, careers in geography, and information about related organizations.

2. National Geographic Society
http://www.nationalgeographic.com/
This site provides links to National Geographic's huge archive of maps, articles, and other documents. Search the site for information about worldwide expeditions of interest to geographers.

3. The New York Times
http://www.nytimes.com/
Browsing through the archives of *The New York Times* will provide you with a wide array of articles and information related to the different subfields of geography.

4. Social Science Internet Resources
http://www.wcsu.ctstateu.edu/library/ss_geography.html
This site is a definitive source for geography-related links to universities, browsers, cartography, associations, and discussion groups.

5. U.S. Geological Survey
http://www.usgs.gov/
This site and its many links are replete with information and resources for geographers, from explanations of El Niño, to mapping, to geography education, to water resources. No geographer's resource list would be complete without frequent mention of the USGS.

Geography in a Changing World

6. Geological Survey of Sweden: Other Geological Surveys
http://www3.sgu.se/links/Surveys_e.shtml
This site provides links to the national geographical surveys of many countries in Europe and elsewhere, including Brazil, South Africa, and the United States. It makes for very interesting and informative browsing.

7. Mission to Planet Earth
http://www.hq.nasa.gov/office/mtpe/
This site will direct you to information about NASA's Mission to Planet Earth program and its Science of the Earth System. Surf here to learn about satellites, El Niño, and even "strategic visions" of interest to geographers.

8. Public Utilities Commission of Ohio
http://www.puc.ohio.gov/consumer/gcc/index.html
PUCO aims for this site to serve as a clearinghouse of information related to global climate change. Its extensive links provide for explanation of the science and chronology of global climate change, acronyms, definitions, and more.

9. Santa Fe Institute
http://acoma.santafe.edu/sfi/research/
This home page of the Santa Fe Institute—a nonprofit, multidisciplinary research and education center—will lead you to a plethora of valuable links related to its primary goal: to create a new kind of scientific research community pursuing emerging science. Such links as Evolution of Language, Ecology, and Local Rules for Global Problems are offered.

Land-Human Relationships

10. Measurement of Air Pollution from Satellites (MAPS)
http://stormy.larc.nasa.gov/overview.html
This page from the National Aeronautics and Space Administration focuses on measurements of carbon monoxide in the atmosphere and provides links to many other sites of interest to geographers. Also click on http://ccf.arc.nasa.gov/dx/basket/storiesetc/satfile.html if you're interested in more Space Links from NASA, including information on satellites.

11. The North-South Institute
http://www.nsi-ins.ca/info.html
Searching this site of the North-South Institute—which works to strengthen international development cooperation and enhance gender and social equity—will help you find information on a variety of development issues.

12. United Nations Environment Programme
http://www.unep.ch/
Consult this home page of UNEP for links to critical topics of concern to geographers, including desertification and the impact of trade on the environment. The site will direct you to useful databases and global resource information.

13. World Health Organization
http://www.who.ch/Welcome.html
This home page of the World Health Organization will provide you with links to a wealth of statistical and analytical information about health in the developing world.

The Region

14. AS at UVA Yellow Pages: Regional Studies
http://xroads.virginia.edu/~YP/regional.html
Those interested in American regional studies will find this site a gold mine. Links to periodicals and other informational resources about the Midwest/Central, Northeast, South, and West regions are provided here.

15. Can Cities Save the Future?
http://pan.cedar.univie.ac.at/habitat/press/press7.html
This press release about the second session of the Preparatory Committee for Habitat II is an excellent discussion of the question of global urbanization.

16. IISDnet
http://iisd1.iisd.ca/
The International Institute for Sustainable Development, a Canadian organization, presents information through gateways entitled Business and Sustainable Development, Developing Ideas, and Hot Topics. Linkages is its multimedia resource for environment and development policymakers.

17. NewsPage
http://pnp1.individual.com/
Individual, Inc., maintains this business-oriented Web site. Geographers will find links to much valuable information about such fields as energy, environmental services, media and communications, and health care.

18. Telecommuting as an Investment: The Big Picture—John Wolf
http://www.svi.org/telework/forums/messages5/48.html
This page deals with the many issues related to telecommuting, including its potential role in reducing environmental pollution. The site discusses such topics as employment law and the impact of telecommuting on businesses and employees.

19. The Urban Environment
http://www.geocities.com/RainForest/Vines/6723/urb/index.html
Global urbanization is discussed fully at this site, which also includes the original 1992 Treaty on Urbanization.

20. Virtual Seminar in Global Political Economy//Global Cities & Social Movements
http://csf.colorado.edu/gpe/gpe95b/resources.html
This Web site is rich in links to subjects of interest in regional studies, such as sustainable cities, megacities, and urban planning. Links to many international nongovernmental organizations are included.

21. WWW-LARCH-LK Archive: Sustainability
http://www.clr.toronto.edu/ARCHIVES/HMAIL/larchl/0737.html
This site gives you the opportunity to read and respond to a discourse on sustainability, with many different opinions and viewpoints represented.

Spatial Interaction and Mapping

22. Edinburgh Geographical Information Systems
http://www.geo.ed.ac.uk/home/gishome.html
This valuable site, hosted by the Department of Geography at the University of Edinburgh, provides information on all aspects of Geographic Information Systems and provides links to other servers worldwide. A GIS reference database as well as a major GIS bibliography is included.

23. GIS Frequently Asked Questions and General Information
http://www.census.gov/geo/gis/faq-index.html
Browse through this site to get answers to FAQs about Geographic Information Systems. It can direct you to general information about GIS as well as guidelines on such specific questions as how to order U.S. Geological Survey maps. Other sources of information are also noted.

24. International Map Trade Association
http://www.maptrade.org/
The International Map Trade Association offers this site for those interested in information on maps, geography, and mapping technology. Lists of map retailers and publishers as well as upcoming IMTA conferences and trade shows are noted.

25. PSC Publications
http://www.psc.lsa.umich.edu/pubs/abs/abs94-319.html
Use this site and its links from the Population Studies Center of the University of Michigan for spatial patterns of immigration and discussion of white and black flight from high-immigration metropolitan areas in the United States.

26. U.S. Geological Survey
http://www.usgs.gov/research/gis/title.html
This site discusses the uses for Geographic Information Systems and explains how GIS works, addressing such topics as data integration, data modeling, and relating information from different sources.

Population, Resources, and Socioeconomic Development

27. African Studies WWW (U.Penn)
http://www.sas.upenn.edu/African_Studies/AS.html
This site will give you access to rich and varied resources that cover such topics as demographics, migration, family planning, and health and nutrition.

28. Human Rights and Humanitarian Assistance
http://info.pitt.edu/~ian/resource/human.htm
Through this site, part of the World Wide Web Virtual Library, you can conduct research into a number of human-rights topics in order to gain a greater understanding of the issues affecting indigenous peoples in the modern era.

29. Hypertext and Ethnography
http://www.umanitoba.ca/faculties/arts/anthropology/tutor/aaa_presentation.new.html
This site, presented by Brian Schwimmer of the University of Manitoba, will be of great value to people who are interested in culture and communication. He addresses such topics as multivocality and complex symbolization, among many others.

30. Research and Reference (Library of Congress)
http://lcweb.loc.gov/rr/
This research and reference site of the Library of Congress will lead you to invaluable information on different countries. It provides links to numerous publications, bibliographies, and guides in area studies that can be of great help to geographers.

31. Space Research Institute
http://arc.iki.rssi.ru/Welcome.html
Browse through this home page of Russia's Space Research Institute for information on its Environment Monitoring Information Systems, the IKI Satellite Situation Center, and its Data Archive.

We highly recommend that you review our Web site for expanded information and our other product lines. We are continually updating and adding links to our Web site in order to offer you the most usable and useful information that will support and expand the value of your Annual Editions. You can reach us at: *http://www.dushkin.com/annualeditions/*.

Unit 1

Unit Selections

1. **The Four Traditions of Geography,** William D. Pattison
2. **The American Geographies,** Barry Lopez
3. **NASA Readies a 'Mission to Planet Earth,'** William K. Stevens
4. **Human Domination of Earth's Ecosystems,** Peter M. Vitousek, Harold A. Mooney, Jane Lubchenco, and Jerry M. Melillo
5. **Counting the Cost: The Growing Role of Economics in Environmental Decisionmaking,** Paul R. Portney
6. **The Coming Climate,** Thomas R. Karl, Neville Nicholls, and Jonathan Gregory
7. **The Emergence of Gasohol: A Renewable Fuel for American Roadways,** William D. Warren and Dennis Roberts
8. **America's Rush to Suburbia,** Kenneth T. Jackson

Key Points to Consider

❖ Why is geography called an integrating discipline?

❖ How is geography related to earth science? Give some examples of these relationships.

❖ What are area studies? Why is the spatial concept so important in geography? What is your definition of geography?

❖ Why do history and geography rely on each other? How have humans affected the weather? The environment in general?

❖ How is individual behavior related to our understanding of environment?

❖ Discuss whether or not change is a good thing. Why is it important to anticipate change?

❖ What does interconnectedness mean in terms of places? Give examples of how you as an individual interact with people in other places. How are you "connected" to the rest of the world?

❖ What will the world be like in the year 2010? Tell why you are pessimistic or optimistic about the future. What, if anything, can you do about the future?

 **Links** www.dushkin.com/online/

6. **Geological Survey of Sweden: Other Geological Surveys**
   http://www3.sgu.se/links/Surveys_e.shtml
7. **Mission to Planet Earth**
   http://www.hq.nasa.gov/office/mtpe/
8. **Public Utilities Commission of Ohio**
   http://www.puc.ohio.gov/consumer/gcc/index.html
9. **Santa Fe Institute**
   http://acoma.santafe.edu/sfi/research/

These sites are annotated on pages 6 and 7.

Geography in a Changing World

What is geography? This question has been asked innumerable times, but it has not elicited a universally accepted answer, even from those who are considered to be members of the geography profession. The reason lies in the very nature of geography as it has evolved through time. Geography is an extremely wide-ranging discipline, one that examines appropriate sets of events or circumstances occurring at specific places. Its goal is to answer certain basic questions.

The first question—Where is it?—establishes the location of the subject under investigation. The concept of location is very important in geography, and its meaning extends beyond the common notion of a specific address or the determination of the latitude and longitude of a place. Geographers are more concerned with the relative location of a place and how that place interacts with other places both far and near. Spatial interaction and the determination of the connections between places are important themes in geography.

Once a place is "located," in the geographer's sense of the word, the next question is, Why is it here? For example, why are people concentrated in high numbers on the North China plain, in the Ganges River Valley in India, and along the eastern seaboard in the United States? Conversely, why are there so few people in the Amazon basin and the Central Siberian lowlands? Generally, the geographer wants to find out why particular distribution patterns occur and why these patterns change over time.

The element of time is another extremely important ingredient in the geographical mix. Geography is most concerned with the activities of human beings, and human beings bring about change. As changes occur, new adjustments and modifications are made in the distribution patterns previously established. Patterns change, for instance, as new technology brings about new forms of communication and transportation and as once-desirable locations decline in favor of new ones. For example, people migrate from once-productive regions such as the Sahel when a disaster such as drought visits the land. Geography, then, is greatly concerned with discovering the underlying processes that can explain the transformation of distribution patterns and interaction forms over time. Geography itself is dynamic, adjusting as a discipline to handle new situations in a changing world.

Geography is truly an integrating discipline. The geographer assembles evidence from many sources in order to explain a particular pattern or ongoing process of change. Some of this evidence may even be in the form of concepts or theories borrowed from other disciplines. The first two articles of this unit provide insight into both the conceptual nature of geography and the development of the discipline over time.

Throughout its history, four main themes have been the focus of research work in geography. These themes or traditions, according to William Pattison in "The Four Traditions of Geography," link geography with earth science, establish it as a field that studies land-human relationships, engage it in area studies, and give it a spatial focus. Although Pattison's article first appeared over 30 years ago, it is still referred to and cited frequently today. Much of the geographical research and analysis engaged in today would fall within one or more of Pattison's traditional areas, but new areas are also opening for geographers. In a particularly thought-provoking essay, the eminent author Barry Lopez discusses local geographies and the importance of a sense of place. William Stevens then reports on a NASA project to study earth systems from space. In "Human Domination of the Earth's Ecosystems," we see how human activity is significantly altering the globe. Then, the growing role of economics in environmental decision making is described. "The Coming Climate" takes its cue from the widely accepted concept of global warming.

The growth of gasohol usage in the United States is traced through map patterns in an article by William Warren and Dennis Roberts. Finally, Kenneth Jackson's article, "America's Rush to Suburbia," points out problems in urban sprawl.

Article 1

The Four Traditions of Geography

William D. Pattison

Late Summer, 1990

To Readers of the *Journal of Geography:*

I am honored to be introducing, for a return to the pages of the *Journal* after more than 25 years, "The Four Traditions of Geography," an article which circulated widely, in this country and others, long after its initial appearance—in reprint, in xerographic copy, and in translation. A second round of life at a level of general interest even approaching that of the first may be too much to expect, but I want you to know in any event that I presented the paper in the beginning as my gift to the geographic community, not as a personal property, and that I re-offer it now in the same spirit.

In my judgment, the article continues to deserve serious attention—perhaps especially so, let me add, among persons aware of the specific problem it was intended to resolve. The background for the paper was my experience as first director of the High School Geography Project (1961–63)—not all of that experience but only the part that found me listening, during numerous conference sessions and associated interviews, to academic geographers as they responded to the project's invitation to locate "basic ideas" representative of them all. I came away with the conclusion that I had been witnessing not a search for consensus but rather a blind struggle for supremacy among honest persons of contrary intellectual commitment. In their dialogue, two or more different terms had been used, often unknowingly, with a single reference, and no less disturbingly, a single term had been used, again often unknowingly, with two or more different references. The article was my attempt to stabilize the discourse. I was proposing a basic nomenclature (with explicitly associated ideas) that would, I trusted, permit the development of mutual comprehension **and** confront all parties concerned with the pluralism inherent in geographic thought.

This intention alone could not have justified my turning to the NCGE as a forum, of course. The fact is that from the onset of my discomfiting realization I had looked forward to larger consequences of a kind consistent with NCGE goals. As finally formulated, my wish was that the article would serve "to greatly expedite the task of maintaining an alliance between professional geography and pedagogical geography and at the same time to promote communication with laymen" (see my fourth paragraph). I must tell you that I have doubts, in 1990, about the acceptability of my word choice, in saying "professional," "pedagogical," and "layman" in this context, but the message otherwise is as expressive of my hope now as it was then.

I can report to you that twice since its appearance in the *Journal,* my interpretation has received more or less official acceptance—both times, as it happens, at the expense of the earth science tradition. The first occasion was Edward Taaffe's delivery of his presidential address at the 1973 meeting of the Association of American Geographers (see *Annals AAG,* March 1974, pp. 1–16). Taaffe's working-through of aspects of an interrelations among the spatial, area studies, and man-land traditions is by far the most thoughtful and thorough of any of which I am aware. Rather than fault him for omission of the fourth tradition, I compliment him on the grace with which he set it aside in conformity to a meta-epistemology of the American university which decrees the integrity of the social sciences as a consortium in their own right. He was sacrificing such holistic claims as geography might be able to muster for a freedom to argue the case for geography as a social science.

The second occasion was the publication in 1984 of *Guidelines for Geographic Education: Elementary and Secondary Schools,* authored by a committee jointly representing the AAG and the NCGE. Thanks to a recently published letter (see *Journal of Geography,* March-April 1990, pp. 85–86), we know that, of five themes commended to teachers in this source,

> The committee lifted the human environmental interaction theme directly from Pattison. The themes of place and location are based on Pattison's spatial or geometric geography, and the theme of region comes from Pattison's area studies or regional geography.

Having thus drawn on my spatial, area studies, and man-land traditions for four of the five themes, the committee could have found the remaining theme, movement, there too—in the spatial tradition (see my sixth paragraph). However that may be, they did not avail themselves of the earth science tradition, their reasons being readily surmised. Peculiar to the elementary and secondary schools is a curriculum category framed as much by theory of citizenship as by theory of knowledge: the social studies. With admiration, I see already in the committee members' adoption of the theme idea a strategy for assimilation of their program to the established repertoire of social studies practice. I see in their exclusion of the earth science tradition an intelligent respect for social studies' purpose.

Here's to the future of education in geography: may it prosper as never before.

W. D. P., 1990

1. Four Traditions of Geography

Reprinted from the Journal of Geography, 1964, pp. 211–216.

In 1905, one year after professional geography in this country achieved full social identity through the founding of the Association of American Geographers, William Morris Davis responded to a familiar suspicion that geography is simply an undisciplined "omnium-gatherum" by describing an approach that as he saw it imparts a "geographical quality" to some knowledge and accounts for the absence of the quality elsewhere.[1] Davis spoke as president of the AAG. He set an example that was followed by more than one president of that organization. An enduring official concern led the AAG to publish, in 1939 and in 1959, monographs exclusively devoted to a critical review of definitions and their implications.[2]

Every one of the well-known definitions of geography advanced since the founding of the AAG has had its measure of success. Tending to displace one another by turns, each definition has said something true of geography.[3] But from the vantage point of 1964, one can see that each one has also failed. All of them adopted in one way or another a monistic view, a singleness of preference, certain to omit if not to alienate numerous professionals who were in good conscience continuing to participate creatively in the broad geographic enterprise.

The thesis of the present paper is that the work of American geographers, although not conforming to the restrictions implied by any one of these definitions, has exhibited a broad consistency, and that this essential unity has been attributable to a small number of distinct but affiliated traditions, operant as binders in the minds of members of the profession. These traditions are all of great age and have passed into American geography as parts of a general legacy of Western thought. They are shared today by geographers of other nations.

There are four traditions whose identification provides an alternative to the competing monistic definitions that have been the geographer's lot. The resulting pluralistic basis for judgment promises, by full accommodation of what geographers do and by plain-spoken representation thereof, to greatly expedite the task of maintaining an alliance between professional geography and pedagogical geography and at the same time to promote communication with laymen. The following discussion treats the traditions in this order: (1) a spatial tradition, (2) an area studies tradition, (3) a man-land tradition and (4) an earth science tradition.

Spatial Tradition
Entrenched in Western thought is a belief in the importance of spatial analysis, of the act of separating from the happenings of experience such aspects as distance, form, direction and position. It was not until the 17th century that philosophers concentrated attention on these aspects by asking whether or not they were properties of things-in-themselves. Later, when the 18th century writings of Immanuel Kant had become generally circulated, the notion of space as a category including all of these aspects came into widespread use. However, it is evident that particular spatial questions were the subject of highly organized answering attempts long before the time of any of these cogitations. To confirm this point, one need only be reminded of the compilation of elaborate records concerning the location of things in ancient Greece. These were records of sailing distances, of coastlines and of landmarks that grew until they formed the raw material for the great *Geographia* of Claudius Ptolemy in the 2nd century A.D.

A review of American professional geography from the time of its formal organization shows that the spatial tradition of thought had made a deep penetration from the very beginning. For Davis, for Henry Gannett and for most if not all of the 44 other men of the original AAG, the determination and display of spatial aspects of reality through mapping were of undoubted importance, whether contemporary definitions of geography happened to acknowledge this fact or not. One can go further and, by probing beneath the art of mapping, recognize in the behavior of geographers of that time an active interest in the true essentials of the spatial tradition—*geometry* and *movement*. One can trace a basic favoring of movement as a subject of study from the turn-of-the-century work of Emory R. Johnson, writing as professor of transportation at the University of Pennsylvania, through the highly influential theoretical and substantive work of Edward L. Ullman during the past 20 years and thence to an article by a younger geographer on railroad freight traffic in the U.S. and Canada in the *Annals* of the AAG for September 1963.[4]

One can trace a deep attachment to geometry, or positioning-and-layout, from articles on boundaries and population densities in early 20th century volumes of the *Bulletin of the American Geographical Society*, through a controversial pronouncement by Joseph Schaefer in 1953 that granted geographical legitimacy only to studies of spatial patterns[5] and so onward to a recent *Annals* report on electronic scanning of cropland patterns in Pennsylvania.[6]

One might inquire, is discussion of the spatial tradition, after the manner of the remarks just made, likely to bring people within geography closer to an understanding of one another and people outside geography closer to an understanding of geographers? There seem to be at least two reasons for being hopeful. First, an appreciation of this tradition allows one to see a bond of fellowship uniting the elementary school teacher, who attempts the most rudimentary instruction in directions and mapping, with the contemporary research geographer, who dedicates himself to an exploration of central-place theory. One cannot only open the eyes of many teachers to the potentialities of their own instruction, through proper exposition of the spatial tradition, but one can also "hang a bell" on research quantifiers in geography, who are often thought to have wandered so far in their intellectual adventures as to have become lost from the rest. Looking outside geography, one may anticipate benefits from the readiness of countless persons to associate the name "geography" with maps. Latent within this readiness is a willingness to recognize as geography, too, what maps are about—and that is the geometry of and the movement of what is mapped.

Area Studies Tradition
The area studies tradition, like the spatial tradition, is quite strikingly represented in classical antiquity by a practitioner to whose surviving work we can point. He is Strabo, celebrated for his *Geography* which is a massive production addressed to the statesmen of Augustan Rome and intended to sum up and regularize knowledge not of the location of places and associated cartographic facts, as in the somewhat later case of Ptolemy, but of the nature of places, their character and their differentiation. Strabo exhibits interesting attributes of the area-studies tradition that can hardly be overemphasized. They are a pronounced tendency toward subscription primarily to literary standards, an almost omnivorous appetite for information and a self-conscious companionship with history.

It is an extreme good fortune to have in the ranks of modern American geography the scholar Richard Hartshorne, who has pondered the meaning of the area-studies tradition with a legal acuteness that few persons would challenge. In his *Nature of Geography*, his 1939 monograph already cited,[7] he scrutinizes exhaustively the implications of the "interesting attributes" identified in connection with Strabo, even though his concern is with quite other and much later authors, largely German. The major literary problem of unities or wholes he considers from every angle. The Gargantuan appetite for miscellaneous information he accepts and rationalizes. The companionship between area studies and history he clarifies by appraising the so-called idiographic content of both and by affirming the tie of both to what he and Sauer have called "naively given reality."

The area-studies tradition (otherwise known as the chorographic tradition) tended to be excluded from early American professional geography. Today it is beset by certain champions of the spatial tradition who would have one believe that somehow the area-studies way of organizing knowledge is only a subdepartment of spatialism. Still, area-studies as a method of presentation lives and prospers in its own right. One can turn today for reassurance on this score to practically any issue of the *Geographical Review*, just as earlier readers could turn at the opening of the century to that magazine's forerunner.

What is gained by singling out this tradition? It helps toward restoring the faith of many teachers who, being accustomed to administering learning in the area-studies style, have begun to wonder if by doing so they really were keeping in touch with professional geography. (Their doubts are owed all too much to the obscuring effect of technical words attributable to the very professionals who have been intent, ironically, upon protecting that tradition.) Among persons outside the classroom the geographer stands to gain greatly in intelligibility. The title "area-studies" itself carries an understood message in the United States today wherever there is contact with the usages of the academic community. The purpose of characterizing a place, be it neighborhood or nation-state, is readily grasped. Furthermore, recognition of the right of a geographer to be unspecialized may be expected to be forthcoming from people generally, if application for such recognition is made on the merits of this tradition, explicitly.

Man-Land Tradition
That geographers are much given to exploring man-land questions is especially evident to anyone who examines geographic output, not only in this country but also abroad. O. H. K. Spate, taking an international view, has felt justified by his observations in nominating as the most significant ancient precursor of today's geography neither Ptolemy nor Strabo nor writers typified in their outlook by the geographies of either of these two men, but rather Hippocrates, Greek physician of the 5th century B.C. who left to posterity an extended essay, *On Airs, Waters and Places*.[8] In this work made up of reflections on human health and conditions of external nature, the questions asked are such as to confine thought almost altogether to presumed influence passing from the latter to the former, questions largely about the effects of winds, drinking water and seasonal changes upon man. Understandable though this uni-directional concern may have been for Hippocrates as medical commentator, and defensible as may be the attraction that this same approach held for students of the condition of man for many, many centuries thereafter, one can only regret that this narrowed version of the man-land tradition, combining all too easily with social Darwinism of the late 19th century, practically overpowered American professional geography in the first generation of its history.[9] The premises of this version governed scores of studies by American geographers in interpreting the rise and fall of nations, the strategy of battles and the construction of public improvements. Eventually this special bias, known as environmentalism, came to be confused with the whole of the man-land tradition in the minds of many people. One can see now, looking back to the years after the ascendancy of environmentalism, that although the spatial tradition was asserting itself with varying degrees of forwardness, and that although the area-studies tradition was also making itself felt, perhaps the most interesting chapters in the story of American professional geography were being written by academicians who were reacting against environmentalism while deliberately remaining within the broad man-land tradition. The rise of culture historians during the last 30 years has meant the dropping of a curtain of culture between land and man, through which it is asserted all influence must pass. Furthermore work of both culture historians and other geographers has exhibited a reversal of the direction of the effects in Hippocrates, man appearing as an independent agent, and the land as a sufferer from action. This trend as presented in published research has reached a high point in the collection of papers titled *Man's Role in Changing the Face of the Earth*. Finally, books and articles can be called to mind that have addressed themselves to the most difficult task of all, a balanced tracing out of interaction between man and environment. Some chapters in the book mentioned above undertake just this. In fact the separateness of this approach is discerned only with difficulty in many places; however, its significance as a general research design that rises above environmentalism, while refusing to abandon the man-land tradition, cannot be mistaken.

The NCGE seems to have associated itself with the man-land tradition, from the time of founding to the present day, more than with any other tradition, although all four of the traditions are amply represented in its official magazine, *The Journal of Geography* and in the proceedings of its annual meetings. This apparent preference on the part of the NCGE members *for defining geography in terms of the man-land tradition* is strong evidence of the appeal that man-land ideas, separately stated, have for persons whose main job is teaching. It should be noted, too, that this inclination reflects a proven acceptance by the general public of learning that centers on resource use and conservation.

Earth Science Tradition
The earth science tradition, embracing study of the earth, the waters of the earth, the atmosphere surrounding the earth and the association between earth and sun, confronts one with a paradox. On the one hand one is assured by professional geographers that their participation in this tradition has declined precipitously in the course of the past few decades, while on the other one knows that college departments of geography across the nation rely substantially, for justification of their role in general education, upon curricular content springing directly from this tradition. From all the reasons that combine to account for this state of affairs, one may, by selecting only two, go far toward achieving an understanding of this tradition. First, there is the fact that American college geography, growing out of departments of geology in many crucial instances, was at one time greatly overweighted in favor of earth science, thus rendering the field unusually liable to a sense of loss as better balance came into being. (This one-time disproportion found reciprocate support for many years in the narrowed, environmentalistic interpretation of the man-land tradition.) Second, here alone in earth science does one encounter subject matter in the normal sense of the term as one reviews geographic traditions. The spatial tradition abstracts certain aspects of reality; area studies is distinguished by a point of view; the man-land tradition dwells upon relationships; but earth science is identifiable through concrete objects. Historians, sociologists and other academicians tend not only to accept but also to ask for help from this part of geography. They readily appreciate earth science as something physically associated with their subjects of study, yet generally beyond their competence to treat. From this appreciation comes strength for geography-as-earth-science in the curriculum.

Only by granting full stature to the earth science tradition can one make sense out of the oft-repeated addage, "Geography is the mother of sciences." This is the tradition that emerged in ancient Greece, most clearly in the work of Aristotle, as a wide-ranging study of natural processes in and near the surface of the earth. This is the tradition that was rejuvenated by Varenius in the 17th century as "Geographia Generalis." This is the tradition that has been subjected to subdivision as the development of science has approached the present day, yielding mineralogy, paleontology, glaciology, meterology and other specialized fields of learning.

Readers who are acquainted with American junior high schools may want to make a challenge at this point, being aware that a current revival of earth sciences is being sponsored in those schools by the field of geology. Belatedly, geography has joined in support of this revival.[10] It may be said that in this connection and in others, American professional geography may have faltered in its adherence to the earth science tradition but not given it up.

In describing geography, there would appear to be some advantages attached to isolating this final tradition. Separation improves the geographer's chances of successfully explaining to educators why geography has extreme difficulty in accommodating itself to social studies programs. Again, separate attention allows one to make understanding contact with members of the American public for whom surrounding nature is known as the geographic environment. And finally, specific reference to the geographer's earth science tradition brings into the open the basis of what is, almost without a doubt, morally the most significant concept in the entire geographic heritage, that of the earth as a unity, the single common habitat of man.

An Overview

The four traditions though distinct in logic are joined in action. One can say of geography that it pursues concurrently all four of them. Taking the traditions in varying combinations, the geographer can explain the conventional divisions of the field. Human or cultural geography turns out to consist of the first three traditions applied to human societies; physical geography, it becomes evident, is the fourth tradition prosecuted under constraints from the first and second traditions. Going further, one can uncover the meanings of "systematic geography," "regional geography," "urban geography," "industrial geography," etc.

It is to be hoped that through a widened willingness to conceive of and discuss the field in terms of these traditions, geography will be better able to secure the inner unity and outer intelligibility to which reference was made at the opening of this paper, and that thereby the effectiveness of geography's contribution to American education and to the general American welfare will be appreciably increased.

Notes

1. William Morris Davis, "An Inductive Study of the Content of Geography," *Bulletin of the American Geographical Society,* Vol. 38, No. 1 (1906), 71.
2. Richard Hartshorne, *The Nature of Geography,* Association of American Geographers (1939), and idem., *Perspective on the Nature of Geography,* Association of American Geographers (1959).
3. The essentials of several of these definitions appear in Barry N. Floyd, "Putting Geography in Its Place," *The Journal of Geography,* Vol. 62, No. 3 (March, 1963). 117–120.
4. William H. Wallace, "Freight Traffic Functions of Anglo-American Railroads," *Annals of the Association of American Geographers,* Vol. 53, No. 3 (September, 1963), 312–331.
5. Fred K. Schaefer, "Exceptionalism in Geography: A Methodological Examination," *Annals of the Association of American Geographers,* Vol. 43, No. 3 (September, 1953), 226–249.
6. James P. Latham, "Methodology for an Instrumental Geographic Analysis," *Annals of the Association of American Geographers,* Vol. 53, No. 2 (June, 1963), 194–209.
7. Hartshorne's 1959 monograph, *Perspective on the Nature of Geography,* was also cited earlier. In this later work, he responds to dissents from geographers whose preferred primary commitment lies outside the area studies tradition.
8. O. H. K. Spate, "Quantity and Quality in Geography," *Annals of the Association of American Geographers,* Vol. 50, No. 4 (December, 1960), 379.
9. Evidence of this dominance may be found in Davis's 1905 declaration: "Any statement is of geographical quality if it contains ... some relation between an element of inorganic control and one of organic response" (Davis, *loc. cit.*).
10. Geography is represented on both the Steering Committee and Advisory Board of the Earth Science Curriculum Project, potentially the most influential organization acting on behalf of earth science in the schools.

The American Geographies

Americans are fast becoming strangers in a strange land, where one roiling river, one scarred patch of desert, is as good as another. America the beautiful exists— a select few still know it intimately—but many of us are settling for a homogenized national geography.

Barry Lopez

Barry Lopez has written The Rediscovery of North America *(Vintage), and his most recent book is* Field Notes *(Knopf).*

It has become commonplace to observe that Americans know little of the geography of their country, that they are innocent of it as a landscape of rivers, mountains, and towns. They do not know, supposedly, the location of the Delaware Water Gap, the Olympic Mountains, or the Piedmont Plateau; and, the indictment continues, they have little conception of the way the individual components of this landscape are imperiled, from a human perspective, by modern farming practices or industrial pollution.

I do not know how true this is, but it is easy to believe that it is truer than most of us would wish. A recent Gallup Organization and National Geographic Society survey found Americans woefully ignorant of world geography. Three out of four couldn't locate the Persian Gulf. The implication was that we knew no more about our own homeland, and that this ignorance undermined the integrity of our political processes and the efficiency of our business enterprises.

As Americans, we profess a sincere and fierce love for the American landscape, for our rolling prairies, freeflowing rivers, and "purple mountains' majesty"; but it is hard to imagine, actually, where this particular landscape is. It is not just that a nostalgic landscape has passed away—Mark Twain's Mississippi is now dammed from Minnesota to Missouri and the prairies have all been sold and fenced. It is that it has always been a romantic's landscape. In the attenuated form in which it is presented on television today, in magazine articles and in calendar photographs, the essential wildness of the American landscape is reduced to attractive scenery. We look out on a familiar, memorized landscape that portends adventure and promises enrichment. There are no distracting people in it and few artifacts of human life. The animals are all beautiful, diligent, one might even say well-behaved. Nature's unruliness, the power

To truly understand geography requires not only time but a kind of local expertise, an intimacy with place few of us ever develop.

of rivers and skies to intimidate, and any evidence of disastrous human land management practices are all but invisible. It is, in short, a magnificent garden, a colonial vision of paradise imposed on a real place that is, at best, only selectively known.

The real American landscape is a face of almost incomprehensible depth and complexity. If one were to sit for a few days, for example, among the ponderosa pine forests and black lava fields of the Cascade Mountains in western Oregon, inhaling the pines' sweet balm on an evening breeze from some point on the barren rock, and then were to step off to the Olympic Peninsula in Washington, to those rain forests with sphagnum moss floors soft as fleece underfoot and Douglas firs too big around for five people to hug, and then head south to walk the ephemeral creeks and sun-blistered playas of the Mojave Desert in southern California, one would be reeling under the sensations. The contrast is not only one of plants and soils, a different array say, of brilliantly colored beetles. The shock to the senses comes from a different shape to the silence, a difference in the very quality of light, in the weight of the air. And this relatively short journey down the West Coast would still leave the traveler with all that lay to the east to explore—the anomalous sand hills of Nebraska, the heat and frog voices of Okefenokee Swamp, the fetch of Chesapeake Bay, the hardwood copses and black bears of the Ozark Mountains.

No one of these places, of course, can be entirely fathomed, biologically or aesthetically. They are mysteries upon which we impose names. Enchantments. We tick the names off glibly but lovingly. We mean no disrespect. Our genuine desire, though we might be skeptical about the time it would take and uncertain of its practical value to us, is to actually know these places. As deeply ingrained in the American psyche as the desire to conquer and control the land is the desire to sojourn in it, to sail up and down Pamlico Sound, to paddle a canoe through Minnesota's boundary waters, to walk on the desert of the Great Salt Lake, to camp in the stony hardwood valleys of Vermont.

To do this well, to really come to an understanding of a specific American geography, requires not only time but a kind of local expertise, an intimacy with place few of us ever develop. There is no way around the former requirement: If you want to know you must take the time. It is not in books. A specific geographical understanding, however, can be sought out and borrowed. It resides with men and women more or less sworn to a place, who abide there, who have a feel for the soil and history, for the turn of leaves and night sounds. Often they are glad to take the outlander in tow.

These local geniuses of American landscape, in my experience, are people in whom geography thrives. They are the antithesis of geographical ignorance. Rarely known outside their own communities, they often seem, at the first encounter, unremarkable and anonymous. They may not be able to recall the name of a particular wildflower—or they may have given it a name known only to them. They might have forgotten the precise circumstances of a local historical event. Or they can't say for certain when the last of the Canada geese passed through in the fall, or can't differentiate between two kinds of trout in the same creek. Like all of us, they have fallen prey to the fallacies of memory and are burdened with ignorance; but they are nearly flawless in the respect they bear these places they love. Their knowledge is intimate rather than encyclopedic, human but not necessarily scholarly. It rings with the concrete details of experience.

America, I believe, teems with such people. The paradox here, between a faulty grasp of geographical knowledge for which Americans are indicted and the intimate, apparently contradictory familiarity of a group of largely anonymous people, is not solely a matter of confused scale. (The local landscape is easier to know than a national geography.) And it is not simply ironic. The paradox is dark. To be succinct: The politics and advertising that seek a national audience must project a national geography; to be broadly useful that geography must, inevitably, be generalized and it is often romantic. It is therefore frequently misleading and imprecise. The same holds true with the entertainment industry, but here the problem might be clearer. The same films, magazines, and television features that honor an imaginary American landscape also tout the worth of the anonymous men and women who interpret it. Their affinity for the land is lauded, their local allegiance admired. But the rigor of their local geographies, taken as a whole, contradicts a patriotic, national vision of unspoiled, untroubled land. These men and women are ultimately forgotten, along with the details of the landscapes they speak for, in the face of more pressing national matters. It is the chilling nature of modern society to find an ignorance of geography, local or national, as excusable as an ignorance of hand tools; and to find the commitment of people to their home places only momentarily entertaining. And finally naive.

If one were to pass time among Basawara people in the Kalahari Desert, or with Kreen-Akrora in the Amazon Basin, or with Pitjantjatjara Aborigines in Australia, the most salient impression they might leave is of an absolutely stunning knowledge of their local geography—geology, hydrology, biology, and weather. In short, the extensive particulars of their intercourse with it.

In 40,000 years of human history, it has only been in the last few hundred years or so that a people could afford to ignore their local geographies as completely as we do and still survive. Technological innovations from refrigerated trucks to artificial fertilizers, from sophisticated cost accounting to mass air transportation, have utterly changed concepts of season, distance, soil productivity, and the real cost of drawing sustenance from the land. It is now possible for a resident of Boston to bite into a fresh strawberry in the dead of winter; for someone in San Francisco to travel to Atlanta in a few hours with no worry of how formidable might be crossing of the Great Basin Desert or the Mississippi River; for an absentee farmer to gain a tax advantage from a farm that leaches poisons into its water table and on which crops are left to rot. The Pitjantjatjara might shake their heads in bewilderment and bemusement, not because they are primitive or ignorant people, not because they have no sense of irony or are incapable of marveling, but because they have not (many would say not yet) realized a world in which such manipulation of the land—surmounting the imperatives of distance it imposes, for example, or turning the large-scale destruction of forests and arable land in wealth—is desirable or plausible.

In the years I have traveled through America, in cars and on horseback, on foot and by raft, I have repeatedly been brought to a sudden state of awe by some gracile or savage movement of animal, some odd wrapping of tree's foliage by the wind, an unimpeded run of dew-laden prairie stretching to a horizon flat as a coin where a pin-dot sun pales the dawn sky pink. I know these things are beyond intellection, that they are the vivid edges of a world that includes but also transcends the human world. In memory, when I dwell on these things, I know that in a truly national literature there should be odes to the Triassic reds of the Colorado Plateau, to the sharp and ghostly light of the Florida Keys, to the aeolian soils of southern Minnesota, and the Palouse in Washington, though the modern mind abjures the literary potential of such subjects. (If the sand and flood water farmers of Arizona and New Mexico were to take the black loams of Louisiana in their hands they would be flabbergasted, and that is the beginning of literature.) I know there should be eloquent evocations of the cobbled beaches of Maine, the plutonic walls of the Sierra Nevada, the orange canyons of the Kaibab Plateau. I have no doubt, in fact, that there are. They are as numerous and diverse as the eyes and fingers that ponder the country—it is that only a handful of them are known. The great majority are to be found in drawers and boxes, in the letters and private journals of millions of workaday people who have regarded their encounters with the land as an engagement bordering on the spiritual, as being fundamentally linked to their state of health.

One cannot acknowledge the extent and the history of this kind of testimony without being forced to the realization that something strange, if not dangerous, is afoot. Year by year, the number of people with firsthand experience in the land dwindles. Rural populations continue to shift to the cities. The family farm is in a state of demise, and government and industry continue to apply pressure on the native peoples of North America to sever their ties with the land. In the wake of this loss of personal and local knowledge from which a real geography is derived, the knowledge on

1 ❖ GEOGRAPHY IN A CHANGING WORLD

which a country must ultimately stand, has [be]come something hard to define but I think sinister and unsettling—the packaging and marketing of land as a form of entertainment. An incipient industry, capitalizing on the nostalgia Americans feel for the imagined virgin landscapes of their fathers, and on a desire for adventure, now offers people a convenient though sometimes incomplete or even spurious geography as an inducement to purchase a unique experience. But the line between authentic experience and a superficial exposure to the elements of experience is blurred. And the real landscape, in all its complexity, is distorted even further in the public imagination. No longer innately mysterious and dignified, a ground from which experience grows, it becomes a curiously generic backdrop on which experience is imposed.

In theme parks the profound, subtle, and protracted experience of running a river is reduced to a loud, quick, safe equivalence, a pleasant distraction. People only able to venture into the countryside on annual vacations are, increasingly, schooled in the belief that wild land will, and should, provide thrills and exceptional scenery on a timely basis. If it does not, something is wrong, either with the land itself or possibly with the company outfitting the trip.

People in America, then, face a convoluted situation. The land itself, vast and differentiated, defies the notion of a national geography. If applied at all it must be applied lightly and it must grow out of the concrete detail of local geographies. Yet Americans are daily presented with, and have become accustomed to talking about, a homogenized national geography. One that seems to operate independently of the land, a collection of objects rather than a continuous bolt of fabric. It appears in advertisements, as a background in movies, and in patriotic calendars. The suggestion is that there can be national geography because the constituent parts are interchangeable and can be treated as commodities. In day-to-day affairs, in other words, one place serves as well as another to convey one's point. On reflection, this is an appalling condescension and a terrible imprecision, the very antithesis of knowledge. The idea that either the Green River in Utah or the Salmon River in Idaho will do, or that the valleys of Kentucky and West Virginia are virtually interchangeable, is not just misleading. For people still dependent on the soil for their sustenance, or for people whose memories tie them to those places, it betrays a numbing casualness, utilitarian, expedient, and commercial frame of mind. It heralds a society in which it is no longer necessary for human beings to know where they live, except as those places are described and fixed by numbers. The truly difficult and lifelong task of discovering where one lives is finally disdained.

If a society forgets or no longer cares where it lives, then anyone with the political power and the will to do so can manipulate the landscape to conform to certain social ideals or nostalgic visions. People may hardly notice that anything has happened, or assume that whatever happens—a mountain stripped of timber and eroding into its creeks—is for the common good. The more superficial a society's knowledge of the real dimensions of the land it occupies becomes, the more vulnerable the land is to exploitation, to manipulation for short-term gain. The land, virtually powerless before political and commercial entities, finds itself finally with no defenders. It finds itself bereft of intimates with indispensable, concrete knowledge. (Oddly, or perhaps not oddly, while American society continues to value local knowledge as a quaint part of its heritage, it continues to cut such people off from any real political power. This is as true for small farmers and illiterate cowboys as it is for American Indians, native Hawaiians, and Eskimos.)

The intense pressure of imagery in America, and the manipulation of images necessary to a society with specific goals, means the land will inevitably be treated like a commodity; and voices that tend to contradict the proffered image will, one way or another, be silenced or discredited by those in power. This is not new to America; the promulgation in America of a false or imposed geography has been the case from the beginning. All local geographies, as they were defined by hundreds of separate, independent native traditions, were denied in the beginning in favor of an imported and unifying vision of America's natural history. The country, the landscape itself, was eventually defined according to dictates of Progress like Manifest Destiny, and laws like the Homestead Act which reflected a poor understanding of the physical lay of the land.

When I was growing up in southern California, I formed the rudiments of a local geography—eucalyptus trees, February rains, Santa Ana winds. I lost much of it when my family moved to New York City, a move typical of the modern, peripatetic style of American life, responding to the exigencies of divorce and employment. As a boy I felt a hunger to know the American landscape that was extreme; when I was finally able to travel on my own, I did so. Eventually I visited most of the United States, living for brief periods of time in Arizona, Indiana, Alabama, Georgia, Wyoming, New Jersey, and Montana before settling 20 years ago in western Oregon.

The astonishing level of my ignorance confronted me everywhere I went. I knew early on that the country could not be held together in a few phrases, that its geography was magnificent and incomprehensible, that a man or woman could devote a lifetime to its elucidation and still feel in the end that he had but sailed many thousands of miles over the surface of the ocean. So I came into the habit of traversing landscapes I wanted to know with local tutors and reading what had previously been written about, and in, those places. I came to value exceedingly novels and essays and works of nonfiction that connected human enterprise to real and specific places, and I grew to be mildly distrustful of work that occurred in no particular place, work so cerebral and detached as to be refutable only in an argument of ideas.

These sojourns in various corners of the country infused me, somewhat to my surprise on thinking about it, with a great sense of hope. Whatever despair I had come to feel at a waning sense of the real land and the emergence of false geographies—elements of the land being manipulated, for example, to create erroneous but useful patterns in advertising—was dispelled by the depth of a single person's local knowledge, by the serenity that seemed to come with that intelligence. Any harm that might be done by people who cared nothing for the land, to whom it was not innately worthy but only something ultimately for sale, I thought, would one day have to meet this kind of integrity, people with the same dignity and transcendence as the land they occupied. So when I traveled, when I rolled my sleeping bag out on the shores of the Beaufort Sea, or in the high pastures of the Absaroka

2. American Geographies

Range in Wyoming, or at the bottom of the Grand Canyon, I absorbed those particular testaments to life, the indigenous color and songbird song, the smell of sun-bleached rock, damp earth, and wild honey, with some crude appreciation of the singular magnificence of each of those places. And the reassurance I felt expanded in the knowledge that there were, and would likely always be, people speaking out whenever they felt the dignity of the Earth imperiled in those places.

The promulgation of false geographies, which threaten the fundamental notion of what it means to live somewhere, is a current with a stable and perhaps growing countercurrent. People living in New York City are familiar with the stone basements, the cratonic geology, of that island and have a feeling for birds migrating through in the fall, their sequence and number. They do not find the city alien but human, its attenuated natural history merely different from that of rural Georgia or Kansas. I find the countermeasure, too, among Eskimos who cannot read but who might engage you for days on the subtleties of sea-ice topography. And among men and women who, though they have followed in the footsteps of their parents, have come to the conclusion that they cannot farm or fish or log in the way their ancestors did; the finite boundaries to this sort of wealth have appeared in their lifetime. Or among young men and women who have taken several decades of book-learned agronomy, zoology, silviculture and horticulture, ecology, ethnobotany, and fluvial geomorphology and turned it into a new kind of local knowledge, who have taken up residence in a place and sought, both because of and in spite of their education, to develop a deep intimacy with it. Or they have gone to work, idealistically, for the National Park Service or the fish and wildlife services or for a private institution like the Nature Conservancy. They are people to whom the land is more than politics and economics. These are people for whom the land is alive. It feeds them, directly, and that is how and why they learn its geography.

In the end, then, if one begins among the blue crabs of Chesapeake Bay and wanders for several years, down through the Smoky Mountains and back to the bluegrass hills, along the drainages of the Ohio and into the hill country of Missouri, where in summer a chorus of cicadas might drown out human conversation, then up the Missouri itself, reading on the way the entries of Meriwether Lewis and William Clark and musing on the demise of the plains grizzly and the sturgeon, crosses west into the drainage of the Platte and spends the evenings with Gene Weltfish's *The Lost Universe,* her book about the Pawnee who once thrived there, then drops south to the Palo Duro Canyon and the irrigated farms of the Llano Estacado in Texas, turns west across the Sangre de Cristo, southernmost of the Rocky Mountain ranges, and moves north and west up onto the slickrock mesas of Utah, those browns and oranges, the ocherous hues reverberating in the deep canyons, then goes north, swinging west to the insular ranges that sit like battleships in the pelagic space of Nevada, camps at the steaming edge of the sulfur springs in the Black Rock desert, where alkaline pans are glazed with a ferocious light, a heat to melt iron, then crosses the northern Sierra Nevada, waist-deep in summer snow in the passes, to descend to the valley of the Sacramento, and rises through groves of the elephantine redwoods in the Coast Range, to arrive at Cape Mendocino, before Balboa's Pacific, cormorants and gulls, gray whales headed north for Unimak Pass in the Aleutians, the winds crashing down on you, facing the ocean over the blue ocean that gives the scene its true vastness, making this crossing, having been so often astonished at the line and the color of the land, the ingenious lives of its plants and animals, the varieties of its darknesses, the intensity of the stars overhead, you would be ashamed to discover, then, in yourself, any capacity to focus on ravages in the land that left you unsettled. You would have seen so much, breathtaking, startling, and outsize, that you might not be able for a long time to break the spell, the sense, especially finishing your journey in the West, that the land had not been as rearranged or quite as compromised as you had first imagined.

After you had slept some nights on the beach, however, with that finite line of the ocean before you and the land stretching out behind you, the wind first battering then cradling you, you would be compelled by memory, obligated by your own involvement, to speak of what left you troubled. To find the rivers dammed and shrunken, the soil washed away, the land fenced, a tracery of pipes and wires and roads laid down everywhere and animals, cutting the eye off repeatedly and confining it—you had expected this. It troubles you no more than your despair over the ruthlessness, the insensitivity, the impetuousness of modern life. What underlies this obvious change, however, is a less noticeable pattern of disruption: acidic lakes, the skies empty of birds, fouled beaches, the poisonous slags of industry, the sun burning like a molten coin in ruined air.

It is a tenet of certain ideologies that man is responsible for all that is ugly, that everything nature creates is beautiful. Nature's darkness goes partly unreported, of course, and human brilliance is often perversely ignored. What is true is that man has a power, literally beyond his comprehension, to destroy. The lethality of some of what he manufactures, the incompetence with which he stores it or seeks to dispose of it, the cavalier way in which he employs in his daily living substances that threaten his health, the leniency of the courts in these matters (as though products as well as people enjoyed the protection of the Fifth Amendment), and the treatment of open land, rivers, and the atmosphere as if, in some medieval way they could still be regarded as disposal sinks of infinite capacity, would make you wonder, standing face to in the wind at Cape Mendocino, if we weren't bent on an errant of madness.

The geographies of North America, the myriad small landscapes that make up the national fabric, are threatened— by ignorance of what makes them unique, by utilitarian attitudes, by failure to include them in the moral universe, and by brutal disregard. A testament of minor voices can clear away an ignorance of any place, can inform us of its special qualities; but no voice, by merely telling a story, can cause the poisonous wastes that saturate some parts of the land to decompose, to evaporate. This responsibility falls ultimately to the national community, a vague and fragile entity to be sure, but one that, in America, can be ferocious in exerting its will.

Geography, the formal way in which we grapple with this areal mystery, is finally knowledge that calls up something in the land we recognize and respond to. It gives us a sense of place and a sense of community. Both are indispensable to a state of well-being, an individual's and a country's.

One afternoon on the Siuslaw River in the Coast Range of Oregon, in January, I hooked a steelhead, a sea-run trout, that told me, through the muscles of my hands and arms and shoulders, something of the nature of the thing I was calling "the Siuslaw River." Years ago I had stood under a pecan tree in Upson Country, Georgia, idly eating the nuts, when slowly it occurred to me that these nuts would taste different from pecans growing somewhere up in South Carolina. I didn't need a sharp sense of taste to know this, only to pay attention at a level no one had ever told me was necessary. One November dawn, long before the sun rose, I began a vigil at the Dumont Dunes in the Mojave Desert in California, which I kept until a few minutes after the sun broke the horizon. During that time I named to myself the colors by which the sky changed and by which the sand itself flowed like a rising tide through grays and silvers and blues into yellows, pinks, washed duns, and fallow beiges.

It is through the power of observation, the gifts of eye and ear, of tongue and nose and finger, that a place first rises up in our mind; afterward, it is memory that carries the place, that allows it to grow in depth and complexity. For as long as our records go back we have held these two things dear, landscape and memory. Each infuses us with a different kind of life. The one feeds us, figuratively and literally. The other protects us from lies and tyranny. To keep landscapes intact and the memory of them, our history in them, alive, seems as imperative a task in modern time as finding the extent to which individual expression can be accommodated, before it threatens to destroy the fabric of society.

If I were now to visit another country, I would ask my local companion, before I saw any museum or library, any factory or fabled town, to walk me in the country of his or her youth, to tell me the names of things and how, traditionally, they have been fitted together in a community. I would ask for the stories, the voice of memory over the land. I would ask about the history of storms there, the age of the trees, the winter color of the hills. Only then would I ask to see the museum. I would want first the sense of a real place, to know that I was not inhabiting an idea. I would want to know the lay of the land first, the real geography, and take some measure of the love of it in my companion before [having] stood before the painting or read works of scholarship. I would want to have something real and remembered against which I might hope to measure their truth.

NASA Readies a 'Mission to Planet Earth'

A goal is to help scientists understand and predict changes in global climate.

By WILLIAM K. STEVENS

VALLEY FORGE, PA.

In a small nondescript room in an equally ordinary low-rise structure just off the Pennsylvania Turnpike a technician named Dave Banks was speaking into a headset.

"Chamber, this is Control," he said. "We are going to start the power-up sequence." With that, Mr. Banks typed a single line of characters into his computer, setting off a cascade of commands to one of the most important scientific space probes ever to be readied for launching.

Twenty-two feet long and 10 feet in diameter, swathed in gold plastic wrap to insulate it in space, the craft is called simply AM-1. Today it is encapsulated in a 37-foot spherical vacuum chamber, where engineers and technicians are testing it in the simulated harshness of the void. But this summer, if all goes well, AM-1 will inaugurate a new era in space science.

The craft's surveillance target is not some distant world, but rather the home planet of those who built it. AM-1 is to be the flagship of a new generation of earth satellites called the Earth Observing System, or EOS, which in turn is the centerpiece of what until recently has been called Mission to Planet Earth: a 15-year effort to subject the interlinked workings of the atmosphere, oceans and land surfaces to detailed scrutiny from space.

The goal of the project, now called simply the Earth Science enterprise, is to fill important gaps in knowledge that prevent scientists from achieving an adequate understanding of the global environment.

At first, the main focus is to be on the earth's changing climate. Without the information that only satellites can glean, many scientists say, researchers will continue to be hobbled in their efforts to predict how the climate will respond to the combination of natural changes and those forced on the planet by human action—emissions of heat-trapping gases from industrial waste, for instance.

The road to this point has been long and contentious. Attacked nearly a decade ago as an egregious example of big science gone wild, of distorted scientific priorities and overblown spending, EOS today has been sweated down to a cheaper, leaner, more flexible program that is perhaps better attuned to scientists' actual needs. AM-1 is about a third the size of earlier conceptions, partly because of advancing miniaturization of parts, and the $6.6 billion that EOS itself will have cost by 2000 is about a third of what was originally to be spent.

Even so, many experts, including top officials of the National Aeronautics and Space Administration, say the $1 billion, five-ton craft, with five instruments aboard, is still too expensive, too big and complex. They say it is carrying too many scientific eggs in one vulnerable basket.

"I can tell you, this is a gripping launch," said Daniel S. Goldin, the Administrator of the space agency. "The loss of that spacecraft would be horrible; I don't want to contemplate the possibilities."

If Mr. Goldin has his way, there will never be another EOS satellite this big and complex. Although it was too late to change AM-1 when he took over the helm of the space agency in 1992, he says, the future will be different: EOS sensors will be distributed among many more smaller, cheaper, simpler satellites so that if one is lost, it will not deal such a big blow to what is widely recognized as a vital scientific undertaking.

That, among other things, has helped the program weather the criticism directed at it in the past by scientists and politicians alike. While EOS "didn't command a consensus five years ago that it was a good program, it does now," said John M. Logsdon, the director of the Space Policy Institute at George Washington University.

So the way ahead appears clear for the launching of AM-1, now scheduled for June 30. And despite the nervousness of Mr. Goldin and others, an air of excitement is building at the Lockheed Martin Astro Space company's spacecraft construction and test site here.

Putting AM-1 to the vacuum-chamber test is "the proof of the pudding," said Eugene D. Keeling, who oversees the assembly and testing of the satellite. "When you hit that, you know everything is coming together," he said.

His boss, Michael Kavka, the director of the AM-1 program, who has been involved in building and launching satellites since the 1960's, bestowed a unique compliment, saying, "This is without a doubt the nicest-looking spacecraft I've ever seen."

Actually, it has lots of angles and protuberances, as a visitor learned firsthand by standing on a ladder at the bottom of the vacuum chamber before it was drained of air and peering upward at AM-1, where technicians in white coveralls were dusting everything to keep it spotless. The angles and protuberances are the satellite's business end: five different remote-sensing instruments that will scan the earth continuously once AM-1 is launched from Vandenberg Air Force Base in California into a pole-to-pole orbit.

The instruments go by their acronymic nicknames, and their functions indicate the scientific breadth of the research program that is about to begin. They include:

- Modis (for Moderate Resolution Imaging Spectroradiometer). This is the flagship instrument aboard the flagship satellite. It is designed to provide accurate surface temperature measurements of the entire globe; ground-based thermometers leave gaps and are subject to error. Modis will also measure biological changes in the ocean, changes in land vegetation, and changes in the properties of clouds and atmospheric aerosols. The latter are tiny droplets of sulfur that reflect sunlight and minute particles that serve as the nuclei for cloud-forming water droplets.
- Ceres (Clouds and the Earth's Radiant Energy System) will measure incoming solar radiation, which drives the climate system, as well as radiation re-emitted by the earth, which, when trapped by greenhouse gases, warms the planet. The instrument will also measure cloud properties.
- Aster (Advanced Spaceborne Thermal Emission and Reflection Radiometer) will make "Zoom" images of terrestrial surface features and take finer readings of surface temperature and cloud cover.
- Misr (Multi-angle Imaging Spectroradiometer) is designed to look at the three-dimensional structure of clouds and aerosol concentrations from several angles. It will also make fine-grained images of surface features—for example, the heights of trees.
- Mopitt (Measurements of Pollution in the Troposphere) will measure atmospheric carbon monoxide, to help better understand the earth's carbon cycle, and methane, a greenhouse gas.

Japan contributed Aster to the mix, and Mopitt is a Canadian contribution. The other three instruments are American-made.

If all the instruments work as designed, scientists will get the clearest picture they have ever had of how the earth-ocean-atmosphere system functions. For instance, two of the instruments will measure aerosols and four are to measure cloud properties. Lack of data about these two features is the biggest impediment faced by researchers who use computer models to simulate and predict the global climate's behavior. Without a realistic rendition of cloud and aerosol behavior, the models' predictions about any human-induced global warming will continue to be shrouded in a frustrating haze of uncertainty.

While virtually everyone concerned agrees that the data to be provided by EOS are essential, there is also a consensus that satellites cannot do the whole job by themselves. Some data can be acquired only by measurements on the ground, at sea or in the air, with sensors aboard aircraft. In the early 1990's, EOS was severely criticized by those who believed it was drawing money away from these more prosaic means of observation. Critics back then also worried that EOS might suck up money that would otherwise support the work of scientists who would use the data.

As conceived at the start of the 1990's, EOS was to consist of an elaborate array of six 15-ton satellites, each carrying 12 sensing instruments—"Battlestar Galacticas," Mr. Goldin calls these behemoths—to be launched over a 12-year period beginning in 1998. A complementary series of smaller satellites was to be sent into orbit starting somewhat earlier. The cost of the program, including operational expenses, was project- ed at $17 billion by 2000 and $30 billion by 2020.

But the project never found solid support. When Mr. Goldin became the NASA Administrator, he set out to make the space agency's programs, "smaller, cheap- er, faster, better," and EOS was a prime target of his intended reforms. Mr. Golden brought in Dr. Charles F. Kennel, a plasma physicist at the University of California at Los Angeles, as associate administrator for Mission to Planet Earth.

Not long after Dr. Kennel arrived, the program came under attack from Congressional Republicans who charged that its purpose was to push a global-warming agenda.

EOS survived that challenge, but was redesigned in accordance with Mr. Goldin's philosophy.

The next EOS satellites will all be smaller than AM-1. One of these, set for launching in 2000, would complement AM-1 by making many of the same measurements (plus some new ones) at a different time of day. (The second satellite is to be called PM-1, because it will cross the Equator in the afternoon instead of the morning, as AM-1 will). Another satellite, to be launched in 2002, is called Chem-1 and will measure atmospheric trace gases, like water vapor, ozone and nitrous oxides, that play critical roles in regulating the earth's climate.

Before the end of this year, the latest in the highly successful series of Landsat satellites, which make images and

3. NASA Readies a 'Mission to Planet Earth'

measure properties of the earth's land areas, is also to be launched. Landsat 7, as it is called, and its successors will be folded into the EOS program. A fifth EOS satellite now scheduled for launching, Ice-sat-1, is to go into orbit in 2001 to measure the growth and shrinkage of global ice sheets.

In the meantime, a number of smaller, complementary satellites, each making a single type of measurement, are to be placed in orbit. The first, launched last November, measures tropical rainfall.

The space agency plans to hold off building EOS spacecraft for the period beyond 2001 as long as possible, both to take advantage of the newest technology and take into account any new research priorities that emerge once the first round of data has been analyzed. Smaller satellites, which can be built faster, make this flexibility possible, Mr. Goldin said.

In further pursuit of flexibility, the whole program now undergoes a full-scale reassessment every two years. As a result of the most recent review, last summer, NASA boiled down its earth-science research targets to five areas: seasonal-to-annual climate variability; climate change over years and decades; atmospheric chemistry and ozone; natural hazards like volcanoes, earthquakes, floods and hurricanes; and, changes in land use and land cover. All told, EOS is ultimately to make 24 separate kinds of measurements.

Mr. Goldin said the scaling-down has freed money to support the research of scientists who will use the EOS data. From about $136 million provided in the fiscal 1997 budget, he said, the total is scheduled to rise gradually until it approaches $200 million in 2002.

Now everyone is totally happy with the balance in resources being allocated to satellites and the equally important conventional measurements. "It is a continuing challenge to make sure we're doing both" adequately, said Mr. Michael MacCracken, an official of the United States Global Change Research Program, the Federal office under which Mission to Planet Earth functions.

Over all, the program "has improved enormously," said Dr. Thomas M. Donahue, a professor of planetary science and physics at the University of Michigan, an early critic of what he called the original "humongous" EOS spacecraft. Criticism from scientists has died down, he said. "I don't hear a lot of complaining now," he said.

Conservative Republicans in Congress say the program is still too big and should rely more on the private sector to perform remote sensing. Some say EOS remains a creature of Democrats and environmentalists.

Its "driving force" is the desire to "prove global warming," said Representative Dana Rohrabacher, a California Republican. Mr. Rohrabacher, who heads the space and aeronautics subcommittee of the House Science Committee, said the space agency should rely more on the private sector and on smaller and cheaper satellites.

NASA officials insist that science alone drives the program. Far from promoting any point of view about global warming, Dr. Kennel said, NASA wants to contribute information and be an "honest broker" in the debate on climate change.

Still, Mr. Goldin said that "we are really listening" to the Congressional critics. "We value criticism at NASA," he said. Mr. Rohrabacher responded, "I think he has heard me."

Dr. Kennel, for his part, said he believed that the worse of EOS's problems were behind it. But while the program is not in mortal danger, he said, it will always be under pressure from Congress and others.

"I think that pressure is healthy," he said.

The Earth Observing System will use satellites to collect data about the earth, like these images showing a short-term shrinking of the Sahara. (NASA)

Human Domination of Earth's Ecosystems

Peter M. Vitousek, Harold A. Mooney, Jane Lubchenco, Jerry M. Melillo

Human alteration of Earth is substantial and growing. Between one-third and one-half of the land surface has been transformed by human action; the carbon dioxide concentration in the atmosphere has increased by nearly 30 percent since the beginning of the Industrial Revolution; more atmospheric nitrogen is fixed by humanity than by all natural terrestrial sources combined; more than half of all accessible surface fresh water is put to use by humanity; and about one-quarter of the bird species on Earth have been driven to extinction. By these and other standards, it is clear that we live on a human-dominated planet.

All organisms modify their environment, and humans are no exception. As the human population has grown and the power of technology has expanded, the scope and nature of this modification has changed drastically. Until recently, the term "human-dominated ecosystems" would have elicited images of agricultural fields, pastures, or urban landscapes; now it applies with greater or lesser force to all of Earth. Many ecosystems are dominated directly by humanity, and no ecosystem on Earth's surface is free of pervasive human influence.

This article provides an overview of human effects on Earth's ecosystems. It is not intended as a litany of environmental disasters, though some disastrous situations are described; nor is it intended either to downplay or to celebrate environmental successes, of which there have been many. Rather, we explore how large humanity looms as a presence on the globe—how, even on the grandest scale, most aspects of the structure and functioning of Earth's ecosystems cannot be understood without accounting for the strong, often dominant influence of humanity.

We view human alterations to the Earth system as operating through the interacting processes summarized in Fig. 1. The growth of the human population, and growth in the resource base used by humanity, is maintained by a suite of human enterprises such as agriculture, industry, fishing, and international commerce. These enterprises transform the land surface (through cropping, forestry, and urbanization), alter the major biogeochemical cycles, and add or remove species and genetically distinct populations in most of Earth's ecosystems. Many of these changes are substantial and reasonably well quantified; all are ongoing. These relatively well-documented changes in turn entrain further alterations to the functioning of the Earth system, most notably by driving global climatic change (1) and causing irreversible losses of biological diversity (2).

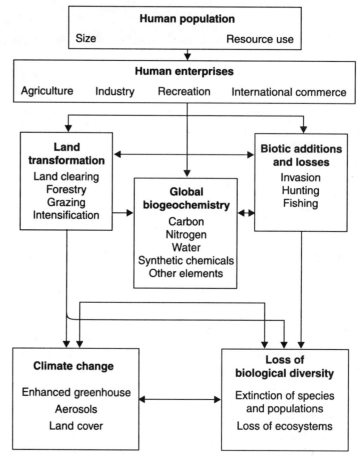

Fig. 1 A conceptual model illustrating humanity's direct and indirect effects on the Earth system [modified from (56)].

P. M. Vitousek and H. A. Mooney are in the Department of Biological Sciences, Stanford University, Stanford, CA 94305, USA. J. Lubchenco is in the Department of Zoology, Oregon State University, Corvallis, OR 97331, USA. J. M. Melillo is at the U.S. Office of Science and Technology Policy, Old Executive Office Building, Room 443, Washington, DC 20502, USA.

4. Human Dominations of Earth's Ecosystems

Land Transformation

The use of land to yield goods and services represents the most substantial human alteration of the Earth system. Human use of land alters the structure and functioning of ecosystems, and it alters how ecosystems interact with the atmosphere, with aquatic systems, and with surrounding land. Moreover, land transformation interacts strongly with most other components of global environmental change.

The measurement of land transformation on a global scale is challenging; changes can be measured more or less straightforwardly at a given site, but it is difficult to aggregate these changes regionally and globally. In contrast to analyses of human alteration of the global carbon cycle, we cannot install instruments on a tropical mountain to collect evidence of land transformation. Remote sensing is a most useful technique, but only recently has there been a serious scientific effort to use high-resolution civilian satellite imagery to evaluate even the more visible forms of land transformation, such as deforestation, on continental to global scales (3).

Land transformation encompasses a wide variety of activities that vary substantially in their intensity and consequences. At one extreme, 10 to 15% of Earth's land surface is occupied by row-crop agriculture or by urban-industrial areas, and another 6 to 8% has been converted to pastureland (4); these systems are wholly changed by human activity. At the other extreme, every terrestrial ecosystem is affected by increased atmospheric carbon dioxide (CO_2), and most ecosystems have a history of hunting and other low-intensity resource extraction. Between these extremes lie grassland and semiarid ecosystems that are grazed (and sometimes degraded) by domestic animals, and forests and woodlands from which wood products have been harvested; together, these represent the majority of Earth's vegetated surface.

The variety of human effects on land makes any attempt to summarize land transformations globally a matter of semantics as well as substantial uncertainty. Estimates of the fraction of land transformed or degraded by humanity (or its corollary, the fraction of the land's biological production that is used or dominated) fall in the range of 39 to 50% (5) (Fig. 2). These numbers have large uncertainties, but the fact that they are large is not at all uncertain. Moreover, if anything these estimates understate the global impact of land transformation, in that land that has not been transformed often has been divided into fragments by human alteration of the surrounding areas. This fragmentation affects the species composition and functioning of otherwise little modified ecosystems (6).

Overall, land transformation represents the primary driving force in the loss of biological diversity worldwide. Moreover, the effects of land transformation extend far beyond the boundaries of transformed lands. Land transformation can affect climate directly at local and even regional scales. It contributes ~20% to current anthropogenic CO_2 emissions, and more substantially to the increasing concentrations of the greenhouse gases methane and nitrous oxide; fires associated with it alter the reactive chemistry of the troposphere, bringing elevated carbon monoxide concentrations and episodes of urban-like photochemical air pollution to remote tropical areas of Africa and South America; and it causes runoff of sediment and nutrients that drive substantial changes in stream, lake, estuarine, and coral reef ecosystems (7–10).

The central importance of land transformation is well recognized within the community of researchers concerned with global environmental change. Several research programs are focused on aspects of it (9, 11); recent and substantial progress toward understanding these aspects has been made (3), and much more progress can be anticipated. Understanding land transformation is a difficult challenge; it requires integrating the social, economic, and cultural causes of land transformation with evaluations of its biophysical nature and consequences. This interdisciplinary approach is essential to predicting the course, and to any hope of affecting the consequences, of human-caused land transformation.

Oceans

Human alterations of marine ecosystems are more difficult to quantify than those of terrestrial ecosystems, but several kinds of information suggest that they are substantial. The human population is concentrated near coasts—about 60% within 100 km—and the oceans' productive coastal margins have been affected strongly by humanity. Coastal wetlands that mediate interactions between land and sea have been altered over large areas; for example, approximately 50% of mangrove ecosystems globally have been transformed or destroyed by human activity (12). Moreover, a recent analysis suggested that although humans use about 8% of the primary production of the oceans, that fraction grows to more than 25% for upwelling areas and to 35% for temperate continental shelf systems (13).

Many of the fisheries that capture marine productivity are focused on top predators, whose removal can alter marine ecosystems out of proportion to their abundance. Moreover, many such fisheries have proved to be unsustainable, at least at our present level of knowledge and control. As of 1995, 22% of recognized marine fisheries were overexploited or already depleted, and 44% more were at their limit of exploitation (14) (Figs. 2 and 3). The consequences of fisheries are not restricted to their target organisms; commercial marine fisheries around the world discard 27 million tons of nontarget animals annually, a quantity nearly one-third as large as total landings (15). Moreover, the dredges and trawls used in some fisheries damage habitats substantially as they are dragged along the sea floor.

A recent increase in the frequency, extent, and duration of harmful algal blooms in coastal areas (16) suggests that human activity has affected the base as well as the top of marine food chains. Harmful algal blooms

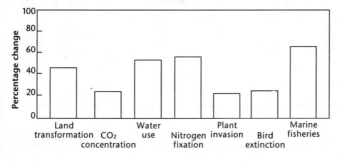

Fig. 2. Human dominance or alteration of several major components of the Earth system, expressed as (from left to right) percentage of the land surface transformed (5); percentage of the current atmospheric CO_2 concentration that results from human action (17); percentage of accessible surface fresh water used (20); percentage of terrestrial N fixation that is human-caused (28); percentage of plant species in Canada that humanity has introduced from elsewhere (48); percentage of bird species on Earth that have become extinct in the past two millennia, almost all of them as a consequence of human activity (42); and percentage of major marine fisheries that are fully exploited, overexploited, or depleted (14).

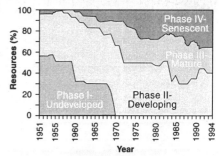

Fig. 3. Percentage of major world marine fish resources in different phases of development, 1951 to 1994 [from (57)]. Undeveloped = a low and relatively constant level of catches; developing = rapidly increasing catches; mature = a high and plateauing level of catches; senescent = catches declining from higher levels.

are sudden increases in the abundance of marine phytoplankton that produce harmful structures or chemicals. Some but not all of these phytoplankton are strongly pigmented (red or brown tides). Algal blooms usually are correlated with changes in temperature, nutrients, or salinity; nutrients in coastal waters, in particular, are much modified by human activity. Algal blooms can cause extensive fish kills through toxins and by causing anoxia; they also lead to paralytic shellfish poisoning and amnesic shellfish poisoning in humans. Although the existence of harmful algal blooms has long been recognized, they have spread widely in the past two decades (16).

Alterations of the Biogeochemical Cycles

Carbon. Life on Earth is based on carbon, and the CO_2 in the atmosphere is the primary resource for photosynthesis. Humanity adds CO_2 to the atmosphere by mining and burning fossil fuels, the residue of life from the distant past, and by converting forests and grasslands to agricultural and other low-biomass ecosystems. The net result of both activities is that organic carbon from rocks, organisms, and soils is released into the atmosphere as CO_2.

The modern increase in CO_2 represents the clearest and best documented signal of human alteration of the Earth system. Thanks to the foresight of Roger Revelle, Charles Keeling, and others who initiated careful and systematic measurements of atmospheric CO_2 in 1957 and sustained them through budget crises and changes in scientific fashions, we have observed the concentration of CO_2 as it has increased steadily from 315 ppm to 362 ppm. Analysis of air bubbles extracted from the Antarctic and Greenland ice caps extends the record back much further; the CO_2 concentration was more or less stable near 280 ppm for thousands of years until about 1800, and has increased exponentially since then (17).

There is no doubt that this increase has been driven by human activity, today primarily by fossil fuel combustion. The sources of CO_2 can be traced isotopically; before the period of extensive nuclear testing in the atmosphere, carbon depleted in ^{14}C was a specific tracer of CO_2 derived from fossil fuel combustion, whereas carbon depleted in ^{13}C characterized CO_2 from both fossil fuels and land transformation. Direct measurements in the atmosphere, and analyses of carbon isotopes in tree rings, show that both ^{13}C and ^{14}C in CO_2 were diluted in the atmosphere relative to ^{12}C as the CO_2 concentration in the atmosphere increased.

Fossil fuel combustion now adds 5.5 ± 0.5 billion metric tons of CO_2-C to the atmosphere annually, mostly in economically developed regions of the temperate zone (18) (Fig. 4). The annual accumulation of CO_2-C has averaged 3.2 ± 0.2 billion metric tons recently (17). The other major terms in the atmospheric carbon balance are net ocean-atmosphere flux, net release of carbon during land transformation, and net storage in terrestrial biomass and soil organic matter. All of these terms are smaller and less certain than fossil fuel combustion or annual atmospheric accumulation; they represent rich areas of current research, analysis, and sometimes contention.

The human-caused increase in atmospheric CO_2 already represents nearly a 30% change relative to the pre-industrial era (Fig. 2), and CO_2 will continue to increase for the foreseeable future. Increased CO_2 represents the most important human enhancement to the greenhouse effect; the consensus of the climate research community is that it probably already affects climate detectably and will drive substantial climate change in the next century (1). The direct effects of increased CO_2 on plants and ecosystems may be even more important. The growth of most plants is enhanced by elevated CO_2, but to very different extents; the tissue chemistry

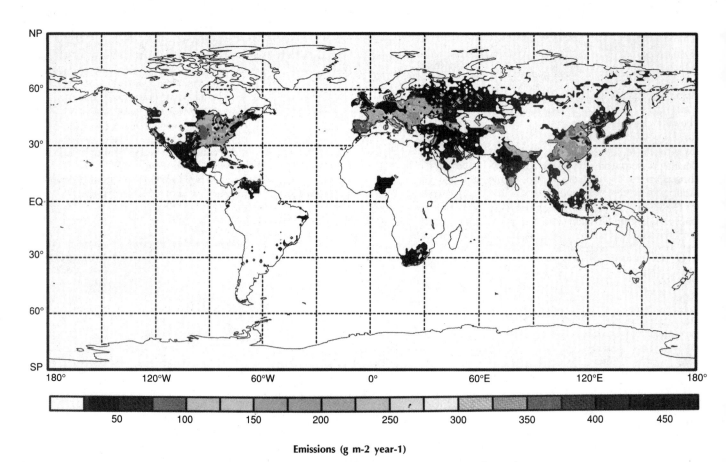

Fig. 4 Goegraphical distribution of fossil fuel sources of CO_2 as of 1990. The global mean is 12.2 g m^{-2} year^{-1}; most emissions occur in economically developed regions of the north temperate zone. EQ, equator; NP, North Pole; SP, South Pole. [Prepared by A.S. Denning, from information in (18)]

of plants that respond to CO_2 is altered in ways that decrease food quality for animals and microbes; and the water use efficiency of plants and ecosystems generally is increased. The fact that increased CO_2 affects species differentially means that it is likely to drive substantial changes in the species composition and dynamics of all terrestrial ecosystems (19).

Water. Water is essential to all life. Its movement by gravity, and through evaporation and condensation, contributes to driving Earth's biogeochemical cycles and to controlling its climate. Very little of the water on Earth is directly usable by humans; most is either saline or frozen. Globally, humanity now uses more than half of the runoff water that is fresh and reasonably accessible, with about 70% of this use in agriculture (20) (Fig. 2). To meet increasing demands for the limited supply of fresh water, humanity has extensively altered river systems through diversions and impoundments. In the United States only 2% of the rivers run unimpeded, and by the end of this century the flow of about two-thirds of all of Earth's rivers will be regulated (21). At present, as much as 6% of Earth's river runoff is evaporated as a consequence of human manipulations (22). Major rivers, including the Colorado, the Nile, and the Ganges, are used so extensively that little water reaches the sea. Massive inland water bodies, including the Aral Sea and Lake Chad, have been greatly reduced in extent by water diversions for agriculture. Reduction in the volume of the Aral Sea resulted in the demise of native fishes and the loss of other biota; the loss of a major fishery; exposure of the salt-laden sea bottom, thereby providing a major source of windblown dust; the production of a drier and more continental local climate and a decrease in water quality in the general region; and an increase in human diseases (23).

Impounding and impeding the flow of rivers provides reservoirs of water that can be used for energy generation as well as for agriculture. Waterways also are managed for transport, for flood control, and for the dilution of chemical wastes. Together, these activities have altered Earth's freshwater ecosystems profoundly, to a greater extent than terrestrial ecosystems have been altered. The construction of dams affects biotic habitats indirectly as well; the damming of the Danube River, for example, has altered the silica chemistry of the entire Black Sea. The large number of operational dams (36,000) in the world, in conjunction with the many that are planned, ensure that humanity's effects on aquatic biological systems will continue (24). Where surface water is sparse or over-exploited, humans use groundwater—and in many areas the groundwater that is drawn upon is nonrenewable, or fossil, water (25). For example, three-quarters of the water supply of Saudi Arabia currently comes from fossil water (26).

Alterations to the hydrological cycle can affect regional climate. Irrigation increases atmospheric humidity in semiarid areas, often increasing precipitation and thunderstorm frequency (27). In contrast, land transformation from forest to agriculture or pasture increases albedo and decreases surface roughness; simulations suggest that the net effect of this transformation is to increase temperature and decrease precipitation regionally (7, 26).

Conflicts arising from the global use of water will be exacerbated in the years ahead, with a growing human population and with the stresses that global changes will impose on water quality and availability. Of all of the environmental security issues facing nations, an adequate supply of clean water will be the most important.

Nitrogen. Nitrogen (N) is unique among the major elements required for life, in that its cycle includes a vast atmospheric reservoir (N_2) that must be fixed (combined with carbon, hydrogen, or oxygen) before it can be used by most organisms. The supply of this fixed N controls (at least in part) the productivity, carbon storage, and species composition of many ecosystems. Before the extensive human alteration of the N cycle, 90 to 130 million metric tons of N (Tg N) were fixed biologically on land each year; rates of biological fixation in marine systems are less certain, but perhaps as much was fixed there (28).

Human activity has altered the global cycle of N substantially by fixing N_2—deliberately for fertilizer and inadvertently during fossil fuel combustion. Industrial fixation of N fertilizer increased from <10 Tg/year in 1950 to 80 Tg/year in 1990; after a brief dip caused by economic dislocations in the former Soviet Union, it is expected to increase to >135 Tg/year by 2030 (29). Cultivation of soybeans, alfalfa, and other legume crops that fix N symbiotically enhances fixation by another ~40 Tg/year, and fossil fuel combustion puts >20 Tg/year of reactive N into the atmosphere globally—some by fixing N_2, more from the mobilization of N in the fuel. Overall, human activity adds at least as much fixed N to terrestrial ecosystems as do all natural sources combined (Fig. 2), and it mobilizes >50 Tg/year more during land transformation (28, 30).

Alteration of the N cycle has multiple consequences. In the atmosphere, these include (i) an increasing concentration of the greenhouse gas nitrous oxide globally; (ii) substantial increases in fluxes of reactive N gases (two-thirds or more of both nitric oxide and ammonia emissions globally are human-caused); and (iii) a substantial contribution to acid rain and to the photochemical smog that afflicts urban and agricultural areas throughout the world (31). Reactive N that is emitted to the atmosphere is deposited downwind, where it can influence the dynamics of recipient ecosystems. In regions where fixed N was in short supply, added N generally increases productivity and C storage within ecosystems, and ultimately increases losses of N and cations from soils, in a set of processes termed "N saturation" (32). Where added N increases the productivity of ecosystems, usually it also decreases their biological diversity (33).

Human-fixed N also can move from agriculture, from sewage systems, and from N-saturated terrestrial systems to streams, rivers, groundwater, and ultimately the oceans. Fluxes of N through streams and rivers have increased markedly as human alteration of the N cycle has accelerated; river nitrate is highly correlated with the human population of river basins and with the sum of human-caused N inputs to those basins (8). Increases in river N drive the eutrophication of most estuaries, causing blooms of nuisance and even toxic algae, and threatening the sustainability of marine fisheries (16, 34).

Other cycles. The cycles of carbon, water, and nitrogen are not alone in being altered by human activity. Humanity is also the largest source of oxidized sulfur gases in the atmosphere; these affect regional air quality, biogeochemistry, and climate. Moreover, mining and mobilization of phosphorus and of many metals exceed their natural fluxes; some of the metals that are concentrated and mobilized are highly toxic (including lead, cadmium, and mercury) (35). Beyond any doubt, humanity is a major biogeochemical force on Earth.

Synthetic organic chemicals. Synthetic organic chemicals have brought humanity many beneficial services. However, many are toxic to humans and other species, and some are hazardous in concentrations as low as 1 part per billion. Many chemicals persist in the environment for decades; some are both toxic and persistent. Long-lived organochlorine compounds provide the clearest examples of environmental consequences of persistent compounds. Insecticides such as DDT and its relatives, and industrial compounds like polychlorinated biphenyls (PCBs), were used widely in North America in the 1950s and 1960s. They were transported globally, accumulated in organisms, and magnified in concentration through food chains; they devastated populations of some predators (notably falcons and eagles) and entered parts of the human food supply in concentrations higher than was prudent. Domestic use of these compounds was phased out in the 1970s in the United States and Canada, and their concentrations declined thereafter. However, PCBs in particular remain readily detectable in many organisms, sometimes approaching thresholds of public health concern (36). They will continue to circulate through organisms for many decades.

Synthetic chemicals need not be toxic to cause environmental problems. The fact that the persistent and volatile chlorofluorocarbons (CFCs) are wholly nontoxic contributed to their widespread use as refrigerants and even aerosol propellants. The subsequent discovery that CFCs drive the breakdown of stratospheric ozone, and especially the later discovery of the Antarctic ozone hole and their role in it, represent great surprises in global environmental science (37). Moreover, the response of the international political system to those discoveries is the best extant illustration that global environmental change can be dealt with effectively (38).

Particular compounds that pose serious health and environmental threats can be and often have been phased out (although PCB production is growing in Asia). Nonetheless, each year the chemical industry produces more than 100 million tons of organic chemicals representing some 70,000 different compounds, with about 1000 new ones being added annually (39). Only a small fraction of the many chemicals produced and released into the environment are tested adequately for health hazards or environmental impact (40).

Biotic Changes

Human modification of Earth's biological resources—its species and genetically distinct populations—is substantial and growing. Extinction is a natural process, but the current rate of loss of genetic variability, of populations, and of species is far above background rates; it is ongoing; and it represents a wholly irreversible global change. At the same time, human transport of species around Earth is homogenizing Earth's biota, introducing many species into new areas where they can disrupt both natural and human systems.

Losses. Rates of extinction are difficult to determine globally, in part because the majority of species on Earth have not yet been identified. Nevertheless, recent calculations suggest that rates of species extinction are now on the order of 100 to 1000 times those before humanity's dominance of Earth (41). For particular well-known groups, rates of loss are even greater; as many as one-quarter of Earth's bird species have been driven to extinction by human activities over the past two millennia, particularly on oceanic islands (42) (Fig. 2). At present, 11% of the remaining birds, 18% of the mammals, 5% of fish, and 8% of plant species on Earth are threatened with extinction (43). There has been a disproportionate loss of large mammal species because of hunting; these species played a dominant role in many ecosystems, and their loss has resulted in a fundamental change in the dynamics of those systems (44), one that could lead to further extinctions. The largest organisms in marine systems have been affected similarly, by fishing and whaling. Land transformation is the single most important cause of extinction, and current rates of land transformation eventually will drive many more species to extinction, although with a time lag that masks the true dimensions of the crisis (45). Moreover, the effects of other components of global environmental change—of altered carbon and nitrogen cycles, and of anthropogenic climate change—are just beginning.

As high as they are, these losses of species understate the magnitude of loss of genetic variation. The loss to land transformation of locally adapted populations within species, and of genetic material within populations, is a human-caused change that reduces the resilience of species and ecosystems while precluding human use of the library of natural products and genetic material that they represent (46).

Although conservation efforts focused on individual endangered species have yielded some successes, they are expensive—and the protection or restoration of whole ecosystems often represents the most effective way to sustain genetic, population, and species diversity. Moreover, ecosystems themselves may play important roles in both natural and human-dominated landscapes. For example, mangrove ecosystems protect coastal areas from erosion and provide nurseries for offshore fisheries, but they are threatened by land transformation in many areas.

Invasions. In addition to extinction, humanity has caused a rearrangement of Earth's biotic systems, through the mixing of floras and faunas that had long been isolated geographically. The magnitude of transport of species, termed "biological invasion," is enormous (47); invading species are present almost everywhere. On many islands, more than half of the plant species are nonindigenous, and in many continental areas the figure is 20% or more (48) (Fig. 2).

As with extinction, biological invasion occurs naturally—and as with extinction, human activity has accelerated its rate by orders of magnitude. Land transformation interacts strongly with biological invasion, in that human-altered ecosystems generally provide the primary foci for invasions, while in some cases land transformation itself is driven by biological invasions (49). International commerce is also a primary cause of the breakdown of biogeographic barriers; trade in live organisms is massive and global, and many other organisms are inadvertently taken along for the ride. In freshwater systems, the combination of upstream land transformation, altered hydrology, and numerous deliberate and accidental species introductions has led to particularly widespread invasion, in continental as well as island ecosystems (50).

In some regions, invasions are becoming more frequent. For example, in the San Francisco Bay of California, an average of one new species has been established every 36 weeks since 1850, every 24 weeks since 1970, and every 12 weeks for the last decade (51). Some introduced species quickly become invasive over large areas (for example, the Asian clam in the San Francisco Bay), whereas others become widespread only after a lag of decades, or even over a century (52).

Many biological invasions are effectively irreversible; once replicating biological material is released into the environment and becomes successful there, calling it back is difficult and expensive at best. Moreover, some species introductions have consequences. Some degrade human health and that of other species; after all, most infectious diseases are invaders over most of their range. Others have caused economic losses amounting to billions of dollars; the recent invasion of North America by the zebra mussel is a well-publicized example. Some disrupt ecosystem processes, altering the structure and functioning of whole ecosystems. Finally, some invasions drive losses in the biological diversity of native species and populations; after land transformation, they are the next most important cause of extinction (53).

Conclusions

The global consequences of human activity are not something to face in the future—as Fig. 2 illustrates, they are with us now. All of these changes are ongoing, and in many cases accelerating; many of them were entrained long before their importance was recognized. Moreover, all of these seemingly disparate phenomena trace to a single cause—the growing scale of the human enterprise. The rates, scales, kinds, and combinations of changes occurring now are fundamentally different from those at any other time in history; we are changing Earth more rapidly than we are understanding it. We live on a human-dominated planet—and the momentum of human population growth, together with the imperative for further economic development in most of the world, ensures that our dominance will increase.

The papers in this special section summarize our knowledge of and provide specific policy recommendations concerning major human-dominated ecosystems. In addition, we suggest that the rate and extent of human alteration of Earth should affect how we think about Earth. It is clear that we control much of Earth, and that our activities affect the rest. In a very real sense, the world is in our hands—and how we handle it will determine its composition and dynamics, and our fate.

Recognition of the global consequences of the human enterprise suggests three complementary directions. First, we can work to reduce the rate at which we alter the Earth system. Humans and human-dominated systems may be able to adapt to slower change, and ecosystems and the species they support may cope more effectively with the changes we impose, if those changes are slow. Our footprint on the planet (54) might then be stabilized at a point where enough space and resources remain to sustain most of the other species on Earth, for their sake and our own. Reducing the rate of growth in human effects on Earth involves slowing human population growth and using resources as efficiently as is practical. Often it is the waste products and by-products of human activity that drive global environmental change.

Second, we can accelerate our efforts to understand Earth's ecosystems and how they interact with the numerous components of human-caused global change. Ecological research is inherently complex and demanding: It requires measurement and monitoring of populations and ecosystems; experimental studies to elucidate the regulation of ecological processes; the development, testing, and validation of regional and global models; and integration with a broad range of biological, earth, atmospheric, and marine sciences. The challenge of understanding a human-dominated planet further requires that the human dimensions of global change—the social, economic, cultural, and other drivers of human actions—be included within our analyses.

Finally, humanity's dominance of Earth means that we cannot escape responsibility for managing the planet. Our activities are causing rapid, novel, and substantial changes to Earth's ecosystems. Maintaining populations, species, and ecosystems in the face of those changes, and maintaining the flow of goods and services they provide humanity (55), will require active management for the foreseeable future. There is no clearer illustration of the extent of human dominance of Earth than the fact that maintaining the diversity of "wild" species and the functioning of "wild" ecosystems will require increasing human involvement.

REFERENCES AND NOTES

1. Intergovernmental Panel on Climate Change, *Climate Change 1995* (Cambridge Univ. Press, Cambridge, 1996), pp. 9–49.
2. United Nations Environment Program, *Global Biodiversity Assessment*, V. H. Heywood, Ed. (Cambridge Univ. Press, Cambridge, 1995).
3. D. Skole and C. J. Tucker, *Science* **260**, 1905 (1993).
4. J. S. Olson, J. A. Watts, L. J. Allison, *Carbon in Live Vegetation of Major World Ecosystems* (Office of Energy Research, U.S. Department of Energy, Washington, DC, 1983).
5. P. M. Vitousek, P. R. Ehrlich, A. H. Ehrlich, P. A. Matson, *Bioscience* **36**, 368 (1986); R. W. Kates, B. L. Turner, W. C. Clark, in (35), pp. 1–17; G. C. Daily, *Science* **269**, 350 (1995).
6. D. A. Saunders, R. J. Hobbs, C. R. Margules, *Conserv. Biol.* **5**, 18 (1991).
7. J. Shukla, C. Nobre, P. Sellers, *Science* **247**, 1322 (1990).
8. R. W. Howarth *et al.*, *Biogeochemistry* **35**, 75 (1996).
9. W. B. Meyer and B. L. Turner II, *Changes in Land Use and Land Cover: A Global Perspective* (Cambridge Univ. Press, Cambridge, 1994).
10. S. R. Carpenter, S. G. Fisher, N. B. Grimm, J. F. Kitchell, *Annu. Rev. Ecol. Syst.* **23**, 119 (1992); S. V. Smith and R. W. Buddemeier, *ibid.*, p. 89; J. M. Melillo, I. C. Prentice, G. D. Farquhar, E.-D. Schulze, O. E. Sala, in (1), pp. 449–481.
11. R. Leemans and G. Zuidema, *Trends Ecol. Evol.* **10**, 76 (1995).
12. World Resources Institute, *World Resources 1996–1997* (Oxford Univ. Press, New York, 1996).
13. D. Pauly and V. Christensen, *Nature* **374**, 257 (1995).
14. Food and Agricultural Organization (FAO), *FAO Fisheries Tech. Pap. 335* (1994).
15. D. L. Alverson, M. H. Freeberg. S. A. Murawski, J. G. Pope, *FAO Fisheries Tech. Pap. 339* (1994).
16. G. M. Hallegraeff, *Phycologia* **32**, 79 (1993).
17. D. S. Schimel *et al.*, in *Climate Change 1994: Radiative Forcing of Climate Change*, J. T. Houghton *et al.*, Eds. (Cambridge Univ. Press, Cambridge, 1995), pp. 39–71.
18. R. J. Andres, G. Marland, I. Y. Fung, E. Matthews, *Global Biogeochem. Cycles* **10**, 419 (1996).
19. G. W. Koch and H. A. Mooney, *Carbon Dioxide and Terrestrial Ecosystems* (Academic Press, San Diego, CA, 1996); C. Körner and F. A. Bazzaz, *Carbon Dioxide, Populations, and Communities* (Academic Press, San Diego, CA, 1996).
20. S. L. Postel, G. C. Daily, P. R. Ehrlich, *Science* **271**, 785 (1996).
21. J. N. Abramovitz, *Imperiled Waters, Impoverished Future: The Decline of Freshwater Ecosystems* (Worldwatch Institute, Washington, DC, 1996).
22. M. I. L'vovich and G. F. White, in (35), pp. 235–252; M. Dynesius and C. Nilsson, *Science* **266**, 753 (1994).
23. P. Micklin, *Science* **241**, 1170 (1988); V. Kotlyakov, *Environment* **33**, 4 (1991).
24. C. Humborg, V. Ittekkot, A. Cociasu, B. Bodungen, *Nature* **386**, 385 (1997).
25. P. H. Gleick, Ed., *Water in Crisis* (Oxford Univ. Press, New York, 1993).
26. V. Gornitz, C. Rosenzweig, D. Hillel, *Global Planet. Change* **14**, 147 (1997).
27. P. C. Milly and K. A. Dunne, *J. Clim.* **7**, 506 (1994).
28. J. N. Galloway, W. H. Schlesinger, H. Levy II, A. Michaels, J. L. Schnoor, *Global Biogeochem. Cycles* **9**, 235 (1995).
29. J. N. Galloway, H. Levy II, P. S. Kasibhatla, *Ambio* **23**, 120 (1994).
30. V. Smil, in (35), pp. 423–436.
31. P. M. Vitousek *et al., Ecol. Appl.*, in press.
32. J. D. Aber, J. M. Melillo, K. J. Nadelhoffer, J. Pastor, R. D. Boone, *ibid.* **1**, 303 (1991).
33. D. Tilman, *Ecol. Monogr.* **57**, 189 (1987).
34. S. W. Nixon et al., *Biogeochemistry* **35**, 141 (1996).
35. B. L. Turner II *et al.*, Eds., *The Earth As Transformed by Human Action* (Cambridge Univ. Press, Cambridge, 1990).
36. C. A. Stow, S. R. Carpenter, C. P. Madenjian, L. A. Eby, L. J. Jackson, *Bioscience* **45**, 752 (1995).
37. F. S. Rowland, *Am. Sci.* **77**, 36 (1989); S. Solomon, *Nature* **347**, 347 (1990).
38. M. K. Tolba *et al.*, Eds., *The World Environment 1972–1992* (Chapman & Hall, London, 1992).
39. S. Postel, *Defusing the Toxics Threat: Controlling Pesticides and Industrial Waste* (Worldwatch Institute, Washington, DC, 1987).
40. United Nations Environment Program (UNEP). *Saving Our Planet—Challenges and Hopes* (UNEP, Nairobi, 1992).
41. J. H. Lawton and R. M. May, Eds., *Extinction Rates* (Oxford Univ. Press, Oxford, 1995); S. L. Pimm, G. J. Russell, J. L. Gittleman, T. Brooks, *Science* **269**, 347 (1995).
42. S. L. Olson, in *Conservation for the Twenty-First Century*, D. Western and M. C. Pearl, Eds. (Oxford Univ. Press, Oxford, 1989), p. 50; D. W. Steadman, *Science* **267**, 1123 (1995).
43. R. Barbault and S. Sastrapradja. in (2), pp. 193–274.
44. R. Dirzo and A. Miranda, in *Plant-Animal Interactions*, P. W. Price, T. M. Lewinsohn, W. Fernandes, W. W. Benson, Eds. (Wiley Interscience, New York, 1991), p. 273.
45. D. Tilman, R. M. May, C. Lehman, M. A. Nowak, *Nature* **371**, 65 (1994).
46. H. A. Mooney, J. Lubchenco, R. Dirzo, O. E. Sala, in (2). pp. 279–325.
47. C. Elton, *The Ecology of Invasions by Animals and Plants* (Methuen, London, 1958); J. A. Drake *et al.*, Eds., *Biological Invasions. A Global Perspective* (Wiley, Chichester, UK, 1989).
48. M. Rejmanek and J. Randall, *Madrono* **41**, 161 (1994).
49. C. M. D'Antonio and P. M. Vitousek, *Annu. Rev. Ecol. Syst.* **23**, 63 (1992).
50. D. M. Lodge, *Trends Ecol. Evol.* **8**, 133 (1993).
51. A. N. Cohen and J. T. Carlton, *Biological Study: Nonindigenous Aquatic Species in a United States Estuary. A Case Study of the Biological Invasions of the San Francisco Bay and Delta* (U.S. Fish and Wildlife Service, Washington, DC, 1995).
52. I. Kowarik, in *Plant Invasions—General Aspects and Special Problems*, P. Pysek, K. Prach, M. Rejmánek, M. Wade, Eds. (SPB Academic, Amsterdam, 1995), p. 15.
53. P. M. Vitousek, C. M. D'Antonio, L. L. Loope, R. Westbrooks, *Am. Sci.* **84**, 468 (1996).
54. W. E. Rees and M. Wackernagel, in *Investing in Natural Capital: The Ecological Economics Approach to Sustainability*, A. M. Jansson, M. Hammer, C. Folke, R. Costanza, Eds. (Island, Washington, DC, 1994).
55. G. C. Daily, Ed., *Nature's Services* (Island, Washington, DC, 1997).
56. J. Lubchenco *et al.*, *Ecology*, **72**, 371 (1991), P. M. Vitousek, *ibid.* **75**, 1861 (1994).
57. S. M. Garcia and R. Grainger. *FAO Fisheries Tech. Pap. 359* (1996).
58. We thank G. C. Daily, C. B. Field, S. Hobbie, D. Gordon, P.A. Matson, and R. L. Naylor for constructive comments on this paper, A. S. Denning and S. M. Garcia for assistance with illustrations, and C. Nakashima and B. Lilley for preparing text and figures for publication.

Counting the Cost

The Growing Role of Economics in Environmental Decisionmaking

By Paul R. Portney

Debates over environmental policy questions have long reflected economic concerns to some extent. When the Clean Air and Clean Water Acts were being debated in the early 1970s, for instance, there were at least perfunctory discussions of the possible effects of regulation on prices, profits, and plant closings. It is only recently, however, that economic issues have come to play a major role in environmental decisionmaking. This article will illustrate this important development and explain why the role of economics has grown. The first step, however, is to briefly examine the history of environmental policy in the United States, along with the reasons why economics initially played such a limited role.

Environmental Policy over Time

The U.S. environmental movement of the 1960s and 1970s had many parents, but Rachel Carson was surely one of them. Her books, particularly *Silent Spring,* alerted people to the unintended but very serious consequences of using the pesticides and herbicides that were such a boon to agriculture. In addition to her writings, three dramatic incidents that occurred in the late 1960s helped to galvanize public opinion about the environment. These influences led to the creation of the Council on Environmental Quality in 1969 and the Environmental Protection Agency (EPA) in 1970, as well as to the passage of numerous laws that still affect businesses, lower levels of government, and ordinary citizens.

The first of the incidents that fostered a new environmental awareness in the United States was the blowout at an oil platform in the Santa Barbara channel, which led to the oiling of considerable areas of beach. As it happened, this accident was one of the first environmental incidents to be widely covered on television. It provoked outrage on the part of the public, who for the first time saw the possible environmental side effects of energy production.

The second important incident was the spontaneous combustion of the Cuyahoga River in Cleveland, Ohio. That river had become an aqueous "oily rag" because of the oil, tanker fuel, and other combustible substances floating on its surface, and on a hot, summer day, it burst into flames. The U.S. public was treated to the sight of the Cleveland Fire Department racing to the banks of the river, hooking up their hoses to fire hydrants, and trying to put the river out. This incident made people realize that water pollution was a more serious problem than they had previously thought. Along with other (admittedly less dramatic) episodes, it led to the passage of the Clean Water Act three years later.

Finally, that same summer, a temperature inversion occurred in Pittsburgh on a day when the steel mills and other industries were going full bore. By noon that day, it was so dark that the city had to turn on its street lights and drivers had to use their headlights. This incident was one of several that led to the passage of the Clean Air Act, the basic law that still governs the control of air pollution at the federal level. Again, a widely reported incident served to underscore the price unwittingly being paid for uncontrolled industrial growth in the United States.

When the U.S. Congress drafted the environmental legislation of the early 1970s, these and other problems were thought to be so obvious and so serious—and, in fact, they were—that members simply could not envision directing EPA to address them only if it was affordable to do so. As a consequence, most of the early laws effectively directed the agency's administrator to disregard cost altogether in setting ambient air quality standards and some water quality standards (the same approach was later taken in other areas, such as standards for drinking water and for the cleanup of hazardous waste sites). While none of the laws explicitly forbade taking economics into account, a series of court tests eventually established the principle that if the laws did not *require* the agency to consider cost (or to balance economic considerations against health and ecological ones), then economic factors were not to be used in setting standards.

At the present time, there are seven major statutes that direct EPA to issue environmental regulations.[1] Thousands of pages of such regulations now govern the behavior of individuals, the actions of the local communities and states in which they live, and the activities of corporations. According to EPA, in 1997, the cost of complying with all these federal regulations is $170 billion—a figure that excludes any costs associated with state regulations that impose standards tighter than those of the federal government.[2] If EPA's estimate is correct (and there are reasons to believe that it is too high), the cost of environmental compliance in the United States amounts to 2.2 percent of the country's gross domestic product (GDP). This is a greater share than any other country devotes to environmental protection. The closest competitors are probably the Netherlands and Germany, which (according to imprecise estimates) devote perhaps 1.8 to 2.0 percent of their respective GDPs to environmental compliance.

As a result of its efforts, the United States has seen fairly substantial—and occasionally breathtaking—improvements in its environmental quality over the past 27 years. Air quality, for example, has improved significantly in virtually every major metropolitan area and with respect to almost every air pollutant of concern to human health and/or the environment. For instance, between 1985 and 1994 the average concentration of lead fell 86 percent, while those of ozone and particulate matter fell 12 percent and 20 percent respectively.[3]

The Growing Role of Economics

Given the magnitude of current U.S. expenditures on environmental quality, it is reasonable to ask whether the improvements such expenditures make possible are large enough and important enough to justify them. This question, of course, lies squarely in the domain of economics, which not only focuses on the efficiency of resource use in general but also provides specific analytical tools for addressing such issues. But as the role of economics in environmental policymaking has grown, it has also become more controversial.

One important area of controversy is "regulatory reform." In the last several sessions of Congress, lawmakers have attempted to pass legislation that would require all federal regulatory agencies dealing with the environment, safety, or health to pay closer attention to the benefits and costs of the regulations they issue. Bills debated in Congress have tried not only to enlarge the role of economics and benefit-cost analysis but also to expand the role of what is known as quantitative risk assessment. The latter would require regulatory agencies to adhere more closely to certain standardized procedures when they are weighing the seriousness of, say, chemicals that might cause cancer or ordinary, garden-variety air pollutants with possible human health or ecological effects.[4]

Despite efforts over the past 20 years or so, no comprehensive legislation of this type has been enacted. The United States Senate recently began debating another such bill (S.981, which was co-sponsored by Senators Fred Thompson (R-Tenn.) and Carl Levin (D-Mich.)) that would require agencies to analyze and compare the benefits and costs of the regulatory actions that they are taking. Importantly—and sensibly—this legislation does not mandate that all benefits and costs be quantified and/or expressed in monetary terms. Nor are agencies required to issue only those regulations for which benefits exceed costs (although they do have to explain why they are taking action in those cases where this is not the case).

Why would any regulator issue a regulation if the benefits do not justify the costs? The answer lies in the origins of the regulatory system. Under Section 109 of the Clean Air Act, for instance, the EPA administrator is not allowed to even look at the costs associated with the alternative standards he or she is considering. As a result, even though both the president and Congress want EPA to do more benefit-cost analysis, a number of environmental statutes effectively preclude this. The regulatory reform legislation currently before Congress would substantially improve the situation, raising the quality of regulation without eliminating the safeguards that Congress put in place to ensure environmental progress.[5]

A second and more contentious issue is EPA's decision in the summer of 1997 to tighten the national ambient air quality standard for ground-level ozone (smog) and to establish a strict new standard for fine particulate matter (tiny soot particles in the air).

> As the role of economics in environmental policymaking has grown, it has also become more controversial.

For at least three years, a debate raged in Washington and elsewhere over the pros and cons of these actions. Much of this debate revolved around how much the new standards would cost. Because many metropolitan areas have not been able to meet the existing ozone standard, while others have just come into compliance after a long struggle, tightening the standard will only increase the number of areas in violation. This, in turn, will mean more frequent inspections of cars in those areas, as well as greater repair bills for those vehicles failing the tougher inspections. A tighter ozone standard and a new standard for fine particles will also mean additional pollution control expenditures by business and local governments. For this reason, a big debate arose about whether the health and other benefits of the tighter standards would justify the costs.

Although the final decision on these two standards focused attention on the role of economics in environmental regulation, it clearly did not settle the question. Congressional Republicans initially considered delaying the new standards or using them as a test case of the powers Congress now has to nullify new regulations, but in the end they took no action (and the "window" for nullification has now closed). However, a number of parties have sued to have the new standards overturned.

In 1970, when the Clean Air Act was written, members of Congress would have found it very hard to envision such wrangling over air quality standards. It was obvious to them that *any* improvement in air quality would be worth whatever it cost, which is no doubt why they wrote the law to exclude cost considerations. But by 1997 the wisdom of that approach was so much less obvious that we were embroiled in a pitched debate about the proper role of economics in environmental policy.

One particularly troubling element of that debate is that neither EPA nor the business community made a reasonable case for their position. One would have thought that EPA, having had a very long time to contemplate and justify the air quality standards it eventually issued, could have done a better job of laying out both the good it expects them to do and the anticipated costs of complying with them. The cost estimates EPA made were both rushed and implausibly low. For its part, the business community reacted in a familiar manner, with some members seeming to suggest that the new standards would result in the closure of every paper mill, petroleum refinery, power plant, and chemical plant in the country—that is, in the veritable end of manufacturing in the United States. They advanced this argument in spite of its obvious absurdity: We have been hearing similar talk since 1970, yet there are still many manufacturing facilities operating profitably within U.S. borders (though fewer than before 1970). The business community might have made a real effort to fill the analytical gap left by EPA by undertaking careful analysis of compliance costs, but (with a few exceptions) they eschewed the high road for invective.

A third issue that illuminates the increasingly important role of economics in environmental policymaking is global climate change. This truly is the "mother of all environmental problems," as well as being one of the richest, most interesting, and most complex public policy issues that one could imagine. Formulating international (or even national) policy in this area requires a great deal of scientific expertise because there is considerable scientific uncertainty about many important aspects of the problem. But skillful policymaking also requires a lot of economic expertise, diplomatic skill, and legal acumen.

The nations of the world took the first step toward reducing the threat of climate change at the Earth Summit in Rio do Janeiro in 1992. There they adopted the Framework Convention on Climate Change, which called for voluntary reductions in carbon emissions (to 1990 levels) by 2000. At a subsequent meeting in Berlin, the developed countries agreed to meet in 1997 and establish binding emissions limits for carbon dioxide as well as a firm timetable for achieving those limits. In accordance with the agreement at Rio, the Clinton administration announced a voluntary emissions reduction program for the United States. Although this program has done some good, it has not accomplished nearly as much as the administration and others had hoped.

As agreed at Berlin, in December 1997, 160 countries from around the world met in Kyoto, Japan, to negotiate the details of a climate accord. The protocol that resulted calls for the industrialized countries (as a group) to reduce their average annual emissions of six greenhouse gases by slightly more than 5 percent (from 1990 levels) by no later than 2012. Developing countries are not required to make any reductions, although they may end up doing so in return for compensation from industrialized countries. The industrialized countries will be permitted to engage in emissions trading under rules to be worked out this coming November in Buenos Aires.

To this point, there has been relatively little public discussion of the economic consequences of the Kyoto protocol. Most of the debate in the United States and around the world has centered on climate science: How quickly is carbon dioxide accumulating in the atmosphere? How much will that accumulation accelerate in the future? How will it affect temperature, both in average terms and in terms of variability? What effects will an appreciably warmer atmosphere have on regional climates, rainfall, drought, and so forth? What will the adverse impacts on human health and the environment be?

To be sure, these are all important questions. But in the wake of the agreement negotiated in Kyoto, it is essential that the United States and other nations have some idea what it might cost them to meet its requirements. For the last year, the U.S. government has been making what can only be described as token efforts to figure out what it might cost to comply with various climate proposals that have been put on the table. Such efforts are not good enough.

What is the range of possible costs? Not surprisingly, it depends on whom one asks. At one extreme, Amory Lovins, a very bright environmental advocate and renewable-energy specialist in Colorado, says that not only will it not cost the United States anything to reduce its emissions, the country will make money doing it. He sees opportunities for profitable energy conservation that people currently are either not perceptive enough to take advantage of or are unable to pursue owing to institutional impediments in the economy.[6] In Lovins's view, therefore, it would be appropriate to set a very ambitious emissions reduction goal.

At the other extreme, and again not surprisingly, some in the business community say that reducing U.S. emissions of carbon dioxide to their 1990 level by 2012 could cost $200 billion a year or more. Under certain assumptions, of course, that estimate could be correct; but it would require the kind of worst-case scenario that those in the business community routinely condemn when used by government regulatory agencies.

> Now the most polluted day in Los Angeles is better than an average or even a good day 27 years ago in most of the country's industrial cities.

5. Counting the Cost

Assuming that neither Lovins nor the more extreme members of the business community is correct, how much would it cost the United States to meet its implied commitment under the Kyoto protocol (i.e., to reduce its emissions by about 7 percent)? Work done by economist James Edmonds and his colleagues at the Pacific Northwest National Laboratory in Richland, Washington, indicates that attaining this goal by 2012 would require a carbon tax of at least $70 per ton (or its equivalent).[7] The eventual "true" cost would depend on several factors, including the costs of reducing greenhouse gases other than carbon dioxide (about which little is known), the scope of emissions trading between industrial and developing countries, and the commitment to R&D on less carbon-intensive fuels by both the public and private sectors between now and 2012. Not surprisingly, this estimate falls squarely between those of the optimists such as Lovins and the pessimists in the business community.

Let us turn briefly to how the United States should go about reducing its emissions of greenhouse gases. The worst way to do this would be for the federal or state governments to start telling the steel, pharmaceutical, chemical, or auto companies how to manufacture their wares; to specify how utilities should generate electricity; and to tell petroleum refiners how to make gasoline or other petroleum products. Nor should we resort to quietly tightening the Corporate Average Fuel Economy (CAFE) standards, even though the United States will need to significantly reduce carbon emissions from motor vehicles if it is to even come close to meeting the goal assigned to it in Kyoto. It would be far better to reduce such emissions through gradual increases in the taxes on gasoline and other carbon-intensive energy sources. Not only would this be a more visible and direct way to deal with emissions, it would also raise revenues that could be used to reduce existing taxes on labor and capital (thus reducing the distortions caused by a new tax). Unfortunately, the United States has tended to use command-and-control approaches in the past, requiring not only specific emissions reductions but also specific types of control technologies to achieve them.

These command-and-control approaches would clearly cost the United States more than is necessary to meet any particular greenhouse gas reduction goal. Indeed, that is the most important lesson to be learned from Title IV of the 1990 Clean Air Act Amendments, which created an emissions trading system for curbing acid deposition. Instead of compelling all utilities to achieve stated reductions in emissions using specified technologies, this system permits those utilities with especially high compliance costs to purchase credits (from those able to effect reductions more economically) instead of actually cutting their own emissions. Analysts estimate that such trading has lowered the annual cost of reducing sulfur dioxide emissions by $1 billion to $3 billion compared with command-and-control regulation. We must remember this lesson as we set about fashioning policies to meet our emissions reduction goals for carbon dioxide.

The three issues just discussed vividly illustrate the increasingly large role that economics is playing in environmental policy debates. It only remains to say a few words about why this is the case.

The Situation Now

Why is it that the environmental legislation of the early 1970s effectively proscribed considering costs in setting most standards, yet economic impacts are now at the center of virtually every environmental legislative and regulatory debate? Answering this question forces one to step back and review briefly the environmental progress that the United States has made during the past 27 years. In traveling around the country today, one can certainly find environmental problems in some localized areas, but rivers are no longer bursting into flames, nor are cities so polluted that midday occasionally resembles nightfall. Back in 1970, our environmental problems were big, obvious, and serious. Today, however, air quality is dramatically better. Now the most polluted day in Los Angeles is better than an average or even a good day 27 years ago in most of the country's industrial cities. Water quality has improved in many places as well, though much less dramatically.

In solving many of the worst problems, we have in effect picked the "low-hanging fruit" of easy control options. Effecting further improvements in environmental quality is vastly more expensive today than it was in 1970, and that is one reason for our current preoccupation with the economics of the environment. By way of illustration, it can now cost as much as $50,000 to prevent the discharge of a single ton of volatile organic compounds in Los Angeles, whereas decades ago one could do so for as little as 50 cents. In this case, at least, the cost of pollution control has grown by as much as five orders of magnitude.

We have also witnessed a significant change in public perceptions of the benefits of pollution control. In the 1970s, environmental problems were regarded—rightly, in many instances—as posing serious and immediate risks to public health. Now many people—and not merely those in the regulated business community—feel that the problems with which the United States is dealing are probably far less serious. To take but one example, reducing the concentration of contaminants in drinking water from 4 parts per billion (ppb) to 3 ppb may lessen health risks somewhat. But even those who dismissed cost considerations in the 1970s are now asking these questions: If each reduction in pollution is much more expensive now, and yet the health benefits are often much smaller than they were when pollution was pervasive, shouldn't we look at how expensive it will be to attain further improvements? And shouldn't we make sure it's worth it to achieve what are in some cases relatively small additional benefits? Many in Congress, academia, the media, and the world at large are increasingly of the view that the nation needs to do a better job of balancing the good that we get from a cleaner environment with the added costs that we incur.

Another reason for the increased attention being paid to the costs of regulation has to do with global economic integration. Back in 1970, the United States was much more isolated economically from the rest of the world than it is today. As a result, any environmental costs imposed on manufacturers then could be passed on to consumers fairly easily. But today, goods manufactured in the United States can often be produced just as easily in Thailand, Brazil, South Korea, Poland, Mexico, or China. If we burden U.S.-based manufacturing or other firms with regulatory costs that companies in other countries do not have to bear, products from those countries will sell more cheaply than U.S. products. And if the price differences are significant, jobs will inevitably migrate to other countries. Nothing so strikes terror in the heart of a member of Congress as the prospect of a plant closing in his or her state or district. While the evidence to date suggests that stringent environmental standards in the United States have *not* adversely affected the competitiveness of U.S. manufacturers in any significant way, much greater attention is now being paid to international differences in the cost of environmental regulations.[8]

There is a fourth reason for increased attention to the costs of environmental regulation. In the 1970s, when the United

States was getting into the environmental business in a big way at the federal level, it was convenient (and made everyone feel good) to think of pollution as a moral issue. According to the logic of the times, we had pollution because companies were unethical and did not care what harm they did to the public. Such an attitude, of course, was not very constructive even then—and it is much less so now.

A rather earthy analogy should make this point clear. People do not think of themselves as immoral because they daily produce—and have to eliminate—some bodily wastes. Rather, each of us recognizes that converting food to energy produces wastes that need to be disposed of. It is not unreasonable to think about the manufacturing system in a similar way. To make the products that people want, manufacturers convert raw materials to both energy and other forms; in the process, they generate residuals that are not unlike the bodily wastes that individuals produce. There is nothing immoral about any of this. Of course, just as society imposes restrictions on how we deal with human wastes, it has every right to prescribe the way companies deal with the wastes that result from industrial production.

In one sense, in fact, it may be immoral *not* to take costs into account in setting environmental standards. Resources are limited in any society, even one as wealthy as the United States. As a consequence, no society can do everything that its citizens might like. On the contrary, every society has to set priorities (just as individuals do), choosing which things to do and which not to do. Failure to consider costs makes it impossible to get the most from the available resources and ultimately means saving fewer lives, preventing fewer illnesses, and protecting fewer species or areas than one otherwise could. Arguably, this comes far closer to acting immorally than does generating pollution by burning fossil fuels or converting materials from one form to another.

To act morally, therefore, we simply must set priorities and weigh both the costs and the benefits of particular environmental and natural resource regulations. That is, we have to ask how much it will cost to attain a certain improvement in environmental quality and whether that is the best way to spend the money. It may well be, for instance, that by spending the same money in some other way we can effect a more substantial improvement in environmental quality or augmentation of our natural resources.

Summary

There are at least four reasons why economic considerations now loom larger in environmental policy debates. First, the cost of additional improvements in environmental quality is generally increasing, largely because we have already exploited the best opportunities. Second, there is growing sentiment that while there are still serious threats to be addressed, the most serious ones have already been dealt with—and thus that the marginal benefits of further regulation are diminishing even as the marginal costs are increasing. Third, economic integration means that environmental standards in the United States need to be set with at least one eye on regulations in other countries, lest we drive manufacturing and other jobs abroad. Finally, while pollution is still seen as something to be avoided, it is no longer widely viewed as a symptom of moral weakness but rather as an unfortunate by-product of our industrial system that must be controlled.

For reasons that should now be clear, the increased attention that is being paid to the economic consequences of regulation is quite healthy. For too long we have concealed the tradeoffs implicit in environmental regulation, thus making it almost impossible for citizens to know when they were going underprotected and when they were paying more for a cleaner environment than they might be willing to if the question were put to them directly and clearly. Environmental decisionmaking could never and should never be reduced to a mere economic calculation. But there is nothing—repeat nothing—wrong with asking how much we are spending and what we are getting for our regulatory dollars. This is a sentiment that extends far beyond the business community and the economics profession.

Paul R. Portney is president of Resources for the Future in Washington, D.C. He may be reached at Resources for the Future, 1616 P Street NW, Washington, DC 20036 (telephone: 202-328-5103; e-mail: portney@rff.org).

NOTES

1. See P. R. Portney and R. N. Stavins, eds., *Public Policies for Environmental Protection*, 2nd ed. (Washington, D.C.: Resources for the Future, forthcoming).

2. U.S. Environmental Protection Agency, *Environmental Investments: The Cost of a Clean Environment*, Report no. EPA-230-11-90-083 (Washington, D.C., 1990), 8–20.

3. U.S. Environmental Protection Agency, *Air Quality Trends*, Report no. EPA-454/F-95-003 (1995), 3.

4. These bills, for instance, would require regulatory agencies to explicitly state the assumptions they made in extrapolating from animal toxicology tests to probable human consequences or from high to low doses.

5. Others, principally those in the environmental advocacy community, have a less sanguine view of this legislation, however.

6. For instance, Lovins argues that consumers could both reduce their energy use and save money by switching from standard incandescent light bulbs to fluorescent light bulbs.

7. See J. Edmonds et al., "The Return to 1990: The Cost of Mitigating United States Carbon Emissions in the Post-2000 Period" (Pacific Northwest National Laboratory working paper, October 1997).

8. See A. Jaffe et al., "Environmental Regulation and the Competitiveness of U.S. Manufacturing: What Does the Evidence Tell Us?," *Journal of Economic Literature* 37 (1995), 132.

The Coming Climate

Meteorological records and computer models permit insights into some of the broad weather patterns of a warmer world

by Thomas R. Karl, Neville Nicholls and Jonathan Gregory

Human beings have in recent years discovered that they may have succeeded in achieving a momentous but rather unwanted accomplishment. Because of our numbers and our technology, it now seems likely that we have begun altering the climate of our planet.

Climatologists are confident that over the past century, the global average temperature has increased by about half a degree Celsius. This warming is thought to be at least partly the result of human activity, such as the burning of fossil fuels in electric power plants and automobiles. Moreover, because populations, national economies and the use of technology are all growing, the global average temperature is expected to continue increasing, by an additional 1.0 to 3.5 degrees C by the year 2100.

Such warming is just one of many consequences that climate change can have. Nevertheless, the ways that warming might affect the planet's environment—and, therefore, its life—are among the most compelling issues in earth science. Unfortunately, they are also among the most difficult to predict. The effects will be complex and vary considerably from place to place. Of particular interest are the changes in regional climate and local weather and especially extreme events—record temperatures, heat waves, very heavy rainfall, or drought, for example—which could very well have staggering effects on societies, agriculture and ecosystems.

Based on studies of how the earth's weather has changed over the past century as global temperatures edged upward, as well as on sophisticated computer models of climate, it now seems probable that warming will accompany changes in regional weather. For example, longer and more intense heat waves—a likely consequence of an increase in either the mean temperature or in the variability of daily temperatures—would result in public health threats and even unprecedented levels of mortality, as well as in such costly inconveniences as road buckling and high cooling loads, the latter possibly leading to electrical brownouts or blackouts.

Climate change would also affect the patterns of rainfall and other precipitation, with some areas getting more and others less, changing global patterns and occurrences of droughts and floods. Similarly, increased variability and extremes in precipitation can exacerbate existing problems in water quality and sewage treatment and in erosion and urban storm-water routing, among others. Such possibilities underscore the need to understand the consequences of humankind's effect on global climate.

Two Prongs

Researchers have two main—and complementary—methods of investigating these climate changes. Detailed meteorological records go back about a century, which coincides with the period during which the global average temperature in-

1 ❖ GEOGRAPHY IN A CHANGING WORLD

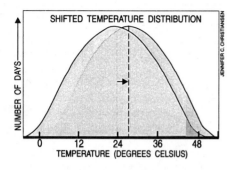

SMALL SHIFTS in the most common daily temperature cause disproportionate increases in the number of extremely hot days. The reason is that temperature distributions are roughly Gaussian. So when the highest point in the Gaussian "bell" moves to the right, the result is a relatively large increase in the probability of exceeding extremely high temperature thresholds. A greater probability of high temperature increases the likelihood of heat waves.

creased by half a degree. By examining these measurements and records, climatologists are beginning to get a picture of how and where extremes of weather and climate have occurred.

It is the relation between these extremes and the overall temperature increase that really interests scientists. This is where another critical research tool—global ocean-atmosphere climate models—comes in.

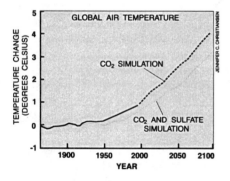

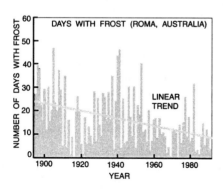

These high-performance computer programs simulate the important processes of the atmosphere and oceans, giving researchers insights into the links between human activities and major weather and climate events.

The combustion of fossil fuels, for example, increases the concentration in the atmosphere of certain greenhouse gases, the fundamental agents of the global warming that may be attributable to humans. These gases, which include carbon dioxide, methane, ozone, halocarbons and nitrous oxide, let in sunlight but tend to insulate the planet against the loss of heat, not unlike the glass of a greenhouse. Thus, a higher concentration means a warmer climate.

Of all the human-caused (anthropogenic) greenhouse gases, carbon dioxide has by far the greatest impact on the global heat budget (calculated as the amount of heat absorbed by the planet less the amount radiated back into space). Contributing to carbon dioxide's greenhouse potency is its persistence: as much as 40 percent of it tends to remain in the atmosphere for centuries. Accumulation of atmospheric carbon dioxide is promoted not only by combustion but also by tropical deforestation.

The second most influential human-caused effect on the earth's radiation budget is probably that of aerosols, which are minute solid particles, sometimes covered by a liquid film, finely dispersed in the atmosphere. They, too, are produced by combustion, but they also come from natural sources, primarily volcanoes. By blocking or reflecting light, aerosols tend to mitigate

GLOBAL AIR TEMPERATURE rise was simulated (*above, left*) by a climate model at the U.K. Meteorological Office's Hanley Center. One line is from a simulation based on carbon dioxide only; the other line also takes into account sulfate. As the global temperature has increased, the number of days with minimums below zero degrees Celsius has gone down. This example (*left*) shows the annual number of days with frost in Roma, Queensland, in Australia.

global warming on regional and global scales. In contrast to carbon dioxide, aerosols have short atmospheric residence times (less than a week) and consequently are concentrated near their sources. At present, scientists are less certain about the radiative effects of aerosols than those of greenhouse gases.

By taking increases in greenhouse gases into account, global ocean-atmosphere climate models can provide some general indications of what we might anticipate regarding changes in weather events and extremes. Unfortunately, however, the capabilities of even the fastest computers and our limited understanding of the linkages among various atmospheric, climatic, terrestrial and oceanic phenomena limit our ability to model important details on the scales at which they occur. For example, clouds are of great significance in the atmospheric heat budget, but the physical processes that form clouds and determine their characteristics operate on scales too small to be accounted for directly in global-scale simulations.

How Hot, and How Often?

The deficiencies in computer models become rather apparent in efforts to reproduce or predict the frequency of climate and weather extremes of all kinds. Of these extremes, temperature is one of the most closely studied, because of its effect on humanity, through health and mortality, as well as cooling loads and other factors. Fortunately, researchers have been able to garner some insights about these extremes by analyzing decades of weather data. For statistical reasons, even slight increases in the average temperature can result in big jumps in the number of very warm days [see *illustration top of page*.]

One of the reasons temperature extremes are so difficult to model is that they are particularly sensitive to unusual circulation patterns and air masses, which can occasionally cause them to follow a trend in the direction opposite that of the mean temperature. For example, in the former Soviet Union, the annual extreme minimum temperature has increased by a degree and a half, whereas the annual extreme maximum showed no change.

The National Climatic Data Center, which is part of the U.S. National Oceanic and Atmospheric Administration (NOAA), has developed a statistical model that simulates the daily maximum and minimum temperatures from three properties of a plot of temperature against time. These three properties are the mean, its daily variance and its day-to-day correlation (the correlation is an indication of how temperatures persist—for example, how often a hot day is followed by another hot day). Given new values of mean, variance and persistence, the model will project the duration and severity of extremes of temperature.

Some of its predictions are surprising. For example, Chicago exhibits considerable variability of temperature from week to week. Even if the mean January temperature went up by four degrees C (an occurrence that may actually take place late in the next century) while the other two properties remained constant, days with minimum temperatures less than –17.8 degrees C (zero degrees Fahrenheit) would still occur. They might even persist for several days in a row. There should also be a significant reduction in the number of early- and late-season freezes. And, not surprisingly, during the summer, uncomfortably hot spells, including so-called killer heat waves, would become more frequent. With just a three degree C increase in the average July temperature, the probability that the heat index (a measure that includes humidity and measures overall discomfort) will exceed 49 degrees C (120 degrees F) sometime during the month increases from one in 20 to one in four.

Because of their effects on agriculture, increases in the minimum are quite significant. Observations over land areas during the latter half of this century indicate that the minimum temperature has increased at a rate more than 50 percent greater than that of the maximum. This increase has lengthened the frost-free season in many parts of the U.S.; in the Northeast, for example, the frost-free season now begins an average of 11 days earlier than it did during the 1950s. A longer frost-free season can be beneficial for many crops grown in places where frost is not very common, but it also affects the growth and development of perennial plants and pests.

The reasons minimum temperatures are going up so much more rapidly than maximums remain somewhat elusive. One possible explanation revolves around cloud cover and evaporative cooling, which have increased in many areas. Clouds tend to keep the days cooler by reflecting sunlight and the nights warmer by inhibiting loss of heat from the surface. Greater amounts of moisture in the soil from additional precipitation and cloudiness inhibit daytime temperature increases because part of the solar energy goes into evaporating this moisture. More conclusive answers, as well as a prediction about whether the asymmetry in daytime and nighttime warming will continue, await better computer models.

Projections of the day-to-day changes in temperature are less certain than those of the mean, but observations have suggested that this variability in much of the Northern Hemisphere's midlatitudes has decreased as the climate has become warmer. Some computer models also project decreases in variability. The variability depends on season and location and is also tied to surface characteristics, such as snow on the ground or moisture in the soil. In midlatitudes, changes in the daily variability of temperature have also been linked to changes in the frequency and intensity of storms and in the location of the paths commonly taken by storms. These storm tracks are, in effect, a succession of eastward-moving midlatitude depressions whose passage dominates the weather.

The relation between these storms and temperature is complex. In a warmer world, the difference of temperature between the tropics and the poles would most likely cover a smaller range, because greater warming is expected near the poles. This factor would tend to weaken storms. On the other hand, high in the atmosphere this difference would be reversed, having the opposite influence. Changes in storms could also happen if anthropogenic aerosols continue to cool the surface regionally, altering the horizontal tempera-

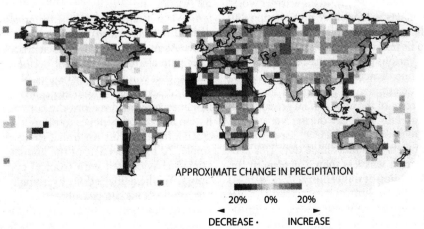

PRECIPITATION TRENDS between 1900 and 1994 reveal a general tendency toward more precipitation at higher latitudes and less precipitation at lower ones.

ture contrasts that control the location of the storm tracks.

More Precipitation

The relation between storms and temperature patterns is one of the reasons it is so difficult to simulate climate changes. The major aspects of climate—temperature, precipitation and storms—are so interrelated that it is impossible to understand one independently of the others. In the global climate system, for example, the familiar cycle of evaporation and precipitation transfers not only water from one place to another but also heat. The heat used at the surface by evaporation of the water is released high in the atmosphere when the water condenses again into clouds and precipitation, warming the surrounding air. The atmosphere then loses this heat by radiating it out into space.

With or without additional greenhouse gases, the earth takes in the same amount of solar energy and radiates the same amount back out into space. With a greater concentration of greenhouse gases, however, the surface is better insulated and can radiate less heat *directly* from the ground to space. The efficiency with which the planet radiates heat to space goes down, which means that the temperature must go up in order for the same amount of heat to be radiated. And as the temperature increases, more evaporation takes place, leading to more precipitation, averaged across the globe.

Precipitation will not increase everywhere and throughout the year, however. (In contrast, all areas of the globe should have warmer temperatures by the end of the next century.) The distribution of precipitation is determined not only by local processes but also by the rates of evaporation and the atmospheric circulations that transport moisture.

For instance, most models predict reduced precipitation in southern Europe in summer as a result of increased greenhouse gases. A significant part of the rainfall in this region comes from local evaporation, with the water not precipitated locally being exported to other areas. Thus, in a warmer climate, increased evaporation in the spring would dry out the soil and lead to less water being available for evaporation and rainfall in the summer.

On a larger scale, most models predict an increase in average precipitation in winter at high latitudes because of greater poleward transport of moisture derived from increased evaporation at low latitudes. Since the turn of the century, precipitation has indeed increased in the high latitudes of the Northern Hemisphere, primarily during the cold season, as temperatures have increased. But for tropical and subtropical land areas, precipitation has actually decreased over the past few decades. This is especially apparent over the Sahel and eastward to Indonesia.

In northernmost North America (north of 55 degrees) and Eurasia, where conditions are normally far below freezing for much of the year, the amount of snowfall has increased over the past several decades. Further increases in snowfall are likely in these areas. Farther south, in southern Canada and the northern U.S., the ratio of snow to rain has decreased, but because of the increase in total precipitation there has been little overall change in the amount of snowfall. In the snow transition belts, where snow is intermittent throughout the cold season, the average snowfall will tend to diminish as the climate warms, before vanishing altogether in some places. Interestingly, areal snow cover during spring and summer abruptly diminished by nearly 10 percent after 1986. This decrease in snow cover has contributed to the rise of spring temperatures in the middle and high latitudes.

Besides the overall amounts of precipitation, scientists are particularly interested in the frequency of heavy downpours or rapid accumulations because of the major practical implications. Intense precipitation can result in flooding, soil erosion and even loss of life. What change do we expect in this frequency?

Whether precipitation occurs is largely determined by the relative humidity, which is the ratio of the concentration of water vapor to its maximum saturation value. When the relative humidity reaches 100 percent, water condenses into clouds, making precipitation possible. Computer models suggest that the distribution of relative humidity will not change much as the climate changes.

The concentration of water vapor needed to reach saturation in the air rises rapidly with temperature, however, at about 6 percent per degree Celsius. So in a warmer climate, the frequency of precipitation (which is related to how often the relative humidity reaches 100 percent) will change less than the amount of precipitation (related to how much water vapor there is in the air). In addition, not only will a warmer world be likely to have more precipitation, but the average precipitation event is likely to be heavier.

Various analyses already support the notion of increased intensity. In the U.S., for example, an average of about 10 percent of the total annual precipitation that falls does so during very heavy downpours in which at least 50 millimeters falls in a single day. This proportion was less than 8 percent at the beginning of this century.

As incredible as it may seem with all this precipitation, the soil in North America, southern Europe and in several other places is actually expected to become drier in the coming decades. Dry soil is of particular concern because of its far-reaching effects, for instance, on crop yields, groundwater resources, lake and river ecosystems and even on down to the foundations of buildings. Higher temperatures dry the soil by boosting the rates of evaporation and transpiration through plants. Several models now project significant increases in the severity of drought. Tempering these predictions, however, are studies of drought frequency and intensity during this century which suggest that at least during the early stages of global warming other factors have overwhelmed the drying effects of warmer weather. For example, in the U.S. and the former U.S.S.R., increases in cloud cover during the past several decades have led to reduced evaporation. In western Russia, in fact, soil moisture has increased.

Stormy Weather

Great as they are, the costs of droughts and heat waves are less obvious than those of another kind of weather extreme: tropical cyclones. These storms, known as hurricanes in the Atlantic and as typhoons in the western North Pacific, can do enormous damage to coastal areas and tropical islands. As the climate warms, scientists anticipate changes in tropical cyclone activity that would vary by region. Not all the consequences would be negative; in some rather arid regions the contribution of tropical cyclones to rainfall is crucial. In northwest Australia, for example, 20 to 50 percent of the annual rainfall is associated with tropical cyclones. Yet the damage done by a single powerful cyclone can be truly spectacular. In August 1992 Hurricane Andrew killed 54 people, left 250,000 homeless and caused $30-billion worth of damage in the Caribbean and in the southeast coastal U.S.

Early discussions of the possible impacts of an enhanced greenhouse effect often suggested more frequent and more intense tropical cyclones. Because these storms depend on a warm surface with unlimited moisture supply, they form only over oceans with a surface temperature of at least 26 degrees C. Therefore, the reasoning goes, global warming will lead to increased ocean temperatures and, presumably, more tropical cyclones.

Yet recent work with climate models and historical data suggests that this scenario is overly simplistic. Other factors, such as atmospheric buoyancy, instabilities in the wind flow, and the differences in wind speed at various heights (vertical wind shear), also play a role in the storms' development. Beyond enabling this rather broad insight, though, climate models have proved of limited use in predicting changes in cyclone activity. Part of the problem is that the simulations are not yet detailed enough to model the very intense inner core of a cyclone.

The historical data are only slightly more useful because they, too, are imperfect. It has been impossible to establish a reliable global record of variability of tropical cyclones through the 20th century because of changes in observing systems (such as the introduction of satellites in the late 1960s) and population changes in tropical areas. Nevertheless, there are good records of cyclone activity in the North Atlantic, where weather aircraft have reconnoitered since the 1940s. Christopher W. Landsea of the NOAA Atlantic Oceanographic and Meteorological Laboratory has documented a decrease in the intensity of hurricanes, and the total number of hurricanes has also followed suit. The years 1991 through 1994 were extremely quiet in terms of the frequency of storms, hurricanes and strong hurricanes; even the unusually intense 1995 season was not enough to reverse this downward trend. It should be noted, too, that the number of typhoons in the northwestern Pacific appears to have gone up.

Overall, it seems unlikely that tropical cyclones will increase significantly on a global scale. In some regions, activity may escalate; in others, it will lessen. And these changes will take place against a backdrop of large, natural variations from year to year and decade to decade.

Midlatitude cyclones accompanied by heavy rainfall, known as extratropical storms, generally extend over a larger area than tropical cyclones and so are more readily modeled. A few studies have been done. A recent one by Ruth Camell and her colleagues at the Hadley Center of the U.K. Meteorological Office found fewer but more intense storms in the North Atlantic under enhanced greenhouse conditions. But the models do not all agree.

Analyses of historical data also do not give a clear conclusion. Some studies suggest that since the late 1980s, North Atlantic winter storm activity has been more extreme than it ever was in the previous century. Over the past few decades, there has also been a trend toward increasing winds and wave heights in the northern half of the North Atlantic Ocean. Other analyses by Hans von Storch and his colleagues at the Max Planck Institute for Meteorology in Hamburg, Germany, found no evidence of changes in storm numbers in the North Sea. In general, as with the tropical cyclones, the available information suggests that there is little cause to anticipate global increases in extratropical storms but that regional changes cannot be ruled out.

The Future

Although these kinds of gaps mean that our understanding of the climate system is incomplete, the balance of evidence suggests that human activities have already had a discernible influence on global climate. In the future, to reduce the uncertainty regarding anthropogenic climate change, especially on the small scales, it will be necessary to improve our computer modeling capabilities, while continuing to make detailed climatic observations.

New initiatives, such as the Global Climate Observing System, and detailed studies of various important climatic processes will help, as will increasingly powerful supercomputers. But the climate system is complex, and the chance always remains that surprises will come about. North Atlantic currents could suddenly change, for example, causing fairly rapid climate change in Europe and eastern North America.

Among the factors affecting our predictions of anthropogenic climate change, and one of our greatest uncertainties, is the amount of future global emissions of greenhouse gases, aerosols and other relevant agents. Determining these emissions is much more than a task for scientists: it is a matter of choice for humankind.

The Authors

THOMAS R. KARL, NEVILLE NICHOLLS and JONATHAN GREGORY were all members of the Intergovernmental Panel on Climate Change, which assessed and reported on the human impact on global climate. Karl is a senior scientist at the National Oceanic and Atmospheric Administration's National Climatic Data Center. Nicholls is a senior principal research scientist at the Australian Bureau of Meteorology Research Center. Gregory is a climate modeler at the Hadley Center of the U.K. Meteorological Office.

Further Reading

CHANGING BY DEGREES: STEPS TO REDUCE GREENHOUSE GASES. U.S. Congress, Office of Technology Assessment, 1991.
POLICY IMPLEMENTATION OF GREENHOUSE WARMING: MITIGATION, ADAPTATION, AND THE SCIENCE BASE. National Academy of Sciences. National Academy Press, 1992.
GLOBAL WARMING DEBATE. Special issue of *Research and Exploration:* A Scholarly Publication of the National Geographic Society, Vol. 9, No. 2; Spring 1993.
GLOBAL WARMING: THE COMPLETE BRIEFING. John T. Houghton. Lion Press, 1994.
CLIMATE CHANGE 1995: THE SCIENCE OF CLIMATE CHANGE. Contribution of Working Group I to the *Second Assessment Report of the Intergovernmental Panel on Climate Change.* Edited by J. T. Houghton, L. G. Meira Filho, B. A. Callendar and N. Harris. Cambridge University Press, 1996.
INDICES OF CLIMATE CHANGE FOR THE UNITED STATES. T. R. Karl, R. W. Knight, D. R. Easterling and R. G. Quayle in *Bulletin of the American Meteorological Society,* Vol. 77, No. 2, pages 279–292; February 1996.

the Emergence of Gasohol:

A Renewable Fuel for American Roadways

Human society has relied on alternative or renewable fuel sources such as alcohol for most of its history.

William D. Warren and Dennis Roberts

The United States of America was the world's largest oil producer at the beginning of the twentieth century. Today, America is the world's largest importer of petroleum. Petroleum provided more than 99 percent of the energy consumed by the transportation sector in 1992. Industrial, commercial and residential sectors rely more on domestic energy sources such as coal-generated electricity and natural gas. Recently, biomass fuels that are produced primarily in the Midwest have captured a small but growing share of the transportation energy market.

Strategic considerations

During the past 20 years concern has risen regarding increasing oil extraction costs in the United States. Recent American oil discoveries continue to be smaller in volume, and thus more expensive to develop. Due to progressively enhanced development costs, domestic oil production is experiencing difficulty competing in the American petroleum market. In addition, petroleum is a non-renewable or stock type of resource. Unlike flow type resources it provides one use and it is consumed. There are additional important reasons why other fuel sources should be developed:

• Dependence on imported oil contributes to conditions of geopolitical instability.

• Reliance on imported oil has massively contributed to the United States' negative trade balance.

• Energy imports have also contributed to the loss of American jobs.

• Petroleum consumption is a primary cause of air pollution in many American cities. A substantial transfer of carbon compounds from the lithosphere to the atmosphere results from the burning of fossil fuels.

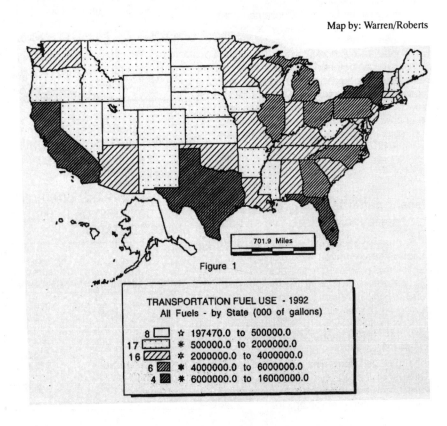

Figure 1

1 ❖ GEOGRAPHY IN A CHANGING WORLD

Alternative fuels: from corn to chicken manure

Human society has relied on alternative or renewable fuel sources, including alcohol, for most of its history. Prior to the beginning of recorded history, humankind discovered the process of manufacturing or distilling alcohol. By the late 1850s, 80 million gallons of alcohol were being produced in the United States. By 1980, U.S. ethanol production had declined to fifty million gallons per year. Since 1980 substantial increases in ethanol production have taken place as production capacity increased to about 1.4 billion gallons per year in 1994. In 1992, 145.7 billion gallons of motor fuel were reported as sold on state taxation reports. More than 8.9 billion gallons were classified as gasohol, and most gasohol is 10 percent ethanol and 90 percent gasoline. Ethanol motor fuel consumption was 893 million gallons or six-tenths of one percent (.6 percent) of United States motor fuel consumption. Thus, despite substantial growth, ethanol's share of the transportation energy fuel market remains small.

Virtually all of the ethanol used in American transportation is manufactured from corn. Soybeans are the basis for perhaps one percent of America's alternative fuel production for transportation. In Brazil, ethanol production is based on sugar cane. Garbage, chicken manure and other organic substances have been used on a very limited basis as source material for ethanol motor fuels.

A study by the Union of Concerned Scientists evaluated biomass energy for electric power generation using switchgrass and hybrid poplar as feedstocks for ethanol plants. Research on potential biomass sources have examined grasses (bahiagrass, bermudagrass, rye, etc.), legumes (alfalfa, crownvetch, clover, etc.), woody species (hybrid poplar and black locust), oilseeds (peanuts, safflower, canola, rapeseed, etc.) Other plants such as comfrey and Jerusalem Artichoke have also been studied.

Recent patterns of energy transportation use

Spatial aspects of motor fuel and gasohol consumption are displayed in the following five figures. Transportation fuel data for 1992 is from the August 1993 issue of *Monthly Motor Fuel Reported by States.* Gasohol data for 1993 is from Federal Highway Administration Table MF-33E, October 1994 titled "Estimated Use of Gasohol—1993." Population data is from The United States Census of Population: 1990. All maps were prepared using the Atlas Pro program of Strategic Mapping, Inc.

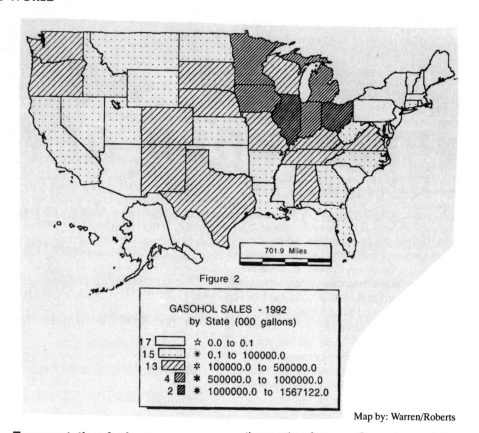

Map by: Warren/Roberts

Transportation fuel use

The distribution of total highway transportation energy consumption for 1992 is presented in Figure 1, the map of "Transportation Fuel Use—All Fuels." California ranked first, consuming 15.6 billion gallons of motor fuel. The four largest states for motor fuel consumption were California, Texas (10.3 billion gallons), Florida (6.7 billion gallons) and New York (6.5 billion gallons). Certainly, a state's population, size and the amount of fuel consumed are closely linked, but features other than population size contribute to this distribution pattern. Table 1, "Transportation Fuel Consumption—Selected States" provides indicators of these other features. State size and trip length considerations contribute to the very high values shown by the large western states of Wyoming, Montana, South Dakota and North Dakota. Also, the availability of electric transit contributes to per capita transportation energy consumption in several states, such as New York state with its extensive subway, electric train and light rail services; California has electric transit operations in five of its metropolitan areas. Per capita income and population density are additional features that influence transportation fuel use.

Table 1

PER CAPITA TRANSPORTATION FUEL CONSUMPTION IN 1992: SELECTED STATES

Number	State	Per Capita Consumption (Gallons)
1	Wyoming	1090.2
2	Montana	749.7
3	South Dakota	743.4
4	North Dakota	724.3
5	Texas	608.3
6	California	525.6
7	Rhode Island	421.2
8	New York	360.9

Gasohol sales

Perhaps the most significant feature in Figure 2, the map of "Gasohol Sales—By State," is the grouping of seventeen states that reported zero gasohol sales in 1992. The consumption values ranged from zero in states such as Maine and New York to 1.6 billion gallons in Illinois and 1.2 billion gallons in Ohio (sales are for 10 percent ethanol and 90 percent gasoline motor fuel). Four additional states produced gasohol sales that exceeded five-hundred million gallons (50 million gallons of ethanol) in 1992. The above six states are all located in the Midwest, center of U.S. corn production. With increasing distance from the Corn Belt, a general decline in gasohol use can be observed. Washington (422,804,000 gallons) and Alabama (280,700,000 gallons) represent anomalies to this pattern. Illinois accounted for 17.7 percent of American gasohol sales, and Ohio's share added an additional 14.1 percent. Eight Midwestern states accounted for 64.9 percent of the gasohol consumed in the United States in 1992. By comparison, two-thirds of America's states consumed less than 100,000,000 gallons (only 1.1 percent of total consumption) of gasohol.

Gasohol market share

The proportion of transportation energy sales that have been captured by gasohol are mapped in Figure 3, "Gasohol Use By State." Figure 2 expressed magnitude values based on total gasohol sales. The data mapped in Figure 3 were calculated by dividing the values in Figure 2 by the values in Figure 1. A similar pattern emerges, but Illinois and Ohio are no longer the most significant states. Nebraska is first with 36 percent of its transportation fuel sales being gasohol. South Dakota was second with 30.8 percent, Iowa was third with 29.8 percent and Illinois was fourth with gasohol obtaining 28.5 percent of the market. Gasohol's market share has increased at a rapid rate in the Midwestern states during the past 10 years. Nevertheless, the region's potential for additional increases in gasohol consumption is large. Gasohol is not available in some areas. A number of petroleum companies refuse to market ethanol products. Also, some of the region's residents continue to have concerns about using gasohol, but people's perceptions are changing. A substantial share of gasohol's future expansion will focus on the Midwest with this region being the primary area for ethanol development.

Per capita consumption of gasohol

In Figure 4, the map of "Gasohol Use—Per Capita," was prepared by dividing the values that were displayed in Figure 2 by population values from the 1990 Census of Population. Per capita consumption values ranged from zero to 235.6 gallons. The five highest states for per capita use were Nebraska (235.6 gallons), South Dakota (229.1 gallons), Iowa (185.2 gallons), Minnesota (148.8 gallons) and Illinois (137.1 gallons). Wyoming, with per capita usage of 113.8 gallons, produced the highest value for a state located outside of the Midwestern core area for gasohol use. Washington, with per capita gasohol consumption of 86.9 gallons in 1992, was the second most important state located outside of the Midwest. As gasohol market shares increase, per capita consumption will also increase. Per capita use values have recorded large increases during the past 10 years. The range of values in the Midwest suggest that large increases are possible in many states.

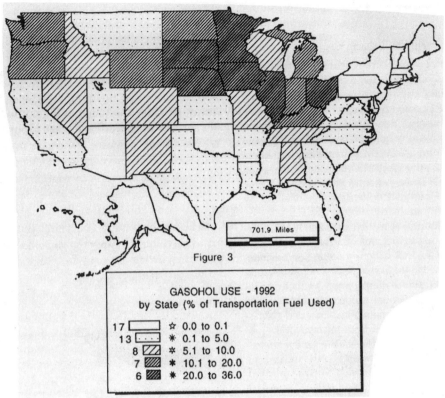

Figure 3

Map by: Warren/Roberts

Gasohol consumption in 1993

The most important change in gasohol consumption data for 1993 resulted from institutional change. Prior to 1993, sales receipts provided the basis for consumption values. Beginning in 1993, the Federal Highway Administration adopted a formula for estimating state values for gasohol consumption. The 1993 data classified gasohol into categories of ten percent and less than 10 percent. Eighty-two percent of U.S. gasohol use was at the 10 percent level. In 1993 Ohio and Illinois ranked first and second in consumption.

For 1993, gasohol consumption declined by 95 million gallons in Illinois. Meanwhile, Ohio increased its consumption by 427 million gallons. California experienced massive growth as consumption increased six-fold from 59.4 million to 360.1 million gallons. Substantial declines in consumption occurred in Alabama, Arkansas, Florida, Kentucky, Nebraska, New Mexico, Texas, Virginia and West Virginia. The number of states with no gasohol consumption was 17 in 1992, and this value declined to eleven in 1993. These patterns can be seen in Figure 5, "Gasohol 1993—Estimated Gallons by State."

Total consumption increased by 15.1

The Midwest is clearly the locus of gasohol development.

percent expanding from 8.9 billion gallons in 1992 to 10.3 billion in 1993. The estimates for the United States' gasohol consumption in 1993 generally conformed to increases in domestic gasohol production. The massive declines that occurred in Illinois and Nebraska are a concern. Both states are very supportive of ethanol fuels. Perhaps the estimating equation is incorrect.

Determinants of gasohol use patterns

The geography of gasohol use is obvious. The Midwest is clearly the locus of gasohol development. Regional outliers were noted with Washington and Alabama providing the most conspicuous examples. Figures 2, 3, 4 and 5 manifest prominent concentrations of gasohol consumption

1 ❖ GEOGRAPHY IN A CHANGING WORLD

RENEWABLE ETHANOL AND ETBE FUEL U.S. EXPORTS

Ethanol processing uses only the starch in the corn. It isolates the protein in the corn in the form of gluten feed and meal. Exporting the 6 million tons of gluten feed creates a larger domestic market for soybean protein. It allows an additional 2.25 million tons of soybean meal to be consumed in the United States. This is a major benefit to the soybean farmers and certainly enhances the price and volume of domestically-used soybeans. The majority of gluten feed is exported to Europe where it is fed to dairy cows and competes only marginally with soybean meal. Corn cannot be imported into Europe because of a high tariff. But when converted to gluten feed, it enters duty free. Therefore, the export of gluten feed represents an important export market for U.S. corn growers.

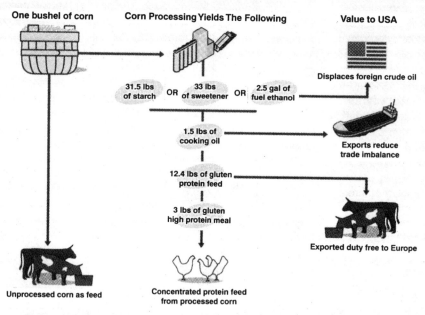

Besides ethanol and high fructose corn syrup, one of the important co-products of corn processing is corn gluten, which contains the same amount of protein as corn used for crude feed. The processing adds about $2 in value to every bushel of corn.

CLEAN ENERGY & THE ENVIRONMENT

with a distinct distance-decay relationship. That is, with increasing distance from the Midwest, the use of ethanol fuels declines. Only one of the six New England states reported any gasohol consumption in 1992.

Gasohol consumption is linked to states that rank high in corn production. Inexpensive surplus corn has provided an important stimulus for the ethanol industry. In the mid–1990s, a bushel of corn sold for slightly more than $2. In 1947, corn prices exceeded $3 per bushel. If corn prices had maintained parity during the past 50 years the existing pattern of gasohol use would not have developed.

A second determinant is the location of ethanol production facilities. Illinois, with 46.1 percent (642.5 million gallons) of the industry's production capacity, is the center of the primary region for gasohol consumption. Other Midwestern states with large production facilities are Iowa (26.8 percent), Indiana (5.4 percent), Ohio (4.7 percent) and Nebraska (4.2 percent). More than ninety percent of the nation's ethanol manufacturing capacity is situated in the Midwest.

Policy and taxation practices make up a third determinant of gasohol use geography. Iowa permits a one cent (US $0.01) reduction, South Dakota a two cent reduction in their motor fuel tax for gasohol. Several other states including Connecticut and New York permitted small motor fuel tax reductions for gasohol. Alaska allowed an eight cent per gallon reduction in gasoline taxes in 1992, but no gasohol was sold in Alaska in that year. Some Midwestern states have allocated funding directly, and through their state universities, for sponsoring research and development activities that have promoted the ethanol industry. In Illinois, the state purchased 45 automobiles which have the ability to operate on gasohol blends that range from 0 to 85 percent (E85 vehicles). The state also supported a grant with the Greater Peoria Transit District for 14 ethanol powered buses (buses are equipped with the E95 engine). Other Midwestern states have sponsored similar programs.

State government can be a negative feature in the promotion of gasohol use. Kansas had rather low values for gasohol use, and the border state of Oklahoma had zero. Oklahoma is clearly an oil state. Kansas contains part of the world's largest natural gas producing field, Panhandle Field. Kansas also contains the large Yates Oil Field. While both states are important producers of corn they play a minor role in the emerging pattern of ethanol fuel use.

The rules of the United States Environmental Protection Agency and of state environmental protection agencies regarding the requirements of the Clean Air Act constitute a fourth determinant. Consumer attitudes and local petroleum prices are additional features that have influenced the regional development of gasohol consumption.

The future: biomass fuel sources and the price of petroleum

During the next decade changes will occur in both the volume and the pattern of gasohol consumption. The Midwest is clearly the core region for the ethanol industry and this feature will continue. States much as Illinois, Iowa and Ohio will experience expansion of their existing patterns of ethanol consumption. States

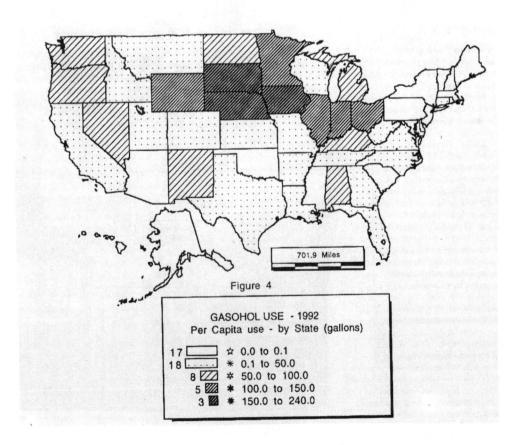

Figure 4

GASOHOL USE - 1992
Per Capita use - by State (gallons)

17 ☆ 0.0 to 0.1
18 ∗ 0.1 to 50.0
 8 ∗ 50.0 to 100.0
 5 ∗ 100.0 to 150.0
 3 ∗ 150.0 to 240.0

Map by: Warren/Roberts

such as Missouri, Kansas and Wisconsin that consume smaller volumes of alcohol fuels could sustain substantial increases. Several of the southeastern states that presently consume miniscule quantities of biomass fuels could experience major expansions of their markets. In 1992, Georgia, Mississippi, North Carolina and South Carolina consumed very little or no gasohol, but they are states with rather large agricultural sectors.

Intensive research activity is occurring in the transportation energy fuels industry. Changes range from new biomass fuel sources to the virtual reinvention of the internal combustion engine. Research and development are progressing on the development of power plants such as the above mentioned E85 and E95 engines. Research also continues on the development of industrial processes that will reduce the cost of manufacturing ethanol.

Government regulations regarding air pollution conditions and federal and state taxation policies will play an important role in determining the future use of renewable fuels and reformulated gasoline. Metropolitan areas that are having problems attaining the air quality mandates of the U.S. Environmental Protection Agency are progressively using more reformulated gasoline. An important issue involves rulings by the U.S. Environmental Protection Agency regarding the renewable oxygen standard (ROS) for reformulated gasoline. At issue is the use of domestically produced Ethyl Tertiary Butyl Ether (ETBE) in reformulated fuels. ETBE is one of several fuel oxygenates that may be used in both oxygenated and reformulated gasoline. ETBE unlike MTBE (methyl tertiary butyl ether) or TAME (tertiary amyl methyl ether) is a renewable fuel oxygenate. Ethanol is used in the production of ETBE. Some multi-national petroleum corporations have devoted extensive resources in order to block standards for reformulated gasoline that would favor renewable fuels. Discussions and revisions of regulations that pertain to the use of renewable fuels are continuing.

Increasing concern regarding the issue of global warming may facilitate the development of renewable fuels. In December 1997 the participants attending the Kyoto Conference on Global Warming agreed to a protocol that called for the reduction of fossil carbon emissions. Subsequently, President Clinton has proposed initiatives that would encourage ethanol development. The authors would like to suggest that it would be appropriate if these initiatives also encouraged the development of other biomass fuel sources.

Petroleum prices will play a critical role regarding the development of biomass transportation fuels. Between 1980 and 1992 the actual price of gasoline declined by almost fifty percent falling from U.S. $2.10 to $1.12 per gallon (prices are adjusted for inflation). Since 1992, fuel prices have continued to fall in response to declining international petroleum prices. In February of 1998 regular gasoline was selling for $0.93.9 per gallon in central Illinois. The development of alternative fuel sources in the United States has been retarded by low prices for imported petroleum.

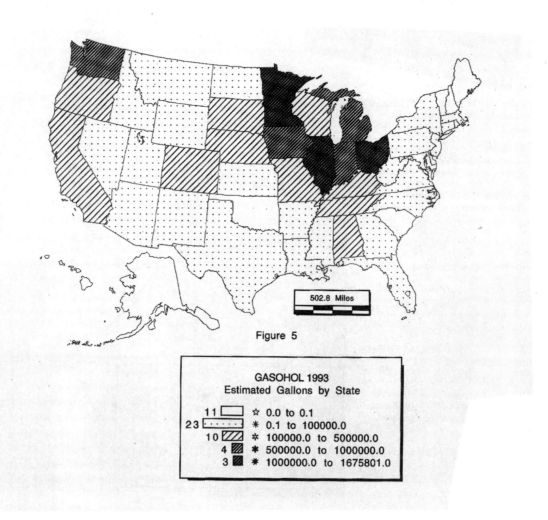

Figure 5

GASOHOL 1993
Estimated Gallons by State

11 ☆ 0.0 to 0.1
23 ✳ 0.1 to 100000.0
10 ✳ 100000.0 to 500000.0
4 ✳ 500000.0 to 1000000.0
3 ✳ 1000000.0 to 1675801.0

Map by: Warren/Roberts

William D. Warren is Professor of Environmental Studies at the University of Illinois-Springfield. His M.A. is from the University of California at Los Angeles and his Ph.D. is from the University of North Carolina. His research has focused on transportation, land use and transportation energy efficiency.

Dennis Roberts is completing an M.A. from the University of Illinois-Springfield. His research interests have focused on geographic information systems and the development of renewable energy resources.

References and further readings

Cohen, Saul B. 1991. The Geopolitical Aftermath of the Gulf War. *FOCUS*, v. 41, (2). Summer. pp. 23–26.

Downstream Alternatives, Inc. 1992. *Changes in Gasoline II*. Bremen, IN.

Green, David L. and Danilo J. Santini, eds. 1994. *Transportation and Global Climatic Change*. Washington, DC: American Council for an Energy-Efficient Economy. pp. 1–357.

Morris, David. 1993. *Ethanol: Bringing Environmental and Economic Benefits to Minnesota*. Minneapolis, MN: Institute for Local Self-Reliance. November.

Shaffer, Shannon, *Oxy-Fuel News*. Potomac, MD: Hart Publications. Published weekly.

Strategic Mapping, Inc. 1990. *Learning Atlas Pro*. San Jose, CA.

Union of Concerned Scientists. 1993. *Powering the Midwest: Renewable Electricity for the Economy and the Environment*. Cambridge, MA: Union of Concerned Scientists. pp. 31–63.

U.S. Department of Transportation. Monthly Motor Fuel Reported by States. 1993. Washington, D.C.: U.S. Department of Transportation. May.

U.S. Department of Agriculture. 1992. New Crops, New Uses, New Markets: Industrial and Commercial Products from U.S. Agriculture. *1992 Yearbook of Agriculture*. Washington, DC: U.S. Department of Agriculture. This source contains a major section on biomass fuels, "Part V. Focus on Renewable Fuels."

U.S. Bureau of the Census. 1992. *1990 Census of Population*. Washington, DC: U.S. Government Printing Office. See CD Rom Diskette Summary Tape File Lc.

America's Rush To Suburbia

By Kenneth T. Jackson

This week in Istanbul, experts from around the globe are attending a United Nations conference on urbanization. The timing is propitious, because in the next few years the world will pass a historic milestone. For the first time, half the earth's population, or more than three billion people, will be living in cities.

At the turn of the century, only 14 percent of us called a city home and just 11 places on the planet had a million inhabitants. Now there are 400 cities with populations of at least one million and 20 megacities of more than 10 million.

But while cities around the world are becoming more dense, those in the United States are moving in the opposite direction. The typical model here is a doughnut—emptiness and desolation at the center and growth on the edges.

Many of the great downtown department stores—including Hudson's in Detroit and Goldsmith's in Memphis—are now closed. Meanwhile, new megamalls, discount centers and factory outlets are springing up every day on the peripheries of America's cities.

Though some cities are still thriving, of the 25 largest cities in 1950, 18 have lost population. For example, from 1950 to 1990, Baltimore lost 22 percent of its population, Philadelphia 23 percent, Chicago 25 percent, Boston 28 percent, Detroit 44 percent and Cleveland 45 percent. (It's true that many cities—Houston, San Diego, Dallas and Phoenix, among them—have grown since 1950, but that is largely because they have annexed their outlying territories. New York City, unique as always, has the same number of people, although its boundaries are unchanged.)

By contrast, during the same period, the suburbs gained more than 75 million people. In 1990, our nation became the first in history to have more suburbanites than city and rural dwellers *combined*.

Why should Americans care whether Portland, Me., or Portland, Ore., is losing inhabitants? Because our system of governance balkanizes social responsibility in our country, a nation divided by race and income.

Only in America are schools, police and fire protection and other services financed largely by local taxes. When middle- and upper-class families flee from the cities, they take with them needed tax revenues.

In Europe, Australia and Japan, such functions are essentially the responsibility of national or at least regional governments. In any of these places, moving from a city to a suburb does not have much impact on a citizen's taxes or on the quality of services.

Americans tend to regard a move to the suburbs as natural—even inevitable—when people are given choices about where to live. But in fact the pattern arises not because land is abundant and cheap (which it is) and not because we have racial and economic divides (which we do) but largely because we have made a series of public policy decisions that other countries have not made.

First, the tax code allows us to deduct mortgage interest and property taxes for both first and second homes. Most other advanced nations do not allow this.

Second, gasoline is essentially not taxed in this country. The 12-country European Union, which has fewer vehicles on the road than the United States does, takes in more than five times as much in gasoline taxes as America does. Our gasoline is cheap compared to that in other advanced industrialized nations, so living in the suburbs, without public transportation, is an attractive option.

Third, the United States has long had a policy, unique in the developed world, of making the provision of public housing voluntary. For the most

---

Kenneth T. Jackson, a professor of history at Columbia University, is editor of "The Encyclopedia of New York City" and author of "Crabgrass Frontier: The Suburbanization of the United States."

part, communities across the country can choose to apply—or not—for public housing. The result of this is that the central cities have become the homes of the poor while the suburbs have become places to escape the poor.

By contrast, the French, British, Germans and Japanese spread public housing around. Indeed, in many countries a demonstrably higher proportion of public housing units go to the periphery than to central city—and this discourages middle-class urban flight.

Finally, in the United States, government at all levels has affected cities by what it has not done. In Europe, land is regarded as a scarce resource that has to be controlled in the public interest rather than exploited for private gain. Thus, governments have acted to preserve open space and deter suburban sprawl.

There are other policies, too, that work against urban areas in the United States, but the larger point is clear: American cities operate under a series of unusual handicaps.

St. Louis offers an extreme example of the consequences of all this. Once the fourth largest city in the nation, the so-called Gateway to the West has become a ghost of its former self. In 1950, it had 857,000 people; by 1990, the population had dwindled to 397,000. Many of its old neighborhoods have become dispiriting collections of eviscerated homes and vacant lots. Aging warehouses and grimy loft factories are now open to the sky; weeds cover once busy railroad sidings.

Will the experience of St. Louis, become typical of other cities in the 21st century?

Tax, housing and gasoline policies doom our cities.

In recent years, such prominent authors as Paul Hawken, John Naisbitt and Alvin Toffler have predicted that cities are doomed and that new telecommunications have made human interaction unnecessary. In the future, they suggest, our journey to work will be from the breakfast table to the home computer. There, in splendid isolation, we will work, shop and play in cyberspace.

Perhaps the futurists are correct, and the cities of our time, like conquered Carthage, will be razed and sowed with salt. But I doubt it. It is more likely that New York, Chicago, Los Angeles, San Francisco, Boston and a dozen or so other places will remain great cities well into the next millennium, despite government policies that cripple them.

That's because the same catalytic mixing of people that creates urban problems and fuels urban conflict also spurs the initiative, innovation and collaboration that taken together move civilization forward. Quite simply, metropolitan centers are the most complex creations of the human mind, and they will not easily yield their role as marketplaces of ideas.

Cities are places where individuals of different bents and pursuits rub shoulders, where most human achievements have been created. Whereas village and rural life, as well as life in the modern shopping mall, is characterized by the endless repetition of similar events, cities remain centers of diversity and opportunity. If they express some of the worst tendencies of modern society, they also represent much of the best.

As Charles E. Merriam, a professor at the University of Chicago, told the United States Conference of Mayors in 1934: "The trouble with Lot's wife was that she looked backward and saw Sodom and Gomorrah. If she had looked forward, she would have seen that heaven is also a pictured as a city."

Unit 2

Unit Selections

9. **The Season of El Niño,** *The Economist*
10. **The Great Climate Flip-Flop,** William H. Calvin
11. **Temperature Rising,** Taras Grescoe
12. **"Dammed If You Do...,"** *World Press Review*
13. **Past and Present Land Use and Land Cover in the USA,** William B. Meyer
14. **China Shoulders the Cost of Environmental Change,** Vaclav Smil

Key Points to Consider

❖ What are the long-range implications of atmospheric pollution? Explain the greenhouse effect.

❖ How can the problem of regional transfer of pollutants be solved?

❖ The manufacture of goods needed by humans produces pollutants that degrade the environment. How can this dilemma be solved?

❖ Where in the world are there serious problems of desertification and drought? Why are these areas increasing in size?

❖ What will be the major forms of energy in the next century?

❖ How are you as an individual related to the land?

❖ Can humankind do anything to ensure the protection of the environment?

❖ What is your attitude toward the environment?

 **Links**   www.dushkin.com/online/

10. **Measurement of Air Pollution from Satellites (MAPS)**
    http://stormy.larc.nasa.gov/overview.html
11. **The North-South Institute**
    http://www.nsi-ins.ca/info.html
12. **United Nations Environment Programme**
    http://www.unep.ch/
13. **World Health Organization**
    http://www.who.ch/Welcome.html

These sites are annotated on pages 6 and 7.

Land-Human Relationships

The home of humankind is Earth's surface and the thin layer of atmosphere enveloping it. Here the human populace has struggled over time to change the physical setting and to create the telltale signs of occupation. Humankind has greatly modified Earth's surface to suit its purposes. At the same time, we have been greatly influenced by the very environment that we have worked to change.

This basic relationship of humans and land is important in geography and, in unit 1, William Pattison identified it as one of the four traditions of geography. Geographers observe, study, and analyze the ways in which human occupants of Earth have interacted with the physical environment. This unit presents a number of articles that illustrate the theme of land-human relationships. In some cases, the association of humans and the physical world has been mutually beneficial; in others, environmental degradation has been the result.

At the present time, the potential for major modifications of Earth's surface and the atmosphere is greater than at any other time in history. It is crucially important that the consequences of these modifications for the environment be clearly understood before such efforts are undertaken.

The first selection in this unit deals with the devastating outcomes in Latin America of the 1997-98 El Niño. "The Great Climate Flip-Flop" suggests that there may be global cooling following the global warming trend and offers ways to prevent it. "Temperature Rising" focuses on a North American hot spot, the Mackenzie Basin. The next piece discusses several major damming projects and their effects on economic growth and environmental degradation. Then, changing land use is examined by William Meyer. Finally, Vaclav Smil's article details the tremendous environmental cost of China's economic growth.

This unit provides a small sample of the many ways in which humans interact with the environment. The outcomes of these interactions may be positive or negative. They may enhance the position of humankind and protect the environment, or they may do just the opposite. We human beings are the guardians of the physical world. We have it in our power to protect, to neglect, or to destroy.

Article 9

THE AMERICAS

The season of El Niño

At last, the El Niño of 1997–98 is returning to its cradle, after scarring Latin America with drought and fire, storm and flood. But it is not over yet, and the fall-out, economic, social and political, will not be cleared up for many months or even, in some places and activities, for years

It swept away roads and bridges, homes and farms, lives and livelihoods. It created a vast lake in a north Peruvian desert, and ruined the fisheries of Chile. It battered Mexico's Pacific coast resort of Acapulco, and lowered water levels, enforcing draught restrictions on ships in the Panama canal. Across stretches of northern Brazil, Guyana and Suriname, even in the island of Trinidad, it dried up vegetation and often with human help—burned up forests. It parched crops too, bringing Brazil's always dry north-east food shortages that had the president scurrying there this week to see relief efforts. Farther south, it swelled the mighty Parana river to eight metres above its normal height, flooding millions of hectares and driving hundreds of thousands of people from their homes in Paraguay, Uruguay and Argentina.

Overall, it killed at least 900 human beings and livestock by the hundreds of thousands. It hit economies for perhaps $20 billion, much of that still to be paid. And, in many countries, it exposed governments to harsh tests and harsher criticism.

And still no one knows just why it happened. Every year, usually around December—hence the name El Niño, the Christ-child—a warm Pacific current flows east to the coasts of Ecuador and Peru. But at random intervals of years, this current becomes a flood. The mechanisms are known. But the root cause is not. And though satellite observations and new weather buoys now enable a fierce El Niño to be foreseen months in advance—as the last big one, in 1982–83, was not—no one can be sure exactly in what form and where the extremes of weather that it brings will strike. Latin America and the Caribbean will not swiftly forget the El Niño of 1997-8.

Drought, flood and uncertainty

Of his two faces, El Niño this time broadly (see the map on the next page) showed drought to the north of the equator, flood—or at least torrential storms—to the south. But little was predictable.

Central America largely escaped the 1982–83 El Niño. This time most of it was hit by drought. So were Mexico's centre and south—but its Pacific coast was struck by hurricanes of wind and rain, rare on that side; the third of these in October smashed Acapulco. The Caribbean has been short of rain: Cuba recently had forest fires, and some east Caribbean islands have had to cut water supplies at night. Colombia has had half its normal rain, and the drought has stretched eastward across Venezuela into northern Brazil, Guyana, Suriname and down into Brazil's north-east.

In contrast, much of Ecuador and Peru have had downpours. It was Peru's Sechura desert, on its far-northern coast, that was converted into a lake, of several thousand square kilometres at its largest, though it is now degenerating into a stagnant swamp. Peru's government had done much to prepare against the rains, building dykes, cleaning waterways and strengthening bridges. El Niño from January to March made light of its efforts, badly damaging 600 kilometres (370 miles) of Peru's main roads and 30 bridges along them. And mocked its forecasts too. The authorities had expected the normally dry south to grow drier still. Not so: a swollen river three months ago flooded the southern city of Ica, near the coast, and it was in the southern mountains that 20 people died when a village was buried by a landslide.

Bolivia for six months has had highland droughts and lowland floods, an unlikely mixture due to unusually high temperatures: what would normally fall as snow on the peaks fell as rain, and ran straight off to drown areas below. Chile had a week of fierce floods last June, midwinter there. Argentina now has floods in the south. But all these pale beside the floods that for four weeks now have swamped northern Argentina and parts of Paraguay and Uruguay, as several big rivers, headed by the giant Parana, have burst their banks. The 250,000 people of Resistencia, capital of Argentina's Chaco province, spent May day wondering whether the Parana, already eight metres above normal, would summon up the extra metre to overwhelm the dyke hurriedly built around their city. It did not; but, in all, some 80,000 square kilometres (31,900 square miles) of Argentina were under water this week—and that excludes areas flooded earlier but now starting to dry out.

Costs, short-term and long

The economic costs imposed by El Niño on the region have been huge. Most of the figures, of course, are heroic estimates, and many are heroic forecasts, at that. Cattle drowned can be crudely counted (140,000, says one estimate for Argentina's Santa Fe province, stretched along 800km of the Parana), and the loss of crops ruined by flood or drought can be reckoned, but the long-term effects may take years to show up.

Not all will be bad. As now in Argentina and Uruguay, one rancher's drowned beast is another's higher beef price, and, more widely helpful, today's disaster can be tomorrow's rebuilding, maybe for the better. There have even been one or two instant winners. Chile's 1997 skiing season was prolonged by abundant snow into October. Tourists have flocked to see

9. Season of El Niño

its Atacama desert in bloom. June's flash floods there made 80,000 people homeless and hit some farmers, especially vegetable growers. Later, excessive rains were too much for some crops. But both also filled reservoirs for the many Chilean farmers who artificially irrigate their land and had suffered three years of low rainfall. So they did for the hydro-power companies, which were within days of enforced electricity rationing. In many countries, road-builders and suppliers of concrete can look forward to a profitable year.

But these are the exceptions. The first to suffer were the fishermen on the Pacific coasts. Peru's boats as early as mid-1997 started finding the shrimps and lobsters usually caught in Ecuador's waters. The Peruvians' usual prey had swum south, looking for food in colder waters off Chile. Peru normally exports $1 billion a year of fishmeal from *anchoveta*, the Pacific pilchard. That business has collapsed, and jobs at sea and in processing plants with it. Chile's jack mackerel in turn went farther south, cutting its catch by 40% in the first quarter of this year. In the fishing port of San Antonio, nine of ten fishmeal plants have closed, and the tenth is handling 20 tonnes a day instead of its usual 100. Only now are Peru's coastal waters starting to cool again.

Farmers' losses have been huge. Drought in Central America has cost Guatamala perhaps 10% of its grain and hard-hit El Salvador 30% of its coffee. Parts of central Panama have been reduced to a dustbowl, and thousands of cattle have had to be slaughtered (not wholly a loss: they were already helping to make the dustbowl).

In Colombia drought has parched 5,000 square kilometres of pasture and 2,000 of maize, sorghum and soya. Overall, says the ministry of agriculture, El Niño has cost 7% of normal output, with the cotton harvest cut by a quarter and milk currently 30% down. Pests have flourished in the coffee plantations east of Bogota. In Brazil's north-east, SUDENE, the regional development agency, talks of drought losses above $4 billion. Cattle have begun to die and the many subsistence farmers have seen their crops wither. Even in this area, not famous for its grain, 2½m tonnes of that are reckoned to have gone.

Argentina's economy ministry this week estimated the total costs of El Niño's floods at $3 billion, almost 1% of GDP. Farming will bear much of that: cattle apart, perhaps 40% of the expected 1½m tonnes of cotton has been lost, 30% of the rice, 50% of the tobacco. This week the wheat and soya crops in the *pampa humida*, Argentina's breadbasket, west of Buenos Aires, were in danger. In the Andean region of Mendoza, home of the country's—rapidly improving and valuable—wineries, vineyards have been damaged by storms. Uruguay's cattle have largely escaped, but 10% of its main grain crop, rice, has gone.

Not all the losses are direct or immediate. Bolivia has seen its soya and cotton hit by floods. But even if their fields were untouched, growers of tropical fruit and vegetables—key crops in the government's efforts to wean farmers off coca-growing—have often seen their roads to market washed away. In the high *altiplano*, poor peasants depend for much of their water on the slow melting of snow. With snowfall lacking, many have lost their crops and will find livestock dying in the coming winter.

Floods can also do long-lasting damage to soil. They wash out nutrients. They may raise salt levels. Irrigation, especially during drought, can also have that effect. And in Guyana and Suriname, coastal farmers irrigating with water from river estuaries have had to stop: because of the rivers' weak flow, it was sea-water.

Roads, power and output

The losses, of course, extend much wider than farming. In Panama, famous for its tropical downpours, the big river Chagres shrank almost to a brook. Lack of rain not only meant drought restrictions on ships (and loss of toll revenue) in the canal. It has also brought power cuts: like most of Central America (except Nicaragua), Panama gets about 70% of its power from hydro-electricity. Several countries, taught by a sharp drought in 1994, have invested in thermal plants, avoiding what they suffered then. But the price, in capital and fuel costs, has been heavy. Colombia too has seen power suppliers struggling to keep up, and river transport, not least of oil, disrupted on its big Magdalena river, which runs north-south deep into the country.

Floods too do more than destroy and kill. Ecuador has been savaged by this El Niño, recording more rain—7,000mm (275 inches)—in the past 14 months than in 17 of the El Niño of 1982–83. It will have to rebuild 19 large bridges and repair 2,500km of roads, at a cost put at $800m. Besides that, farmers have lost $1 billion in output. But on top come the unquantifiable costs of disruption to transport. Lorries stand in interminable queues where roads and bridges have been washed away. So do pedestrians, waiting to pay to cross makeshift plank-and-rope bridges. As in much of Latin America, especially its many mountain areas, a road blocked is movement blocked: without an enormous detour, if at all, traffic simply has no way round.

That has been convenient for Bolivia's weekend travellers to the tropical valleys east of La Paz, who have found "landslide" a neat excuse for not being at the office on Monday morning. No complaint came—except from her father—when a young Ecuadorean was forced by a sudden flood to pass the night at the house of her boyfriend. But the economy has suffered, as raw materials or components (or workers) did not arrive, and goods (or crops or ores) could not be shipped out. Service industries too have lost, as telephones have gone dead and tourist trips been cancelled. The storm that smashed Acapulco spared its glossy hotels, but—even arriving in low season—menaced its $1 billion a year and 150,000 jobs based on tourism.

The human cost

The heaviest cost, though not the most noticed, has been to human happiness, health and lives. Amid all the averages, many people, mostly poor ones, have lost everything: crops, jobs, homes and often hope. In Ecuador, some stand begging at the roadside, crying from shame as much as grief.

Peru has recorded nearly 350,000 people driven from their homes, Argentina 150,000. Some, though by no means all, have had their houses simply swept away. Many of Ecuador's 50,000 have taken refuge in schools left empty during the holidays. But many in all these countries have clung on to their homes, come what might, surrounded by polluted floodwaters; drinking and cooking with them too, with the predictable results—diarrhoea and other intestinal diseases, leptospirosis (caused by animal urine in the water), even cholera. Mosquito-borne malaria and dengue fever have multiplied.

Just how much is El Niño's fault is uncertain: dengue was already alarming Brazil. The death toll too is uncertain; a rough count says around 1,000. Peru blames 300 deaths on El Niño, Ecuador 250, with another 150 missing. Chile lost 20 in the June floods, Argentina about as many in the floods now going on, which were well signalled in advance. One landslip in a coastal resort killed 17 Ecuadoreans last weekend. The Acapulco hurricane cost Mexico 200–400 lives in a day, and at least 19 people died in a single tragedy there this week, when a forest fire—one of thousands that have burned more than 2,000 square kilometres—suddenly overwhelmed those fighting it. But South America's worst reported disaster happily never occurred: a Bolivian goldmine was indeed buried in a mudslide, but earlier flooding had already driven the 70–80 supposed victims to leave.

Yet the true toll is surely far worse. Drought and malnutrition, added to poverty, must have killed far greater numbers than any disaster, but leaving no record.

The natural world

Wildlife too has paid a price. Nature will quickly replace the starving sea-lions and gulls found on Peru's beaches. She will take longer with the burnt savannahs on the Venezuelan-Brazilian border, and far longer with the forests that have burned in so many drought-hit areas, not least the Amazon rainforest destroyed in Roraima state, in northern Brazil, when fires—often lit by farmers clearing land—spread from the savannahs. Even a swamp "of international importance" has burned in Trinidad.

It was the Amazon blaze, engulfing some 3,250 square kilometres of forest, that aroused the outside world, largely indifferent to human suffering, to the fact of El Niño. There may be more to come: the southern edge of the forest too is dry; and this is the area subject to logging and, from June on, to the yearly burning by migrant would-be farmers desperate for land. Yet big fires in any hilly region—as in southern Mexico, in Colombia, even in Trinidad—can do enduring damage. Today's treeless hillside is tomorrow's one lashed by tropical rains, and tomorrow's landslips, instant run-off of rain and flash floods.

Leaders put to the test

The authorities' response to warnings of El Niño, and to its arrival, has varied widely—and been widely, but not always fairly, criticised. Mexicans were quick to denounce the lack of precise forewarning of the Acapulco storm, and then the local theft of relief supplies. Yet action had been fast.

Brazil's rulers, local and federal, were accused of slow and inept response to the Roraima fires; and, more damningly, of ignoring months of drought warnings. President Fernando Henrique Cardoso rushed to the north-east this week, having (correctly) blamed churchmen and the MST landless movement for encouraging raids on food stores (and being told, as correctly, that it was not they who invented hunger). The leading newsweekly, Veja, castigated him—the Veja that in a December story recommended this as just the year for a north-eastern holiday, with El Niño promising "an exceptional season, with lots of sun, blue sky, warm water and soft breezes".

Colombia's government—it has other worries—did little to prepare for El Niño, and even let farmers set fires across huge areas of land, supposedly to enrich the soil, before it banned the practice. Governments in Ecuador, Peru, Bolivia and Chile declared states of emergency in parts of their territory; not so Colombia's, which has been much blamed for that. Paraguay's electioneering politicians have largely ignored its floods; Ecuador's are vying in plans for reconstruction. Argentine politicians—and the public, truly Latin American in this, whatever else—have shown solidarity with victims of the floods, and President Carlos Menem is promising $1 billion to clear up. But critics like Domingo Cavallo, once his economy minister, have damned the lack of advance precautions.

The true artist of El Niño has been Peru's President Alberto Fujimori. His government spent $300m in advance (not all in the right places, but at least the ones that looked right at the time); and since El Niño struck he has rushed about frenetically taking personal charge of relief efforts, even rescue attempts. Too frenetically, say some critics, who claim presidential efforts are muddling those of people on the spot. Maybe, maybe not; but his poll ratings, 30% in mid–1997, now stand at 45%.

The Great Climate Flip-flop

by WILLIAM H. CALVIN

ONE of the most shocking scientific realizations of all time has slowly been dawning on us: the earth's climate does great flip-flops every few thousand years, and with breathtaking speed. We could go back to ice-age temperatures within a decade—and judging from recent discoveries, an abrupt cooling could be triggered by our current global-warming trend. Europe's climate could become more like Siberia's. Because such a cooling would occur too quickly for us to make readjustments in agricultural productivity and supply, it would be a potentially civilization-shattering affair, likely to cause an unprecedented population crash. What paleoclimate and oceanography researchers know of the mechanisms underlying such a climate flip suggests that global warming could start one in several different ways.

For a quarter century global-warming theorists have predicted that climate creep is going to occur and that we need to prevent greenhouse gases from warming things up, thereby raising the sea level, destroying habitats, intensifying storms, and forcing agricultural rearrangements. Now we know—and from an entirely different group of scientists exploring separate lines of reasoning and data—that the most catastrophic result of global warming could be an abrupt cooling.

We are in a warm period now. Scientists have known for some time that the previous warm period started 130,000 years ago and ended 117,000 years ago, with the return of cold temperatures that led to an ice age. But the ice ages aren't what they used to be. They were formerly thought to be very gradual, with both air temperature and ice sheets changing in a slow, 100,000-year cycle tied to changes in the earth's orbit around the sun. But our current warm-up, which started about 15,000 years ago, began abruptly, with the temperature rising sharply while most of the ice was still present. We now know that there's nothing "glacially slow" about temperature change: superimposed on the gradual, long-term cycle have been dozens of abrupt warmings and coolings that lasted only centuries.

The back and forth of the ice started 2.5 million years ago, which is also when the ape-sized hominid brain began to develop into a fully human one, four times as large and reorganized for language, music, and chains of inference. Ours is now a brain able to anticipate outcomes well enough to practice ethical behavior, able to head off disasters in the making by extrapolating trends. Our civilizations began to emerge right after the continental ice sheets melted about 10,000 years ago. Civilizations accumulate knowledge, so we now know a lot about what has been going on, what has

> "Climate change" is popularly understood to mean greenhouse warming, which, it is predicted, will cause flooding, severe windstorms, and killer heat waves. But warming could lead, paradoxically, to drastic cooling—a catastrophe that could threaten the survival of civilization

---

William H. Calvin is a theoretical neurophysiologist at the University of Washington at Seattle.

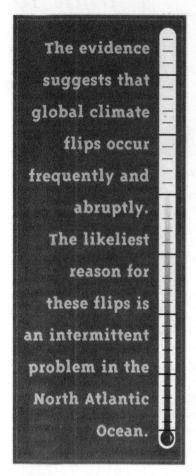

The evidence suggests that global climate flips occur frequently and abruptly. The likeliest reason for these flips is an intermittent problem in the North Atlantic Ocean.

made us what we are. We puzzle over oddities, such as the climate of Europe.

Keeping Europe Warm

EUROPE is an anomaly. The populous parts of the United States and Canada are mostly between the latitudes of 30° and 45°, whereas the populous parts of Europe are ten to fifteen degrees farther north. "Southerly" Rome lies near the same latitude, 42°N, as "northerly" Chicago—and the most northerly major city in Asia is Beijing, near 40°N. London and Paris are close to the 49°N line that, west of the Great Lakes, separates the United States from Canada. Berlin is up at about 52°, Copenhagen and Moscow at about 56°. Oslo is nearly at 60°N, as are Stockholm, Helsinki, and St. Petersburg; continue due east and you'll encounter Anchorage.

Europe's climate, obviously, is not like that of North America or Asia at the same latitudes. For Europe to be as agriculturally productive as it is (it supports more than twice the population of the United States and Canada), all those cold, dry winds that blow eastward across the North Atlantic from Canada must somehow be warmed up. The job is done by warm water flowing north from the tropics, as the eastbound Gulf Stream merges into the North Atlantic Current. This warm water then flows up the Norwegian coast, with a westward branch warming Greenland's tip, 60°N. It keeps northern Europe about nine to eighteen degrees warmer in the winter than comparable latitudes elsewhere—except when it fails. Then not only Europe but also, to everyone's surprise, the rest of the world gets chilled. Tropical swamps decrease their production of methane at the same time that Europe cools, and the Gobi Desert whips much more dust into the air. When this happens, something big, with worldwide connections, must be switching into a new mode of operation.

The North Atlantic Current is certainly something big, with the flow of about a hundred Amazon Rivers. And it sometimes changes its route dramatically, much as a bus route can be truncated into a shorter loop. Its effects are clearly global too, inasmuch as it is part of a long "salt conveyor" current that extends through the southern oceans into the Pacific.

I hope never to see a failure of the northernmost loop of the North Atlantic Current, because the result would be a population crash that would take much of civilization with it, all within a decade. Ways to postpone such a climatic shift are conceivable, however—old-fashioned dam-and-ditch construction in critical locations might even work. Although we can't do much about everyday weather, we may nonetheless be able to stabilize the climate enough to prevent an abrupt cooling.

Abrupt Temperature Jumps

THE discovery of abrupt climate changes has been spread out over the past fifteen years, and is well known to readers of major scientific journals such as *Science* and *Nature*. The abruptness data are convincing. Within the ice sheets of Greenland are annual layers that provide a record of the gases present in the atmosphere and indicate the changes in air temperature over the past 250,000 years—the period of the last two major ice ages. By 250,000 years ago *Homo erectus* had died out, after a run of almost two million years. By 125,000 years ago *Homo sapiens* had evolved from our ancestor species—so the whiplash climate changes of the last ice age affected people much like us.

In Greenland a given year's snowfall is compacted into ice during the ensuing years, trapping air bubbles, and so paleoclimate researchers have been able to glimpse ancient climates in some detail. Water falling as snow on Greenland carries an isotopic "fingerprint" of what the temperature was like en route. Counting those tree-ring-like layers in the ice cores shows that cooling came on as quickly as droughts. Indeed, were another climate flip to begin next year, we'd probably complain first about the drought, along with unusually cold winters in Europe. In the first few years the climate could cool as much as it did during the misnamed Little Ice Age (a gradual cooling that lasted from the early Renaissance until the end of the nineteenth century), with tenfold greater changes over the next decade or two.

The most recent big cooling started about 12,700 years ago, right in the midst of our last global warming. This cold period, known as the Younger Dryas, is named for the pollen of a tundra flower that turned up in a lake bed in Denmark when it shouldn't have. Things had been warming up, and half the ice sheets covering Europe and Canada had already melted. The return to ice-age temperatures lasted 1,300 years. Then, about 11,400 years ago, things suddenly warmed up again, and the earliest agricultural villages were established in the Middle East. An abrupt cooling got started 8,200 years ago, but it aborted within a century, and the temperature changes since then have been gradual in compari-

10. Great Climate Flip-Flop

son. Indeed, we've had an unprecedented period of climate stability.

Coring old lake beds and examining the types of pollen trapped in sediment layers led to the discovery, early in the twentieth century, of the Younger Dryas. Pollen cores are still a primary means of seeing what regional climates were doing, even though they suffer from poorer resolution than ice cores (worms churn the sediment, obscuring records of all but the longest-lasting temperature changes). When the ice cores demonstrated the abrupt onset of the Younger Dryas, researchers wanted to know how widespread this event was. The U.S. Geological Survey took old lake-bed cores out of storage and re-examined them.

Ancient lakes near the Pacific coast of the United States, it turned out, show a shift to cold-weather plant species at roughly the time when the Younger Dryas was changing German pine forests into scrublands like those of modern Siberia. Subarctic ocean currents were reaching the southern California coastline, and Santa Barbara must have been as cold as Juneau is now. (But the regional record is poorly understood, and I know at least one reason why. These days when one goes to hear a talk on ancient climates of North America, one is likely to learn that the speaker was forced into early retirement from the U.S. Geological Survey by budget cuts. Rather than a vigorous program of studying regional climatic change, we see the shortsighted preaching of cheaper government at any cost.)

In 1984, when I first heard about the startling news from the ice cores, the implications were unclear—there seemed to be other ways of interpreting the data from Greenland. It was initially hoped that the abrupt warmings and coolings were just an oddity of Greenland's weather—but they have now been detected on a worldwide scale, and at about the same time. Then it was hoped that the abrupt flips were somehow caused by continental ice sheets, and thus would be unlikely to recur, because we now lack huge ice sheets over Canada and Northern Europe. Though some abrupt coolings are likely to have been associated with events in the Canadian ice sheet, the abrupt cooling in the previous warm period, 122,000 years ago, which has now been detected

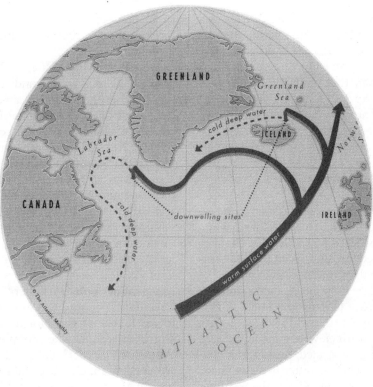

THE NORTHERN LOOP OF THE NORTH ATLANTIC CURRENT

even in the tropics, shows that flips are not restricted to icy periods; they can also interrupt warm periods like the present one.

There seems to be no way of escaping the conclusion that global climate flips occur frequently and abruptly. An abrupt cooling could happen now, and the world might not warm up again for a long time: it looks as if the last warm period, having lasted 13,000 years, came to an end with an abrupt, prolonged cooling. That's how our warm period might end too.

Sudden onset, sudden recovery—this is why I use the word "flip-flop" to describe these climate changes. They are utterly unlike the changes that one would expect from accumulating carbon dioxide or the setting adrift of ice shelves from Antarctica. Change arising from some sources, such as volcanic eruptions, can be abrupt—but the climate doesn't flip back just as quickly centuries later.

Temperature records suggest that there is some grand mechanism underlying all of this, and that it has two major states. Again, the difference between them amounts to nine to eighteen degrees—a range that may depend on how much ice there is to slow the responses. I call the colder one the "low state." In discussing the ice ages there is a tendency to think of warm as good—and therefore of warming as better. Alas, further warming might well kick us out of the "high state." It's the high state that's good, and we may need to help prevent any sudden transition to the cold low state.

Although the sun's energy output does flicker slightly, the likeliest reason for these abrupt flips is an intermittent problem in the North Atlantic Ocean, one that seems to trigger a major rearrangement of atmospheric circulation. North-south ocean currents help to redistribute equatorial heat into the temperate zones, supplementing the heat transfer by winds. When the warm currents penetrate farther than usual into the northern seas, they help to melt the sea ice that is reflecting a lot of sunlight back into space, and so the earth becomes warmer. Eventually that helps to melt ice sheets elsewhere.

The high state of climate seems to involve ocean currents that deliver an extraordinary amount of heat to the vicinity of Iceland and Norway. Like bus routes or conveyor belts, ocean currents must have a return loop. Unlike most ocean currents, the North Atlantic Current has a return loop that runs deep beneath the ocean surface. Huge amounts of seawater sink at known downwelling sites every winter, with the water heading south when it reaches the bottom. When that annual flushing fails for some years, the conveyor belt stops moving and so heat stops flowing so far north—and apparently we're popped back into the low state.

Flushing Cold Surface Water

SURFACE waters are flushed regularly, even in lakes. Twice a year they sink, carrying their load of atmospheric gases downward. That's because water density changes with temperature. Water is densest at about 39°F (a typical refrigerator setting—anything that you take out of the refrigerator, whether you place it on the kitchen counter or move it to the freezer, is going to expand a little). A lake surface cooling down in the autumn will eventually sink into the less-dense-because-warmer waters below, mixing things up. Seawater is more complicated, because salt content also helps to determine whether water floats or sinks. Water that evaporates leaves its salt behind; the resulting saltier water is heavier and thus sinks.

The fact that excess salt is flushed from surface waters has global implications, some of them recognized two centuries ago. Salt circulates, because evaporation up north causes it to sink and be carried south by deep currents. This was posited in 1797 by the Anglo-American physicist Sir Benjamin Thompson (later known, after he moved to Bavaria, as Count Rumford of the Holy Roman Empire), who also posited that, if merely to compensate, there would have to be a warmer northbound current as well. By 1961 the oceanographer Henry Stommel, of the Woods Hole Oceanographic Institution, in Massachusetts, was beginning to worry that these warming currents might stop flowing if too much fresh water was added to the surface of the northern seas. By 1987 the geochemist Wallace Broecker, of Columbia University, was piecing together the paleoclimatic flip-flops with the salt-circulation story and warning that small nudges to our climate might produce "unpleasant surprises in the greenhouse."

Oceans are not well mixed at any time. Like a half-beaten cake mix, with strands of egg still visible, the ocean has a lot of blobs and streams within it. When there has been a lot of evaporation, surface waters are saltier than usual. Sometimes they sink to considerable depths without mixing. The Mediterranean waters flowing out of the bottom of the Strait of Gibraltar into the Atlantic Ocean are about 10 percent saltier than the ocean's average, and so they sink into the depths of the Atlantic. A nice little Amazon-sized waterfall flows over the ridge that connects Spain with Morocco, 800 feet below the surface of the strait.

Another underwater ridge line stretches from Greenland to Iceland and on to the Faeroe Islands and Scotland. It, too, has a salty waterfall, which pours the hypersaline bottom waters of the Nordic Seas (the Greenland Sea and the Norwegian Sea) south into the lower levels of the North Atlantic Ocean. This salty waterfall is more like thirty Amazon Rivers combined. Why does it exist? The cold, dry winds blowing eastward off Canada evaporate the surface waters of the North Atlantic Current, and leave behind all their salt. In late winter the heavy surface waters sink en masse. These blobs, pushed down by annual repetitions of these late-winter events, flow south, down near the bottom of the Atlantic. The same thing happens in the Labrador Sea between Canada and the southern tip of Greenland.

Salt sinking on such a grand scale in the Nordic Seas causes warm water to flow much farther north than it might otherwise do. This produces a heat bonus of perhaps 30 percent beyond the heat provided by direct sunlight to these seas, accounting for the mild winters downwind, in northern Europe. It has been called the Nordic Seas heat pump.

Nothing like this happens in the Pacific Ocean, but the Pacific is nonetheless affected, because the sink in the Nordic Seas is part of a vast worldwide salt-conveyor belt. Such a conveyor is needed because the Atlantic is saltier than the Pacific (the Pacific has twice as much water with which to dilute the salt carried in from rivers). The Atlantic would be even saltier if it didn't mix with the Pacific, in long, loopy currents. These carry the North Atlantic's excess salt southward from the bottom of the Atlantic, around the tip of Africa, through the Indian Ocean, and up around the Pacific Ocean.

There used to be a tropical shortcut, an express route from Atlantic to Pacific, but continental drift connected North America to South America about three million years ago, damming up the easy route for disposing of excess salt. The dam, known as the Isthmus of Panama, may have been what caused the ice ages to begin a short time later, simply because of the forced detour. This major change in ocean circulation, along with a climate that had already been slowly cooling for millions of years, led not

> Huge amounts of seawater sink every winter in the vicinity of Iceland and Norway. When that flushing fails for some years, apparently we're popped into an abrupt cooling.

only to ice accumulation most of the time but also to climatic instability, with flips every few thousand years or so.

Failures of Flushing

FLYING above the clouds often presents an interesting picture when there are mountains below. Out of the sea of undulating white clouds mountain peaks stick up like islands.

Greenland looks like that, even on a cloudless day—but the great white mass between the occasional punctuations is an ice sheet. In places this frozen fresh water descends from the highlands in a wavy staircase.

Twenty thousand years ago a similar ice sheet lay atop the Baltic Sea and the land surrounding it. Another sat on Hudson's Bay, and reached as far west as the foothills of the Rocky Mountains—where it pushed, head to head, against ice coming down from the Rockies. These northern ice sheets were as high as Greenland's mountains, obstacles sufficient to force the jet stream to make a detour.

Now only Greenland's ice remains, but the abrupt cooling in the last warm period shows that a flip can occur in situations much like the present one. What could possibly halt the salt-conveyor belt that brings tropical heat so much farther north and limits the formation of ice sheets? Oceanographers are busy studying present-day failures of annual flushing, which give some perspective on the catastrophic failures of the past.

In the Labrador Sea, flushing failed during the 1970s, was strong again by 1990, and is now declining. In the Greenland Sea over the 1980s salt sinking declined by 80 percent. Obviously, local failures can occur without catastrophe—it's a question of how often and how widespread the failures are—but the present state of decline is not very reassuring. Large-scale flushing at both those sites is certainly a highly variable process, and perhaps a somewhat fragile one as well. And in the absence of a flushing mechanism to sink cooled surface waters and send them southward in the Atlantic, additional warm waters do not flow as far north to replenish the supply.

There are a few obvious precursors to flushing failure. One is diminished wind chill, when winds aren't as strong as usual, or as cold, or as dry—as is the case in the Labrador Sea during the North Atlantic Oscillation. This El Niño-like shift in the atmospheric-circulation pattern over the North Atlantic, from the Azores to Greenland, often lasts a decade. At the same time that the Labrador Sea gets a lessening of the strong winds that aid salt sinking, Europe gets particularly cold winters. It's happening right now: a North Atlantic Oscillation started in 1996.

Another precursor is more floating ice than usual, which reduces the amount of ocean surface exposed to the winds, in turn reducing evaporation. Retained heat eventually melts the ice, in a cycle that recurs about every five years.

Yet another precursor, as Henry Stommel suggested in 1961, would be the addition of fresh water to the ocean surface, diluting the salt-heavy surface waters before they became unstable enough to start sinking. More rain falling in the northern oceans—exactly what is predicted as a result of global warming—could stop salt flushing. So could ice carried south out of the Arctic Ocean.

There is also a great deal of unsalted water in Greenland's glaciers, just uphill from the major salt sinks. The last time an abrupt cooling occurred was in the midst of global warming. Many ice sheets had already half melted, dumping a lot of fresh water into the ocean.

The Nordic Seas sink is part of a worldwide conveyor belt. There used to be a shortcut, but it was dammed up by the Isthmus of Panama, which may have begun the ice ages.

A brief, large flood of fresh water might nudge us toward an abrupt cooling even if the dilution were insignificant when averaged over time. The fjords of Greenland offer some dramatic examples of the possibilities for freshwater floods. Fjords are long, narrow canyons, little arms of the sea reaching many miles inland; they were carved by great glaciers when the sea level was lower. Greenland's east coast has a profusion of fjords between 70°N and 80°N, including one that is the world's biggest. If blocked by ice dams, fjords make perfect reservoirs for meltwater.

Glaciers pushing out into the ocean usually break off in chunks. Whole sections of a glacier, lifted up by the tides, may snap off at the "hinge" and become icebergs. But sometimes a glacial surge will act like an avalanche that blocks a road, as happened when Alaska's Hubbard glacier surged into the Russell fjord in May of 1986. Its snout ran into the opposite side, blocking the fjord with an ice dam. Any meltwater coming in behind the dam stayed there. A lake formed, rising higher and higher—up to the height of an eight-story building.

Eventually such ice dams break, with spectacular results. Once the dam is breached, the rushing waters erode an ever wider and deeper path. Thus the entire lake can empty quickly. Five months after the ice dam at the Russell fjord formed, it broke, dumping a cubic mile of fresh water in only twenty-four hours.

The Great Salinity Anomaly, a pool of semi-salty water derived from about 500 times as much unsalted water as that released by Russell Lake, was tracked from 1968 to 1982 as it moved south from Greenland's east coast. In 1970 it arrived in the Labrador Sea, where it prevented the usual salt sinking. By 1971–1972 the semi-salty blob was off Newfoundland. It then crossed the Atlantic and passed near the Shetland Islands around 1976. From there it was carried northward by the warm Norwegian Current, whereupon some of it swung west again to arrive off Greenland's east coast—where it had started its inch-per-second journey. So freshwater blobs drift, sometimes causing major trouble, and Greenland floods thus have the potential to stop the enormous heat transfer that keeps the North Atlantic Current going strong.

The Greenhouse Connection

OF this much we're sure: global climate flip-flops have frequently happened in the past, and they're likely to happen again. It's also clear that sufficient global warming could trigger an abrupt cooling in at least two ways—by increasing high-latitude rainfall or by melting Greenland's ice, both of which could put enough fresh water into the ocean surface to suppress flushing.

Further investigation might lead to revisions in such mechanistic explanations, but the result of adding fresh water to the ocean surface is pretty standard physics. In almost four decades of subsequent research Henry Stommel's theory has only been enhanced, not seriously challenged.

Up to this point in the story none of the broad conclusions is particularly speculative. But to address how all these nonlinear mechanisms fit together—and what we might do to stabilize the climate—will require some speculation.

Even the tropics cool down by about nine degrees during an abrupt cooling, and it is hard to imagine what in the past could have disturbed the whole earth's climate on this scale. We must look at arriving sunlight and departing light and heat, not merely regional shifts on earth, to account for changes in the temperature balance. Increasing amounts of sea ice and clouds could reflect more sunlight back into space, but the geochemist Wallace Broecker suggests that a major greenhouse gas is disturbed by the failure of the salt conveyor, and that this affects the amount of heat retained.

In Broecker's view, failures of salt flushing cause a worldwide rearrangement of ocean currents, resulting in—and this is the speculative part—less evaporation from the tropics. That, in turn, makes the air drier. Because water vapor is the most powerful greenhouse gas, this decrease in average humidity would cool things globally. Broecker has written, "If you wanted to cool the planet by 5°C [9°F] and could magically alter the water-vapor content of the atmosphere, a 30 percent decrease would do the job."

Just as an El Niño produces a hotter Equator in the Pacific Ocean and generates more atmospheric convection, so there might be a subnormal mode that decreases heat, convection, and evaporation. For example, I can imagine that ocean currents carrying more warm surface waters north or south from the equatorial regions might, in consequence, cool the Equator somewhat. That might result in less evaporation, creating lower-than-normal levels of greenhouse gases and thus a global cooling.

To see how ocean circulation might affect greenhouse gases, we must try to account quantitatively for important nonlinearities, ones in which little nudges provoke great responses. The modern world is full of objects and systems that exhibit "bistable" modes, with thresholds for flipping. Light switches abruptly change mode when nudged hard enough. Door latches suddenly give way. A gentle pull on a trigger may be ineffective, but there comes a pressure that will suddenly fire the gun. Thermostats tend to activate heating or cooling mechanisms abruptly—also an example of a system that pushes back.

We must be careful not to think of an abrupt cooling in response to global warming as just another self-regulatory device, a control system for cooling things down when it gets too hot. The scale of the response will be far beyond the bounds of regulation—more like when excess warming triggers fire extinguishers in the ceiling, ruining the contents of the room while cooling them down.

Preventing Climate Flips

THOUGH combating global warming is obviously on the agenda for preventing a cold flip, we could easily be blindsided by stability problems if we allow global warming per se to remain the main focus of our climate-change efforts. To stabilize our flip-flopping climate we'll need to identify all the important feedbacks that control climate and ocean currents—evaporation, the reflection of sunlight back into space, and so on—and then estimate their relative strengths and interactions in computer models.

Feedbacks are what determine thresholds, where one mode flips into another. Near a threshold one can sometimes observe abortive responses, rather like the act of stepping back onto a curb several times before finally running across a busy street. Abortive responses and rapid chattering between modes are common problems in nonlinear systems with not quite enough oomph—the reason that old fluorescent lights flicker. To keep a bistable system firmly in one state or the other, it should be kept away from the transition threshold.

We need to make sure that no business-as-usual climate variation, such as an El Niño or the North Atlantic Oscillation, can push our climate onto the slippery slope and into an abrupt cooling. Of particular importance are combinations of climate variations—this winter, for ex-

10. Great Climate Flip-Flop

ample, we are experiencing both an El Niño and a North Atlantic Oscillation—because such combinations can add up to much more than the sum of their parts.

We are near the end of a warm period in any event; ice ages return even without human influences on climate. The last warm period abruptly terminated 13,000 years after the abrupt warming that initiated it, and we've already gone 15,000 years from a similar starting point. But we may be able to do something to delay an abrupt cooling.

Do something? This tends to stagger the imagination, immediately conjuring up visions of terraforming on a science-fiction scale—and so we shake our heads and say, "Better to fight global warming by consuming less," and so forth.

Surprisingly, it may prove possible to prevent flip-flops in the climate—even by means of low-tech schemes. Keeping the present climate from falling back into the low state will in any case be a lot easier than trying to reverse such a change after it has occurred. Were fjord floods causing flushing to fail, because the downwelling sites were fairly close to the fjords, it is obvious that we could solve the problem. All we would need to do is open a channel through the ice dam with explosives before dangerous levels of water built up.

Timing could be everything, given the delayed effects from inch-per-second circulation patterns, but that, too, potentially has a low-tech solution: build dams across the major fjord systems and hold back the meltwater at critical times. Or divert eastern-Greenland meltwater to the less sensitive north and west coasts.

Fortunately, big parallel computers have proved useful for both global climate modeling and detailed modeling of ocean circulation. They even show the flips. Computer models might not yet be able to predict what will happen if we tamper with downwelling sites, but this problem doesn't seem insoluble. We need more well-trained people, bigger computers, more coring of the ocean floor and silted-up lakes, more ships to drag instrument packages through the depths, more instrumented buoys to study critical sites in detail, more satellites measuring regional variations in the sea surface, and perhaps some small-scale trial runs of interventions.

It would be especially nice to see another dozen major groups of scientists doing climate simulations, discovering the intervention mistakes as quickly as possible and learning from them. Medieval cathedral builders learned from their design mistakes over the centuries, and their undertakings were a far larger drain on the economic resources and people power of their day than anything yet discussed for stabilizing the climate in the twenty-first century. We may not have centuries to spare, but any economy in which two percent of the population produces all the food, as is the case in the United States today, has lots of resources and many options for reordering priorities.

Three Scenarios

FUTURISTS have learned to bracket the future with alternative scenarios, each of which captures important features that cluster together, each of which is compact enough to be seen as a narrative on a human scale. Three scenarios for the next climatic phase might be called population crash, cheap fix, and muddling through.

The population-crash scenario is surely the most appalling. Plummeting crop yields would cause some powerful countries to try to take over their neighbors or distant lands—if only because their armies, unpaid and lacking food, would go marauding, both at home and across the borders. The better-organized countries would attempt to use their armies, before they fell apart entirely, to take over countries with significant remaining resources, driving out or starving their inhabitants if not using modern weapons to accomplish the same end: eliminating competitors for the remaining food.

This would be a worldwide problem—and could lead to a Third World War—but Europe's vulnerability is particularly easy to analyze. The last abrupt cooling, the Younger Dryas, drastically altered Europe's climate as far east as Ukraine. Present-day Europe has more than 50 million people. It has excellent soils, and largely grows its own food. It could no longer do so if it lost the extra warming from the North Atlantic.

There is another part of the world with the same good soil, within the same latitudinal band, which we can use for a quick comparison. Canada lacks Europe's winter warmth and rainfall, because it has no equivalent of the North Atlantic Current to preheat its eastbound weather systems. Canada's agriculture supports about 28 million people. If Europe had weather like Canada's, it could feed only one out of twenty-three present-day Europeans.

Any abrupt switch in climate would also disrupt food supply routes. The only reason that two percent of our population can feed the other 98 percent is that we have a well-developed system of transportation and middlemen—but it is not very robust. The system allows for large urban populations in

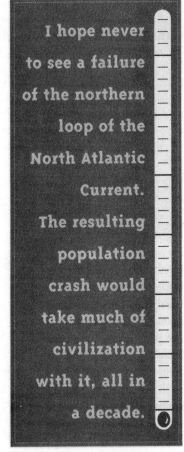

I hope never to see a failure of the northern loop of the North Atlantic Current. The resulting population crash would take much of civilization with it, all in a decade.

the best of times, but not in the case of widespread disruptions.

Natural disasters such as hurricanes and earthquakes are less troubling than abrupt coolings for two reasons: they're short (the recovery period starts the next day) and they're local or regional (unaffected citizens can help the overwhelmed). There is, increasingly, international cooperation in response to catastrophe—but no country is going to be able to rely on a stored agricultural surplus for even a year, and any country will be reluctant to give away part of its surplus.

In an abrupt cooling the problem would get worse for decades, and much of the earth would be affected. A meteor strike that killed most of the population in a month would not be as serious as an abrupt cooling that eventually killed just as many. With the population crash spread out over a decade, there would be ample opportunity for civilization's institutions to be torn apart and for hatreds to build, as armies tried to grab remaining resources simply to feed the people in their own countries. The effects of an abrupt cold last for centuries. They might not be the end of *Homo sapiens*—written knowledge and elementary education might well endure—but the world after such a population crash would certainly be full of despotic governments that hated their neighbors because of recent atrocities. Recovery would be very slow.

A slightly exaggerated version of our present know-something-do-nothing state of affairs is know-nothing-do-nothing: a reduction in science as usual, further limiting our chances of discovering a way out. History is full of withdrawals from knowledge-seeking, whether for reasons of fundamentalism, fatalism, or "government lite" economics. This scenario does not require that the shortsighted be in charge, only that they have enough influence to put the relevant science agencies on starvation budgets and to send recommendations back for yet another commission report due five years hence.

A cheap-fix scenario, such as building or bombing a dam, presumes that we know enough to prevent trouble, or to nip a developing problem in the bud. But just as vaccines and antibiotics presume much knowledge about diseases, their climatic equivalents presume much knowledge about oceans, atmospheres, and past climates. Suppose we had reports that winter salt flushing was confined to certain areas, that abrupt shifts in the past were associated with localized flushing failures, *and* that one computer model after another suggested a solution that was likely to work even under a wide range of weather extremes. A quick fix, such as bombing an ice dam, might then be possible. Although I don't consider this scenario to be the most likely one, it is possible that solutions could turn out to be cheap and easy, and that another abrupt cooling isn't inevitable. Fatalism, in other words, might well be foolish.

A muddle-through scenario assumes that we would mobilize our scientific and technological resources well in advance of any abrupt cooling problem, but that the solution wouldn't be simple. Instead we would try one thing after another, creating a patchwork of solutions that might hold for another few decades, allowing the search for a better stabilizing mechanism to continue.

We might, for example, anchor bargeloads of evaporation-enhancing surfactants (used in the southwest corner of the Dead Sea to speed potash production) upwind from critical downwelling sites, letting winds spread them over the ocean surface all winter, just to ensure later flushing. We might create a rain shadow, seeding clouds so that they dropped their unsalted water well upwind of a given year's critical flushing sites—a strategy that might be particularly important in view of the increased rainfall expected from global warming. We might undertake to regulate the Mediterranean's salty outflow, which is also thought to disrupt the North Atlantic Current.

Perhaps computer simulations will tell us that the only robust solutions are those that re-create the ocean currents of three million years ago, before the Isthmus of Panama closed off the express route for excess-salt disposal. Thus we might dig a wide sea-level Panama Canal in stages, carefully managing the changeover.

Staying in the "Comfort Zone"

STABILIZING our flip-flopping climate is not a simple matter. We need heat in the right places, such as the Greenland Sea, and not in others right next door, such as Greenland itself. Man-made global warming is likely to achieve exactly the opposite—warming Greenland and cooling the Greenland Sea.

A remarkable amount of specious reasoning is often encountered when we contemplate reducing carbon-dioxide emissions. That increased quantities of greenhouse gases will lead to global warming is as solid a scientific prediction as can be found, but other things influence climate too, and some people try to escape confronting the consequences of our pumping more and more greenhouse gases into the atmosphere by supposing that something will come along miraculously to counteract them. Volcanos spew sulfates, as do our own smokestacks, and these reflect some sunlight back into space, particularly over the North Atlantic and Europe. But we can't assume that anything like this will counteract our longer-term flurry of carbon-dioxide emissions. Only the most naive gamblers bet against physics, and only the most irresponsible bet with their grandchildren's resources.

To the long list of predicted consequences of global warming—stronger storms, methane release, habitat changes, ice-sheet melting, rising seas, stronger El Niños, killer heat waves—we must now add an abrupt, catastrophic cooling. Whereas the familiar consequences of global warming will force expensive but gradual adjustments, the abrupt cooling promoted by man-made warming looks like a particularly efficient means of committing mass suicide.

We cannot avoid trouble by merely cutting down on our present warming trend, though that's an excellent place to start. Paleoclimatic records reveal that any notion we may once have had that the climate will remain the same unless pollution changes it is wishful thinking. Judging from the duration of the last warm period, we are probably near the end of the current one. Our goal must be to stabilize the climate in its favorable mode and ensure that enough equatorial heat continues to flow into the waters around Greenland and Norway. A stabilized climate must have a wide "comfort zone," and be able to survive the El Niños of the short term. We can design for that in computer models of climate, just as architects design earthquake-resistant skyscrapers. Implementing it might cost no more, in relative terms, than building a medieval cathedral. But we may not have centuries for acquiring wisdom, and it would be wise to compress our learning into the years immediately ahead. We have to discover what has made the climate of the past 8,000 years relatively stable, and then figure out how to prop it up.

Those who will not reason
Perish in the act:
Those who will not act
Perish for that reason.

—W. H. Auden

Temperature rising

The Mackenzie Basin—one of three climate hot spots worldwide—is a harbinger of climate change globally, say scientists. A landmark study evaluates the impact on humans and wildlife as they adapt to melting permafrost, landslides, forest fires and floods

By Taras Grescoe

SPRING CAME EARLY to the Arctic this year and in Old Crow, the Yukon's most northern settlement, people are worried. "Normally the leaves start budding in the middle of June," says Joe Tetlichi, from this Vuntut Gwich'in community of 250 poised on the permafrost between the Arctic Circle and the tree line, "but largely things have been greening up in May." A few more weeks of summer might be fine, says Tetlichi, without the other changes he and others have witnessed: sudden, short-lived cloudbursts, accompanied by thunder, rather than the usual days of light drizzle. Drier summers with more lightning have ignited spruce forests and drastically lowered water levels.

"In the 1960s and '70s, we could go just about anywhere with outboard motors," says Tetlichi. "Now we have to stay within the main waterways." Elders have told Tedichi about changes on the flats where the Porcupine and Old Crow rivers meet, how higher summer temperatures are causing berries to dry up, encouraging grizzly bears to stay out longer to forage for food. "They were still seeing bear tracks in December, which is very unusual. Normally the grizzlies start hibernating in late September or early October," says Tetlichi. "They told us about a lot of changes with the plants, with the berries, with the animals. People here are very concerned about what's going on with the weather."

11. Temperature Rising

The warming watershed

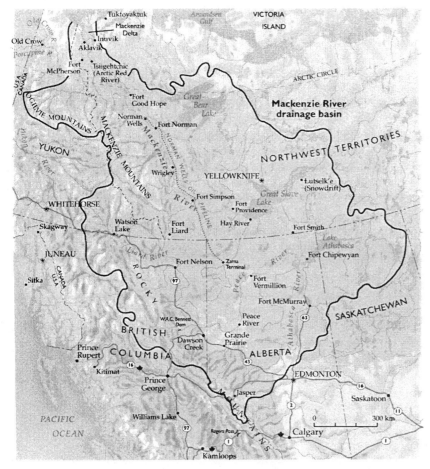

STEVEN FICK, TANYA VANDERHAM/CANADIAN GEOGRAPHIC; BASE: CANADA BASE MAP SERIES, GEOMATICS CANADA, NATURAL RESOURCES CANADA

average of half a degree. Increasingly, people have been asking whether these trends are likely to continue. Now, an extensive Canadian study is providing some of the most detailed answers yet to that question.

The Mackenzie Basin Impact Study (MBIS) focusses on the 1.8-million-square-kilometre drainage basin of the continent's largest northerly flowing river. The six-year, $950,000 effort documents how rising temperatures have already melted permafrost, contributed to hundreds of new landslides and increased the area burned by forest fires. Then, using the most sophisticated computer technology available, its authors project the future impact of continued global warming on the area's wildlife, vegetation and human inhabitants. The information is being used to help understand other observed phenomena—the significant retreat of glaciers worldwide, serious floods, droughts, and changes in wildlife behaviour and habitat.

Physicists were the first to theorize that the world's climate was warming due to changes in the composition of the atmosphere caused by human activity. Since the 1980s, climate models—the computer programs used to simulate what the climate would be like under various atmospheric conditions—have become significantly more accurate at representing worldwide weather patterns. The MBIS took that work still further by bringing together forest ecologists, permafrost geologists, wildlife biologists, and the observations of scientists and residents in northern communities. Supported by, among others, the federal government, the Northwest Territories government, B.C. Hydro, the University of Victoria and Esso Resources Ltd. (the company contributed a one-time grant in 1991), the study has already been recognized worldwide as one of the most significant bodies of work in the climate field. The long-awaited report, released in August, provides unprecedented

Such observations have become commonplace around the world and, in particular, in and around three northern regions termed climate "hot spots"—Lake Baikal in Siberia, the northwestern region of Alaska and, southeast of Old Crow, the Mackenzie Basin—where, over the last century, temperatures have risen by three times the global

Acts of emission

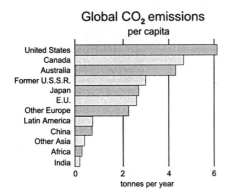

Canada has 0.56% of the world's population and emits 2% of its carbon oxide.
CO_2 comprises 80% of greenhouse gases.

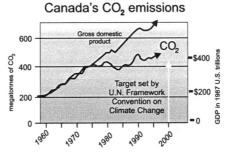

Since the beginning of the Industrial Revolution, the amount of carbon dioxide in the atmosphere has risen by 30%. The growing gap between GDP and carbon dioxide emissions is a result of conservation efforts, less energy-intensive industry, and the use of alternative energy sources.

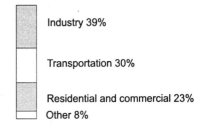

Sources of Canada's CO_2

Industry 39%

Transportation 30%

Residential and commercial 23%

Other 8%

STEVEN FICK/CANADIAN GEOGRAPHIC; SOURCES: M. GRUBB, INTERGOVERNMENTAL PANEL ON CLIMATE CHANGE (GLOBAL EMISSIONS DATA); POLLUTION DATA BRANCH, ENVIRONMENTAL PROTECTION SERVICE, ENVIRONMENT CANADA (CANADIAN EMMISSIONS DATA); STATISTICS CANADA (GDP DATA)

2 ❖ LAND-HUMAN RELATIONSHIPS

Rising temperatures have already melted permafrost, contributed to hundreds of new landslides and increased the area burned by forest fires

documentation of changes wrought by global warming and gives us a glimpse of what the future might look like if greenhouse gas emissions are not curtailed.

THE LIVING LABORATORY that is the Mackenzie Basin offered scientists a remarkably varied cross-section of geography to study. "This is an area that not only includes glaciers and snow-capped Rocky Mountains," the study's leader Stewart Cohen points out, "but also barley fields and pasture lands along the Peace River, freshwater deltas, wide expanses of boreal forest, wetland environment for bison and migratory birds, peatlands and, in the very far north, it fringes on the tundra." Since 1789, when Alexander Mackenzie explored the full length of the river that bears his name, the Mackenzie has been the highway to the western Canadian Arctic. The 4,241-kilometre-long river draws its waters from as far south as the Alberta Rockies near Jasper and its current sweeps Douglas fir logs all the way from British Columbia to the Arctic Ocean. It is fed by the Liard, Peace and Athabasca rivers, as well as by three bodies of fresh water, Great Bear, Athabasca and Great Slave lakes.

These ecosystems have undergone significant changes. "Over the last 35 years, there's been a substantial increase in temperatures—about a degree a decade—in the centre of the basin," says Cohen, an Environment Canada climatologist. "There has been glacial retreat, landslides, and a lowering of water levels. It's real warming." In addition, the southern edge of permafrost in the prairie provinces has receded northward by about 100 kilometres during the past century. In 1994, forest fires swept through the boreal forest in the Northwest Territories, affecting three million hectares—three times the annual average. A year later, fires forced the evacuation of about 900 people from Fort Norman and Norman Wells, through which Exxon's oil passes on its journey south via the Norman Wells pipeline. The cause? Warmer than usual spring water temperatures which increase run-off before the ground has thawed, leaving little opportunity for moisture to be absorbed.

Not surprisingly, such changes have had a striking impact on the Arctic landscape. "The vast majority of pines actually need fires to open up their cones—they don't reproduce otherwise," says Canadian Forest Service climatologist Ross Benton. "But when fires happen every five years, that's well beyond the forest's capacity to regenerate." Larry Dyke, a Geological Survey of Canada research scientist who has been studying the Mackenzie Basin since the late 1970s, says fires have a profound effect at ground level. "One year you've got forest and other vegetation covering the ground; the next year this is destroyed." With this insulating layer gone, the ground—two-thirds of which is underlain by permafrost—is suddenly exposed to much higher temperatures than normal. This in turn deepens the active layer, the metre-or-so of ground above the permafrost that thaws every summer, melting ground ice and promoting landslides. With more forest fires, landslides are more common. "These can be very rapid," says Jan Aylsworth, a surficial geologist who has catalogued 3,400 landslides in the area. "A thin layer of land will come down, and that failure in turn can induce one behind it. This will expose more icy ground and the landslide can grow in area and depth."

Joe Benoit, the 38-year-old president of the Aklavik Renewable Resources Council, has witnessed remarkable changes on the Mackenzie Delta, some of which he described to a MBIS round table. "You see these forests of willow trees sticking out of the water," says the Gwich'in native from Alilavik, N.W.T., "and willows aren't supposed to grow like that." Increased flows (some caused by British Columbia's W. A. C. Bennett Dam, thousands of kilometres upstream on the Peace River) have eroded the bank, laying bare the permafrost at bends in the river; warmer summer air in turn melts the ground, causing vast sections of what was once the bank to collapse. The additional silt in the water also has an impact on the ecology of the river. "Before, we might have lost a few feet of bank in a season," says Benoit. Lately, he has watched hectares at a time disappear into the river. "I've not seen it happening like that in my lifetime."

PROJECTIONS MADE by the MBIS scientists show that collapsing river banks may be the least of the changes northern inhabitants will face in their lifetimes. The study incorporated protections of climate change undertaken by the Canadian Centre for Climate Modelling and Analysis in Victoria. The centre's model, widely used by the international climate research community, was first developed in the 1980s and has since been refined. It creates a three-dimensional picture of how the atmosphere, and hence climate, is likely to react to changes in greenhouse gas levels. Using the Atmospheric Environment Service's supercomputer in Dorval, Que., the centre's scientists created a portrait of how the global climate would react if carbon dioxide (which accounts for 80 percent of greenhouse emissions) continues to increase until it is double

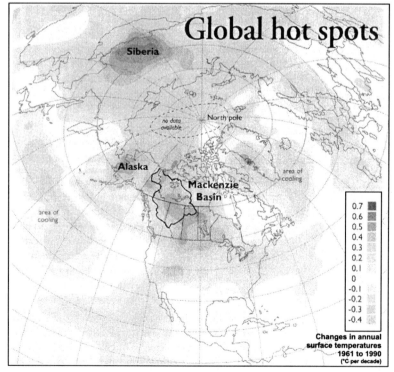

CLIMATE RESEARCH BRANCH, ENVIRONMENT CANADA

what it was before the Industrial Revolution. Scientists suggest increases in greenhouse gas concentrations, equivalent to a doubling of carbon dioxide, could occur as soon as 2040.

For most of Southern Canada, the model suggests an average yearly warming of about 4°C in summer and 4° to 8° in winter. In the Mackenzie Basin, the prediction is a warming of 2° to 5° in summer and 4° to 8° in winter. The increase in the annual average temperature could continue at the rate of a degree a decade over the next 50 years. Observations of recent trends show the Mackenzie Basin has warmed more quickly this century than any other part of Canada. The researchers who studied the permafrost, wildlife and forests of the Mackenzie Basin relied on these figures—as well as the model's predictions of changes in precipitation—to hypothesize what the changing Canadian Arctic might look like.

According to Michael Brklacich, a Carleton University geographer who contributed to the report, wheat farmers in the Peace River region may benefit from longer and warmer growing seasons but those benefits may be offset somewhat by drier summers. Irrigation might provide some relief, but the economic feasibility of this is unknown. Warmer weather may allow economically useful tree species such as the Douglas fir to move north if the trees can seed on new terrain rapidly enough, but it will also enlarge the domain of fire and pests, which move even more rapidly. Increase the number of hot summer days even slightly, says climatologist Benton, and insects such as the white pine weevil, which currently inhabits southern parts of the basin, could exist throughout. "The larvae burrow under the bark and kill the leader of the spruce," says Benton. "You end up with some stunted, weird-looking trees that are pretty useless for timber."

Whether or not humans and wildlife can adapt to the changes being predicted is a hot topic in the MBIS report. Will advances in fire fighting technology and forest management practices, for example, be enough to counter the impact on human activity and wildlife populations from increased forest fires? (The area burned annually could increase by up to two and a half times.) How can one measure the impact of a loss of wildlife on native communities that rely increasingly on a wage economy and imported goods from the south?

Some scenarios are more fully explored. For scientists studying wildlife, where predicting the effects of changing temperatures is as complex as the intricate interaction between animals and their habitat, the Mackenzie study offers good and bad news.

Of the 40 species of shorebirds breeding regularly in Canada, 26 species breed almost exclusively in the arctic or subarctic, including the long-billed dowitcher, which is known only to breed in the Mackenzie delta, and the near-extinct Eskimo curlew. Most shorebirds favour shallow wetland habitats, making them "an excellent indicator group for modelling the effects of climate change on wildlife populations," notes the report. With warmer summers causing greater evaporation and a loss of marshes, predictions are that there will be a decline in the most common and most aquatic species—the red-necked phalarope. A detrimental impact is also predicted for northern-nesting geese due to a disruption in feeding areas. But the picture is not completely bleak: shore-birds that prefer drier conditions—common snipe, semipalmated sandpiper and lesser golden plover—are expected to increase in number.

The effects of longer, warmer summers on the caribou "appear more severe," although questions remain. Three herds enter the basin, each still serving as a significant food source for nearby native communities. The 140,000-strong Bluenose herd's migration takes it across the northern reaches of the basin. To the east is the 300,000-plus Bathurst herd. In the west, the 160,000-strong Porcupine herd. Each year, the Porcupine herd travels 500 kilometres between the Yukon's Ogilvie Mountains and summer calving grounds on the Beaufort Sea. "When the Porcupine herd comes through Old Crow it's something to see," says Joe Tedichi, who chairs the Porcupine Caribou Management Board. "The only thing attaching us to our tradition and culture is the Porcupine Herd," he says.

Wildlife biologist Don Russell says that if winter temperatures increase by 4° to 8°, "precipitation will also increase, and you're going to see years of incredibly high snow."

Cold snaps and hot spells

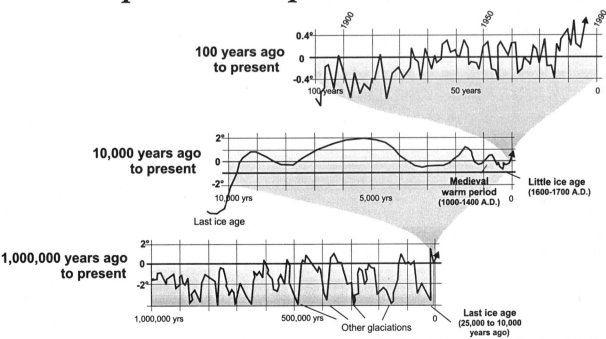

Climate change is a fact whether viewed over 100, 10,000, or a million years. Despite some cool periods, the trend over this century has been toward higher temperatures.

'The recipe of our air is being altered by 5.5 billion cooks, most of them not realizing that they are part of the kitchen staff'

Caribou will expend more energy pawing through deeper snow to get at lichens, the mainstay of their winter diet. As temperatures increase in the summer, insects become more active: at 6°, clouds of mosquitoes descend on the herd; at 13°, the parasitic nasal bot fly appears, spraying its living larvae up the caribou's nostrils. More warm days allow these insects to harass the caribou longer. Avoiding the pests takes caribou to windy but often barren ground. Between 1989 and 1992, the Porcupine herd's population dropped by 18,000, with 1990 being a particularly bad year. "Snow depth reached 80 centimetres, even in the good areas," says Russell. "When spring came, the snow melted all at once; and then they had a really warm July, which meant more insect harassment. By the time the fall came, those animals were bone racks." These are the conditions—snow-heavy winters and rain-free Julys—the Canadian Climate Centre's model indicates will be the norm in the Mackenzie Basin in decades to come.

Despite all of that, Russell says the caribou could adapt to the new conditions. While insects would be a problem, longer summers would provide fresh vegetation earlier and early springs could improve calf survival rates. Winters would be more difficult but caribou could move to areas south of the Porcupine River where there would be less snow accumulation, he says.

For the half-million human inhabitants of the Mackenzie Basin, the most important variable may well be the pace of change. At a study workshop in Yellowknife last year, a representative of the Inuvialuit—most of whom still rely on fish, caribou and whale to survive—emphasized that they could adapt if changes took place over 500 years. "Changes of nature have been slow," Dean Arey of Inuvik, N.W.T., told the first of six round-table discussions, "and the Inuvialuit have been able to alter their lifestyles." But, as Cohen notes, significant warming in the Arctic is taking place on a scale of decades, not centuries. A study by University of Saskatchewan geographer Robert Bone suggests that the predominantly native communities of Tsiigehtchic, Fort Good Hope, Fort McPherson and Tuktoyaktuk could experience major problems as the ground gives way under buildings (especially those with shallow foundations), roads and power lines. In Tuktoyaktuk, several buildings have already been abandoned, due to rising sea levels and shore erosion which global warming could be accelerating. A model by Geological Survey of Canada geologist Fred Wright shows that, given a 2° warming in mean annual temperature—an increase that climate models suggest could take place over the next two decades—virtually all of the permafrost around Fort Simpson, which sits in the middle of the basin, would slowly disappear. However, with enough warning, threatened communities could be relocated.

For Jim Bruce, a meteorologist and former assistant deputy minister of Environment Canada, the MBIS is an important piece of work as Canada prepares to attend the international summit on climate change in Kyoto, Japan, in December, where participating nations will work to further a United Nations' agreement to limit greenhouse gas emissions.

"The Mackenzie Basin was one of the regions in the world where the warming trend showed first, and so it gives us some real clues to what might be in store for many other parts of the world in the future," says Bruce, now head of the Canadian Climate Program, which co-ordinates federal and provincial agencies, universities and private sector groups with an interest in climate-related issues. "It's also one of the few places in which there has been a real attempt to engage the local people in assessing what the impacts might be on their lives."

In 1992, Canada was one of 150 countries to sign the Rio de Janeiro agreement promising to stabilize greenhouse gas emissions at 1990 levels by the year 2000. It was an ambitious commitment considering our rate of emissions. Canadians—by heating their homes, running their appliances, and driving their cars—generate 5.4 tonnes of greenhouse gases per capita each year. Add emissions from industry and energy production and that number jumps to 21 tonnes per capita, putting us second only to the United States among developed countries. And those emissions are expected to increase by 8.2 percent by the year 2000. "The recipe of our air is being altered by 5.5 billion cooks, most of them not realizing that they are part of the kitchen staff," writes Cohen in the MBIS report.

Some European nations, including Holland, England and Germany, are expected to meet their year 2000 commitments by raising taxes on oil and investing in energy alternatives. While environmentalists in Canada have lobbied for a carbon tax, Prime Minister Jean Chrétien told Parliament last year that he was against the idea despite its success in limiting the consumption of fossil fuels elsewhere.

When Canada appears in Japan next month, says John Robinson, director of the University of British Columbia Sustainable Development Research Institute, "it should be a major embarrassment. We've gone from being white hats on this issue to being in the back of the pack." The problem, however, is that Canada is unlikely to be embarrassed because the majority of Rio signatories have not met their emissions targets either. In fact, annual emissions of carbon dioxide worldwide are expected to increase by three billion tonnes by the year 2000.

There are solutions: more efficient automobiles, decreased subsidies to the coal and oil industries, setting emission reduction targets for industries, switching from coal and oil to natural gas, and investing in alternative energies such as solar and wind power. The price tag for all this might not be as onerous as industry leaders suggest. "Most studies predict that the economic costs of reducing greenhouse gas emissions are effectively negligible for Canada," says Bruce. "The coal and oil industries will suffer, but others will gain. And there's a clear benefit. If we invest in solar and wind power development and energy efficiency, which are far more labour intensive than fossil fuels, we would create more jobs than we had with coal and oil." Unfortunately, Canada's record suggests that there is little political will to take strong action on any of these fronts.

Meanwhile, in the Mackenzie Basin, landslides, melting permafrost, and forest fires show that climate change is already having an impact on the Canadian landscape. "Governments have simply not come to grips with the fact that we can't keep on burning large amounts of fossil fuels forever," says Bruce. "The atmosphere is beginning to kick back at us, and will do so, more vigorously, as time goes on."

Taras Grescoe is a freelance writer living in Montreal.

The battle over big water projects

Dammed if You Do . . .

■ *In the 1940s, Woody Guthrie sang the praises of the great Grand Coulee Dam on the "King Columbia River." Today, plans for dams and other large water projects, such as massive dredging operations or river diversions, are more likely to launch protests than laudatory verses. Our cover package looks at a burgeoning global controversy encompassing concerns about the environment, energy, economics, and development.*

The Economist

It must once have seemed like power for free, a gift from the heavens. A dam creates electricity from nothing other than the falling of water through its turbines. The supply is replenished in perpetuity by the munificence of nature. Nothing is contaminated. No pollutants are belched into the atmosphere.

The lake that forms behind a dam, meanwhile, should allow a country to make the best use of what is often its most precious resource. A dam lets people have water when they want it (for irrigation) and holds it back when they do not (by preventing floods). Often the lake is a tourist attraction, a rich source of fish, and a fallback in times of drought.

Dams are seen, by many poor countries in particular, as the only cost-effective way to generate and often export power. They also create jobs and can be status symbols. However, dams have well-publicized vices. They are notorious for causing great environmental change. And they force massive human resettlement, mostly of people who live where the lake is due to appear. The World Bank estimated in 1994 that the 300 "large" dams built every year—those more than 50 feet high and some of those only 33 feet high—force some 4 million people to leave their homes.

The real impact that dams have on the environment and on people has rarely if ever been included in a dam's cost-benefit analysis. The benefits are generally taken for granted and are not properly measured against the outcome of refraining from building a dam.

"Only in a handful of cases have people displaced by a dam ended up better off."

But in recent years, protests against dams have grown more common. Having stopped the building of nuclear reactors in rich countries, public opposition is starting now to alter the predilection for dams in poor ones. Some big dam projects have been abandoned halfway through; funding for new ones is not so easy to find. Even the World Bank, formerly a fan of dams and still their single biggest financier, has become cautious. Hence its request for a meeting of interested parties, from dam builders and financiers to conservationists and dam protesters, which took place in April [1997] in Gland, Switzerland.

[Stephanie Flanders writes in the centrist *Financial Times* of London that "the depth of disagreement between some of the participants" was so deep that it was surprising "they had all managed to stay in one room, let alone have a constructive debate."

[According to Flanders, discussions focused on the practices of the World Bank, the largest provider of financing for dams in developing countries. "Participants agreed to set up an international group to assess the past experience of planning, building, and living with large dams," she writes. "The group would have two years to complete the review and come up with the guidelines on avoiding the social and environmental pitfalls of building dams."—WPR]

Many of the worst environmental effects of dams stem from their supposed benefits. For instance, the constant and reliable irrigation a dam provides can waterlog the ground. The water brings underground salt to the surface, then evaporates and leaves it behind. Eventually, the soil becomes too salty for crops to survive. Even the prevention of floods is a mixed blessing. The silt that was once carried downstream by a swollen river, replenishing the soil and nutrients, no longer makes its journey. Instead it clogs up the reservoir.

To these and other problems, there are some partial fixes. Underground drainage may prevent salinization; channels in the dam can allow some of the silt through; fish ladders allow certain migrating fish to continue their journeys.

But even with these in place, the effect of a dam is unpredictable. The Aswan Dam and the Aswan High Dam, both on the Nile, are commonly held up as exemplars of planning. But the arable land downstream is being eroded away, partly because it is not getting enough silt.

The protection from floods that dams offer turns out to be one of their most troubling drawbacks. Traditionally, much of the land near a river is irrigated by floods and planted as the waters recede. A dam can stop this from happening and so rob millions of people downstream of their livelihood. It is usually assumed that the dam's irrigation of other land will make up for the economic loss, but a study of the Kainji Dam on the Niger River reckoned that the dam reduced rice production downstream by 18 percent and the fish catch by 60 to 70 percent. And a study of the Bakolori Dam on a tributary of the Niger found that its economic benefits were outweighed by the loss of crops, fish, and livestock by villagers downstream.

The thorniest problem is the uprooting and resettlement of people. Those evicted by a dam often have to change their way of life as well as their location. The World Bank itself reckons that only in a handful of cases have most of the people displaced by a dam ended up better off.

The social penalties that dams impose are much better understood nowadays. What is still lacking is a mechanism that enables those costs to be taken into account when a dam is planned. Governments keen on a dam being built are often reluctant to force up expenses by insisting on too many environmental or resettlement measures. And when an external agency such as the World Bank tries to attach such conditions, governments (and the bank) are often bad at enforcing them.

All this has conspired to give dams a bad name. Even the World Bank has shifted its emphasis toward coal as a source of power. But burning fossil fuels has unpleasant side effects of its own. That is why the bank remains interested in dams and keen to find a way to make them acceptable. But in the long run, maybe only when the world runs out of profitable places to put dams will the building of harmful ones stop.

—*"The Economist" (conservative newsmagazine), London, April 19, 1997.*

The Human Debris of China's Mega-Dams

RELOCATION'S LINGERING EFFECTS

South China Morning Post

■ *In recent years, the Three Gorges Dam project has been widely covered in the international press. Beijing's government-owned "China Daily" predictably says the project has not hurt the environment. But the dam on the Yangtze will create a huge inland sea and force the relocation of more than 1 million people. These people are far from the only Chinese who were ever evicted to make way for a dam.*

When a group of foreign journalists recently paid a rare visit to the impoverished mountains of northern Guangdong province in southern China, local peasants tracked them down and asked them to smuggle out a petition bewailing their fate. In 1958, Mao Zedong ordered the construction of a dam on the Xinfeng River that flooded the land where 100,000 peasants lived. Forty years later the displaced population has doubled, and many are still living in dire poverty.

"During the past 10 years no one has cared for our lives. Only one person in 14 has found a job," says the petition, written on behalf of 300 peasants.

During the Mao era, China built 84,000 dams, including 2,700 large and medium-size reservoirs that displaced more than 10 million people. A report released by the Chinese government in 1984 revealed that one third of these involuntary resettlers were living in abject poverty 30 years later, while another third were just getting by.

Like the others in northern Guangdong, Li Nienkun lost his land in 1958 when, at the height of the Great Leap Forward, his house and land were submerged. His family trekked into the surrounding hills. The authorities provided inadequate grain rations. "Quite a few people starved to death," Li says.

Few details of the hardships suffered by victims of huge dams had been published in China until April, 1996, when the magazine *Chinese Writer* described how "refugees" were moved from the Yellow River to arid regions in western Shaanxi and Ningxia Hui after cadres set fire to their houses. Compensation was about $43 per household, and peasants were exhorted to display "self-reliance." When grain rations fell to 7.5 pounds a month, people tried to flee but were stopped along the Yellow River. "It was as bitter as the exodus of the Jews out of Egypt," the magazine said.

Conditions in Guangdong may not have been so terrible as those in north China, but officials in Guangdong's Heyuan prefecture admit the large influx of landless peasants plunged the district into such poverty that, decades later, the raising of living standards is still a challenge. Two years ago, the Guangdong government began a renewed effort to end the poverty created by the damming of the Xinfeng River. Thousands of peasants who were scratching out an existence in the steep mountains around the reservoir were brought into government-run settlement camps.

Chen Heling, a resettlement officer at one camp, Le Yuan (Happiness Garden), said 20,000 people had been encouraged to leave the region and move in with relatives working in the province's big cities. Another 20,000 still lived in the mountains, while 10,000 were resettled in camps.

As in other mountainous areas of southern China, the local authorities have ambitious employment schemes. Such measures have won China praise from international resettlement experts. In 1994, the World Bank reviewed 200 dam projects around the world. "We found China has had a remarkably satisfactory resettlement performance during the 1980s and '90s," says Michael Cernea, the World Bank's top expert on resettlement.

Yet, many of those interviewed at Le Yuan camp complained bitterly about their new conditions. "They forced us to come here but there is nothing to do, and we are all poor now," says Li Nienkun.

Those peasants who dared to draft the petition originally belonged to a commune brigade resettled on land close to Heyuan city, but in 1988, when Heyuan was allowed to expand, their land was appropriated. Critics of China's resettlement policies believe such cases illustrate the lack of democratic consultation in China and the weakness of villagers in defending their interests against those of the state.

—*Jasper Becker, "South China Morning Post" (independent), Hong Kong, May 3, 1997.*

An Unholy Alliance In East Malaysia

LINKING DAMS AND LOGGING

THE AUSTRALIAN

In the east Malaysian state of Sarawak, the confluence of the Rajang and Balui rivers is marked by the Bakun Rapids. Here, churning whirlpools and crashing white water have undone many an experienced boatman. No sane boatman would even attempt crossing it today—20 yards above the cataract, heavy excavating machinery is gouging earth from the hillside and dumping it into the river below. Diversion tunnels, marking the initial stage of the ambitious Bakun River Dam, are taking shape.

In the first months of construction, on-site witnesses reported 10 workers crushed in accidental cave-ins. The deaths were not reported in the Malaysian press. And when freelance journalist M.G.G. Pillai wrote them on his Web site, he was promptly sued for libel by Tan Sri Ting Pek Khiing, the flamboyant millionaire executive chairman of the project development company, Ekran Berhad. Few facets of the $6.2-billion Bakun Hydroelectric Project are not secret. There has been no public debate on the project.

Following construction of the diversion tunnels, the main project will see a concrete dam wall as high as a 60-story building rise from the rapids to create a catchment area of about 5,700 square miles in the center of Borneo's tropical rain forest. It will be the largest hydroelectric dam in Southeast Asia, and it will inundate hundreds of species of rain-forest flora and fauna, many of them protected and endangered. But contrary to statements by environmental groups and critics of the project, the affected area is far from pristine. Large sections have already been slashed and burned by shifting cultivators and mercilessly logged by East Malaysian timber companies.

The rising waters of the catchment will displace nearly 10,000 indigenous people who inhabit long houses along the hillsides overlooking the present rivers. Although "fair and reasonable" compensation is promised for their losses, they will leave behind not only their rambling homes but also a traditional culture based on farming, fishing, and rain-forest hunting and gathering. Further, says Tom Jalong of Malaysia's Friends of the Earth, "The people have not received any detailed information as to what compensation they will get for their homes or how much per acre."

The Malaysian government says the Bakun hydroelectric plant will generate 2,400 megawatts of power, 2,000 of which will be transferred 416 miles by submarine cable to peninsular Malaysia. While the country's economy has grown by more than 7 percent each year for the past decade, energy consumption has risen at a rate of 10 percent a year. The Bakun project will allow the government to reduce its current dependence on coal burning.

However, even the government acknowledges that power companies generated an energy surplus of some 64 percent last year. As Kua Kia Soong, spokesman for a coalition of nongovernmental organizations opposed to the project, points out in his 1996 book, *Malaysia's Energy Crisis,* the country's wealth of clean-burning natural gas reserves is currently exported to Japan.

But the Malaysian government, and in particular Prime Minister Mahathir Mohamad, believes the project will create low-cost renewal power. Mahathir was succinct in dismissing the dam's critics last year: "Malaysia wants to develop, and I say to the so-called environmentalists, 'Mind your own business.'"

To be sure they do just that, 60 security police are stationed at the remote Bakun site. Journalists, in particular, are prohibited from entering the area. The government says the site is security controlled on safety grounds.

I traveled upriver from Sarawak's coastal port of Sibu into the dam-affected area. At the project site, I witnessed the extraordinary scale of logging. The *Business Times* of Singapore estimated the value of rain-forest hardwoods in the catchment area at about $400 million in 1994; that figure has now climbed to at least $507 million as hardwood prices soar. These logging spoils belong to project developer Ekran.

In Sarawak, it is difficult to distinguish the loggers from the government. Conflicts of interest and nepotism stain the project. The two sons of Sarawak's chief minister and resources minister, Taib Mahmud, hold more than 4 million shares in Ekran Berhad. Environment Minister James Wong owns one of the state's largest timber operations. It's little wonder Ekran has won tax concessions that will see the conglomerate pay no taxes on profits for the project's 10-year development period.

Still, the project's greatest human impact is on the local tribes—the Kenyah, Penan, Ukit, and Lahanan peoples living in 15 long houses earmarked for submersion by the dammed rivers. The government plans to resettle them near the present outpost of Belaga. Each family will be housed in an apartment block and will receive about 3 acres of land for farming. Most residents believe the resettlement will result in their working as plantation coolies for subsistence wages, divorced from their river/forest culture.

The underlying reason that the government wants complete control of the catchment probably lies with the Sarawak economy's insatiable dependence upon logging and timber exports to Japan, Taiwan, and other countries. Since the early 1980s, logging has flourished along those waterways, taking first the established primary forest and later attacking secondary forest and regrowth. Replanting is unheard of. Extracted logs have left a barren surface on the forest floor; this has caused massive erosion for miles along the area's riverbanks.

The rise of the dammed rivers to the new height will allow loggers access to a previously inaccessible level of forest, free from the interference of natives. It will extend the life of Sarawak's timber industry, if not by much. Even the International Tropical Timber Organization, which environmentalists regard as pro-logging, warned in 1991 that Sarawak would be denuded in 13 years.

While many see the Bakun Dam construction as a *fait accompli*, opposition to it continues. Opponents see their last opportunity to stop it in the difficulties being experienced by Ekran in arranging financial backing. A report written by former Chase Manhattan Bank director Mark Mansley last year casts doubts over official estimates of the dam's capabilities, as well as its potential investment prospects. It concludes that the hydro project will almost certainly be incapable of reaching capacity energy-production targets.

Mansley's report was not published in Malaysia, but it has given dam opponents like Kua Kia Soong fresh resolve. "We know that prospective investors are very dubious," he says. "As long as the fate of the natives is still uncertain, and as long as the forests have not been completely logged—we have faith we can stop the project."

—Dennis Schulz, *"The Australian"* (centrist), Sydney, May 10–11, 1997.

Creating a River, Destroying a Swamp

BIRDS OR BARGES?

VEJA

One of the largest and richest ecosystems in the world, southwestern Brazil's swampy Pantanal, has escaped destruction because it is one of the most forbidding places on Earth. A half-million people live in an area the size of England dominated by alligators, fish, birds, and clouds of insects. But now engineers, businessmen, diplomats, and consultants are at work on the Paraguay-Paraná Waterway, a 2,140-mile shipping corridor that would link the cities of Cáceres in Mato Grosso state and the Uruguayan port of Nueva Palmira. If the waterway is built according to plan, the swamp could be destroyed.

According to the plan under consideration by the Inter-American Development Bank, which would provide funding for the $102-million project, it would be necessary to double or, in some places, even triple the depth of the Paraguay River, the main river in the Pantanal. Bends and turns in the river would be straightened out, and some 29 million cubic yards—4 million cartloads—of sediment would have to be removed.

In the last several months, the planned waterway has begun to generate international interest—and the concern is not unwarranted. Brazil's four neighbors in the Rio de la Plata basin—Argentina, Paraguay, Uruguay, and Bolivia—all have compelling arguments for the project. By making way for year-round barge traffic, it would, they say, break the isolation of the area encompassing central western Brazil, eastern Bolivia, and northern Paraguay.

The waterway is vital to Argentina because 75 percent of its population lives in one area affected by the project. For Paraguay, the waterway would reduce its dependence on Brazil. Today, almost all Paraguay's agricultural production is carried by truck and exported through the port of Paranaguá, in Paraná state. And the waterway would provide Bolivia's long-dreamed-of exit to the sea. "This project would be a bridge

12. Dammed If You Do...

Paraguay-Paraná Waterway: *Shipping route.*

between old Bolivia, a poor producer of cocaine, and the Bolivia of the future, which would produce grains," says Bolivian businessman Miguel Aguirre.

For Brazil, the benefits are far less obvious. In fact, according to Israel Klabin, president of the Brazilian Foundation for Sustainable Development, Brazil will benefit least from the project, but will suffer the most environmental damage.

Studies show that the waterway would interfere with the flow of the Paraguay River, which is responsible for periodically turning the Pantanal into the world's largest flood plain. The Pantanal is a giant hydrographic funnel. The quantity of water that enters it is so large, and the land is so flat, that the rivers are not able to drain away all the water.

Carlos Eduardo Morelli Tucci, an engineer with the Institute for Water Resources Research at the Federal University of Rio Grande do Sul, has found that only 40 percent of the water entering the swamp is drained out by the rivers. "The other 60 percent evaporates after remaining for months in basins and flooded areas, which operate as natural nurseries for flora and fauna," he says.

Dredging the Paraguay River means widening the mouth of that funnel. That would increase the flow and speed of the river and diminish the amount of water stored in the swamp. The result of that process would be a reduction in swampy areas of from 9 percent to 50 percent, according to different estimates.

The swamp is teeming with life, bursting with the sounds of wind and birds and other animals. One cannot be there for more than a minute before hearing something move. Dawn breaks to the chatter of birds, millions of them in all imaginable colors, sizes, and shapes. Key to the biological chain of life in the swamp are the fish. Some 262 species have been catalogued, some as long as six feet and weighing as much as 220 pounds. They feed on nutrients deposited in the flood areas and serve as food for thousands of other species of birds, reptiles, and mammals. And scientists have only begun to venture into this marshy vastness.

"This is an extraordinary laboratory of wildlife," says Emiko Kawakami, an ichthyologist. "Animals that elsewhere are at risk of extinction are living here in a state of preservation so primitive that it is as if human civilization had never existed."

In a recent study, officials from Embrapa, the Brazilian Agricultural Research Corp., surveyed every square mile of the swamp, calculating the population of some species. The numbers are striking: 32 million alligators, 2.5 million capybara, 71,000 wild deer, 35,000 swamp stag, and 9,800 broods of wild boar.

Water Fight

NewAfrican

Namibia does not have the water it needs for its growing economy and population. Namibia's landlocked eastern neighbor, Botswana, plays host to large parts of the Kalahari Desert and is also not a water-rich nation.

But the northern part of Botswana is home to the Okavango Basin. The Okavango, which rises in central Angola, flows into Botswana, where it creates the world's largest inland wetland. The basin is a robust and resilient ecosystem that could probably survive almost any mishap—except having the water turned off.

But now, the Namibian government, in its increasingly desperate search for water, is planning massive unilateral diversion of Okavango River water without consulting Angola or Botswana. The water would be transported hundreds of miles across desert land by a combination of pipeline and aqueduct to the central region and the city of Windhoek. Funding for the project—$227 million—has been pledged by the Chinese government.

The plan's impact on the basin could be catastrophic. The Botswanan government could see the entire north of the country turn to Kalahari dust.

And there is another option. Experts believe that only half of Namibia's available ground-water resources are being used, and some suggest that could provide an alternative to taking Okavango River water.

The Namibian government seems, however, to have been swayed by the promise of easy money from the Chinese. Also, Owens Corning, a U.S. corporation that could supply the $50 million in pipes, has suggested that the pipeline would accelerate the rate of Namibian industrialization.

Current expectations are that the Namibian Parliament will debate, and perhaps authorize, the project before an environmental impact study has been fully completed.

Meanwhile the Botswanan government looks on with increasing alarm and frustration. In some quarters, the belief is growing that if Namibia continues to decline meaningful negotiation on the Okavango issue, a military response will become Botswana's only option. The prospect of two of the most responsible governments in Southern Africa coming into armed conflict over water is a poor omen for the international cooperation needed to resolve water-rights issues.

—*Adrian Jenvey, "New African" (monthly), London, April, 1997.*

The swamp acts like a sponge, retaining the waters and sediments that descend from the Andes ice melt and the flood waters from the Brazilian highland plain. Geologists imagine that millions of years from now, a considerable part of the soils now in the Andes and the high plains will have been brought down into the swamp by the Paraguay River and its tributaries. The area will then become a large forest, with a topography similar to today's Amazon.

The waterway would mean ever larger movements of freight in the vicinity of the Pantanal, and some Bolivian businessmen have even proposed moving fuel on barges on the waterway. A collision involving one of those barges could be "an unimaginable disaster," says Angelo Rabelo, secretary of the environment in Corumbá. "An oil slick would take months to cross the swamp from north to south and would seep into thousands of square miles of flooded areas. It would be virtually impossible to clean up what was left."

The original waterway plan called for cutting a canal from Cáceres to Corumbá at a total cost of $1.2 billion for the first phase alone. Reaction to that was so negative that the Inter-American Development Bank and the United Nations Development Program, the two funding agencies for the project, rejected it. Now, an environmental impact study has just been finished on the new, more moderate version of the project. It says that the harm done to the swamp area would be "small to moderate."

But the very scientists contracted to do the study do not seem to support its conclusions. "We had a month to do the whole job," says Brazilian biologist Alvaro de Almeida, who was hired by the consortium to do the biological inventory of the swamp. "It is impossible to evaluate such a large ecosystem in so short a time."

Today, the Pantanal contains Brazil's largest cattle herds—nearly 4 million cows and oxen, spread over 2,500 plantations. The numerous birds help prevent the spread of pests that affect cattle herds in other parts of Brazil. Herds of thousands of head are controlled by cowboys on horseback. This is old-fashioned farming, with the cattle kept in a semi-wild state.

In recent years, ranch owners and environmentalists have begun worrying about soy production, which silts up the rivers, and gold prospecting, which contaminates the fish with mercury. These, though, are small concerns compared with the risks of a waterway that could change the face of the swamp forever.

—Laurentino Gomes, "Veja" (centrist newsmagazine),
São Paulo, April 23, 1997.

Past and Present Land Use and Land Cover in the USA

WILLIAM B. MEYER

Dr. William B. Meyer is a geographer currently employed on the research faculty of the George Perkins Marsh Institute at Clark University in Worcester, Massachusetts. His principal interests lie in the areas of global environmental change with particular emphasis on land use and land cover change, in land use conflict, and in American environmental history.

"Land of many uses," runs a motto used to describe the National Forests, and it describes the United States as a whole just as well. "Land of many covers" would be an equally apt, but distinct, description. *Land use* is the way in which, and the purposes for which, human beings employ the land and its resources: for example, farming, mining, or lumbering. *Land cover* describes the physical state of the land surface: as in cropland, mountains, or forests. The term land cover originally referred to the kind and state of vegetation (such as forest or grass cover), but it has broadened in subsequent usage to include human structures such as buildings or pavement and other aspects of the natural environment, such as soil type, biodiversity, and surface and groundwater. A vast array of physical characteristics—climate, physiography, soil, biota—and the varieties of past and present human utilization combine to make every parcel of land on the nation's surface unique in the cover it possesses and the opportunities for use that it offers. For most practical purposes, land units must be aggregated into quite broad categories, but the frequent use of such simplified classes should not be allowed to dull one's sense of the variation that is contained in any one of them.

Land cover is affected by natural events, including climate variation, flooding, vegetation succession, and fire, all of which can sometimes be affected in character and magnitude by human activities. Both globally and in the United States, though, land cover today is altered principally by direct human use: by agriculture and livestock raising, forest harvesting and management, and construction. There are also incidental impacts from other human activities such as forests damaged by acid rain from fossil fuel combustion and crops near cities damaged by tropospheric ozone resulting from automobile exhaust.

Changes in land cover by land use do not necessarily imply a degradation of the land. Indeed, it might be presumed that any change produced by human use is an improvement, until demonstrated otherwise, because someone has gone to the trouble of making it. And indeed, this has been the dominant attitude around the world through time. There are, of course, many reasons why it might be otherwise. Damage may be done with the best of intentions when the harm inflicted is too subtle to be perceived by the land user. It may also be done when losses produced by a change in land use spill over the boundaries of the parcel involved, while the gains accrue largely to the land user. Economists refer to harmful effects of this sort as *negative externalities*, to mean secondary or unexpected consequences that may reduce the net value of production of an activity and displace some of its costs upon other parties. Land use changes can be undertaken because they return a net profit to the land user, while the impacts of negative externalities such as air and water pollution, biodiversity loss, and increased flooding are borne by others. Conversely, activities that result in secondary benefits (or *positive externalities*) may not be undertaken by landowners if direct benefits to them would not reward the costs.

Over the years, concerns regarding land degradation have taken several overlapping (and occasionally conflicting) forms. *Conservationism* emphasized the need for careful and efficient management to guarantee a sustained supply of productive land resources for future generations. *Preservationism* has sought to protect scenery and ecosystems in a state as little human-altered as possible. Modern *environmentalism* subsumes many of these goals and adds new concerns that cover the varied secondary effects of land use both on land cover and on other related aspects of the global environment. By and large, American attitudes in the past century have shifted from a tendency to interpret human use as improving the condition of the land towards a tendency to see human impact as primarily destructive. The term "land reclamation" long denoted the conversion of land from its natural cover; today it is more often used to describe the restoration and repair of land damaged by human use. It would be easy, though, to exaggerate the shift in attitudes. In truth, calculating the balance of costs and benefits from many land use and land cover changes is enormously difficult. The full extent and consequences of proposed changes are often less than certain, as is their possible irreversibility and thus their lasting significance for future generations.

WHERE ARE WE?

The United States, exclusive of Alaska and Hawaii, assumed its present size and shape around the middle of the 19th century. Hawaii is relatively small, ecologically distinctive, and profoundly affected by a long and distinctive history of human use; Alaska is huge and little affected to date by direct land use. In this review assessment we therefore survey land use and land cover change, focusing on the past century and a half, only in the conterminous or lower 48 states. Those states cover an area

"The adjustments that are made in land use and land cover in coming years will in some way alter the life of nearly every living thing on Earth."

of almost 1900 million acres, or about 3 million square miles.

How land is *used*, and thus how *land cover* is altered, depends on who owns or controls the land and on the pressures and incentives shaping the behavior of the owner. Some 400 million acres in the conterminous 48 states—about 21% of the total—are federally owned. The two largest chunks are the 170 million acres of western rangeland controlled by the Bureau of Land Management and the approximately equal area of the National Forest System. Federal land represents 45% of the area of the twelve western states, but is not a large share of any other regional total. There are also significant land holdings by state governments throughout the country.

Most of the land in the United States is privately owned, but under federal, state, and local restrictions on its use that have increased over time. The difference between public and private land is important in explaining and forecasting land use and land coverage change, but the division is not absolute, and each sector is influenced by the other. Private land use is heavily influenced by public policies, not only by regulation of certain uses but through incentives that encourage others. Public lands are used for many private activities; grazing on federal rangelands and timber extraction from the national forests by private operators are the most important and have become the most controversial. The large government role in land use on both government and private land means that policy, as well as economic forces, must be considered in explaining and projecting changes in the land. Economic forces are of course significant determinants of policy—perhaps the most significant—but policy remains to some degree an independent variable.

There is no standard, universally accepted set of categories for classifying land by either use or cover, and the most commonly used, moreover, are hybrids of land cover and land use. Those employed here, which are by and large those of the U.S. National Resources Inventory conducted every five years by relevant federal agencies, are cropland, forest, grassland (pasture and rangeland), wetlands, and developed land.

- *Cropland* is land in farms that is devoted to crop production; it is not to be confused with total farmland, a broad land use or land ownership category that can incorporate many forms of land cover.
- *Forest land* is characterized by a predominance of tree cover and is further divided by the U.S. Census into timberland and non-timberland. By definition, the former must be capable of producing 20 cubic feet of industrial wood per acre per year and remain legally open to timber production.
- *Grassland* as a category of land cover embraces two contrasting Census categories of use: pasture (enclosed and what is called improved grassland, often closely tied to cropland and used for intensive livestock raising), and range (often unenclosed or unimproved grazing land with sparser grass cover and utilized for more extensive production).
- *Wetlands* are not a separate Census or National Resources Inventory category and are included within other categories: swamp, for example, is wetland forest. They are defined by federal agencies as lands covered all or part of the year with water, but not so deeply or permanently as to be classified as water surface *per se*.
- The U.S. government classifies as *developed* land urban and built-up parcels that exceed certain size thresholds. "Developed" or "urban" land is clearly a use rather than a cover category. Cities and suburbs as they are politically defined have rarely more than half of their area, and often much less, taken up by distinctively "urban" land cover such as buildings and pavement. Trees and grass cover

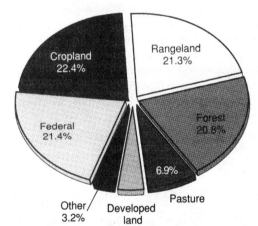

Land Class	Area in Million Acres	Fraction of Total Area
Land Use and Cover in the Conterminous U.S.		
Privately Owned (shown in diagram below)		
Cropland	422	22.4%
Rangeland	401	21.3
Forest	391	20.8
Pasture	129	6.9
Developed	77	4.1
Other Catagories[1]	60	3.2
Federally Owned[2]	404	21.4
TOTAL[3]	1884	100%

[1] Other minor covers and surface water
[2] Federal land is approximately half forest and half rangeland
[3] Included in various catagories is about 100 million acres of wetland, covering about 5% of the national area

Table 1 Source: U.S. 1987 National Resources Inventory, published in 1989. U.S. Government Printing Office.

substantial areas of the metropolitan United States; indeed, tree cover is greater in some settlements than in the rural areas surrounding them.

By the 1987 U.S. National Resources Inventory, non-federal lands were divided by major land use and land cover classes as follows: cropland, about 420 million acres (22% of the entire area of the 48 states); rangeland about 400 million (21%); forest, 390 million (21%);

pasture, 130 million (7%); and developed land, 80 million (4%). Minor covers and uses, including surface water, make up another 60 million acres (Table 1). The 401 million acres of federal land are about half forest and half range. Wetlands, which fall within these other Census classes, represent approximately 100 million acres or about five percent of the national area; 95 percent of them are freshwater and five percent are coastal.

These figures, for even a single period, represent not a static but a dynamic total, with constant exchanges among uses. Changes in the area and the location of cropland, for example, are the result of the *addition* of new cropland from conversion of grassland, forest, and wetland and its *subtraction* either by abandonment of cropping and reversion to one of these less intensive use/cover forms or by conversion to developed land. The main causes of forest *loss* are clearing for agriculture, logging, and clearing for development; the main cause of forest *gain* is abandonment of cropland followed by either passive or active reforestation. Grassland is converted by the creation of pasture from forest, the interchange of pasture and cropland, and the conversion of rangeland to cropland, often through irrigation.

Change in wetland is predominantly loss through drainage for agriculture and construction. It also includes natural gain and loss, and the growing possibilities for wetland creation and restoration are implicit in the Environmental Protection Agency's "no *net* loss" policy (emphasis added). Change in developed land runs in only one direction: it expands and is not, to any significant extent, converted to any other category.

Comparison of the American figures with those for some other countries sets them in useful perspective. The United States has a greater relative share of forest and a smaller relative share of cropland than does Europe as a whole and the United Kingdom in particular.

Though Japan is comparable in population density and level of development to Western Europe, fully two-thirds of its area is classified as forest and woodland, as opposed to ten percent in the United Kingdom; it preserves its largely mountainous forest area by maintaining a vast surplus of timber imports over exports, largely from the Americas and Southeast Asia.

Regional patterns within the U.S. (using the four standard government regions of Northeast, Midwest, South, and West) display further variety. The Northeast, though the most densely populated region, is the most heavily wooded, with three-fifths of its area in forest cover. It is also the only region of the four in which "developed" land, by the Census definition, amounts to more than a minuscule share of the total; it covers about eight percent of the Northeast and more than a quarter of the state of New Jersey. Cropland, not surprisingly, is by far the dominant use/cover in the Midwest, accounting for just under half of its expanse. The South as a whole presents the most balanced mix of land types: about 40 percent forest, 20 percent each of cropland and rangeland, and a little more than ten percent pasture. Western land is predominantly rangeland, with forest following and cropland a distant third. Wetlands are concentrated along the Atlantic seaboard, in the Southeast, and in the upper Midwest. Within each region, of course, there is further variety at and below the state level.

WHERE HAVE WE BEEN?

The public domain, which in 1850 included almost two-thirds of the area of the present conterminous states, has gone through two overlapping phases of management goals. During the first, dominant in 1850 and long thereafter, the principal goal of management was to transfer public land into private hands, both to raise revenue and to encourage settlement and land improvements. The government often attached conditions (which were sometimes complied with) to fulfill other national goals, such as swamp drainage, timber planting, and railroad construction in support of economic development.

The second phase, that of federal retention and management of land, began with the creation of the world's first national park, Yellowstone, shortly after the Civil War. It did not begin to be a significant force, however, until the 1890s, when 40 million acres in the West were designated as federal forest reserves, the beginning of a system that subsequently expanded into other regions of the country as well. Several statutory vestiges of the first, disposal era remain (as in mining laws, for example), but the federal domain is unlikely to shrink noticeably in coming decades, in spite of repeated challenges to the government retention of public land and its regulation of private land. In recent years, such challenges have included the "Sagebrush Rebellion" in the rangelands of the West in the 1970s and 1980s calling for the withdrawal of federal control, and legal efforts to have many

13. Past and Present Land Use

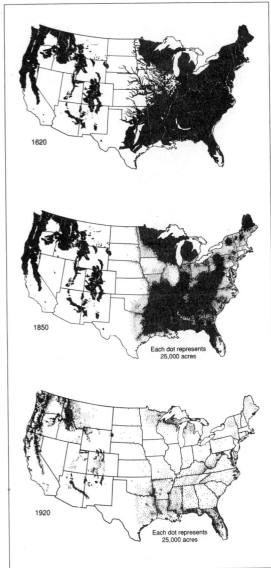

Figure 1 Area of virgin forest: top to bottom 1620, 1850, and 1920 as published by William B. Greeley, "The Relation of Geography to Timber Supply," Economic Geography, vol. 1, pp. 1–11 (1925). The depiction of U.S. forests in the later maps may be misleading in that they show only old-growth forest and not total tree cover.

land use regulations classified as "takings," or as exercises of the power of eminent domain. This classification, where it is granted, requires the government to compensate owners for the value of development rights lost as a result of the regulation.

Cropland

Total cropland rose steadily at the expense of other land covers throughout most of American history. It reached a peak during the 1940s and has subsequently fluctuated in the neighborhood of 400 million acres, though the precise figure depends on the definition of cropland used. Long-term regional patterns have displayed more variety. Cropland abandonment in some areas of New England began to be significant in some areas by the middle of the nineteenth century. Although total farmland peaked in the region as late as 1880 (at 50%) and did not decline sharply until the turn of the century, a steady decline in the subcategory of cropland and an increase in other farmland covers such as woodland and unimproved pasture was already strongly apparent. The Middle Atlantic followed a similar trajectory, as, more recently, has the South. Competition from other, more fertile sections of the country in agricultural production and within the East from other demands on land and labor have been factors; a long-term rise in agricultural productivity caused by technological advances has also exerted a steady downward pressure on total crop acreage even though population, income, and demand have all risen.

Irrigated cropland on a significant scale in the United States extends back only to the 1890s and the early activities in the West of the Bureau of Reclamation. Growing rapidly through about 1920, the amount of irrigated land remained relatively constant between the wars, but rose again rapidly after 1945 with institutional and technological developments such as the use of center-pivot irrigation drawing on the Ogallala Aquifer on the High Plains. It reached 25 million acres by 1950 and doubled to include about an eighth of all cropland by about 1980. Since then the amount of irrigated land has experienced a modest decline, in part through the decline of aquifers such as the Ogallala and through competition from cities for water in dry areas.

Forests

At the time of European settlement, forest covered about half of the present 48 states. The greater part lay in the eastern part of the country, and most of it had already been significantly altered by Native American land use practices that left a mosaic of different covers, including substantial areas of open land.

Forest area began a continuous decline with the onset of European settlement that would not be halted until the early twentieth century. Clearance for farmland and harvesting for fuel, timber, and other wood products represented the principal sources of pressure. From an estimated 900 million acres in 1850, the wooded area of the entire U.S. reached a low point of 600 million acres around 1920 (Fig. 1).

It then rose slowly through the postwar decades, largely through abandonment of cropland and regrowth on cutover areas, but around 1960 began again a modest decline, the result of settlement expansion and of higher rates of timber extraction through mechanization. The agricultural censuses recorded a drop of 17 million acres in U.S. forest cover between 1970 and 1987 (though data uncertainties and the small size of the changes relative to the total forest area make a precise dating of the reversals difficult). At the same time, if the U.S. forests have been shrinking in area they have been growing in density and volume. The trend in forest biomass has been consistently upward; timber stock measured in the agricultural censuses from 1952 to 1987 grew by about 30%.

National totals of forested area again represent the aggregation of varied regional experiences. Farm abandonment in much of the East has translated directly into forest recovery, beginning in the mid- to late-nineteenth century (Fig. 2). Historically, lumbering followed a regular pattern of harvesting one region's resources and moving on to the next; the once extensive old-growth forest of the Great Lakes, the South, and the Pacific Northwest represented successive and overlapping frontiers. After about 1930, frontier-type exploitation gave way to a greater emphasis on permanence and management of stands by timber companies. Wood itself has declined in importance as a natural resource, but forests have been increasingly valued and protected for a range of other services, including wildlife habitat, recreation, and streamflow regulation.

Grassland

The most significant changes in grassland have involved impacts of grazing on the western range. Though data for many periods are scanty or suspect, it is clear that rangelands have often been seriously overgrazed, with deleterious consequences including soil erosion and compaction, increased streamflow variability, and floral and faunal biodiversity loss as well as reduced value for production. The net value of grazing use on the western range is nationally small, though significant locally, and pressures for tighter management have increasingly been guided by ecological and preservationist as well as production concerns.

Wetland

According to the most recent estimates, 53% of American wetlands were lost between the 1780s and the 1980s, principally to drainage for agriculture. Most of the conversion presumably took place during the twentieth century; between the 1950s and the 1970s alone, about 11 million acres were lost. Unassisted private action was long thought to drain too little; since mid-century, it has become apparent that the opposite is true, that unfettered private action tends to drain too much, i.e., at the expense of now-valued wetland. The positive externalities once expected from drainage—improved public health and beautification of an unappealing natural landscape—carry less weight today than the negative ones that it produces. These include the decline of wildlife, greater extremes of streamflow, and loss of a natural landscape that is now seen as more attractive than a human-modified one. The rate of wetland loss has now been cut significantly by regulation and by the removal of incentives for drainage once offered by many government programs.

Developed land

As the American population has grown and become more urbanized, the land devoted to settlement has increased in at least the same degree. Like the rest of the developed world, the United States now has an overwhelmingly non-farm population residing in cities, suburbs, and towns and villages. Surrounding urban areas is a classical frontier of rapid and sometimes chaotic land use and land cover change. Urban impacts go beyond the mere subtraction of land from other land uses and land covers for settlement and infrastructure; they also involve the mining of building materials, the disposal of wastes, the creation of parks and water supply reservoirs, and the introduction of pollutants in air, water, and soil. Long-term data on urban use and cover trends are unfortunately not available. But the trend in American cities has undeniably been one of residential dispersal and lessened settlement densities as transportation technologies have improved; settlement has thus required higher amounts of land per person over time.

WHERE ARE WE GOING?

The most credible projections of changes in land use and land cover in the United States over the next fifty years have come from recent assessments produced under the federal laws that now mandate regular national inventories of resource stocks and prospects. The most recent inquiry into land resources, completed by the Department of Agriculture in 1989 (and cited at the end of this article), sought to project their likely extent and con-

13. Past and Present Land Use

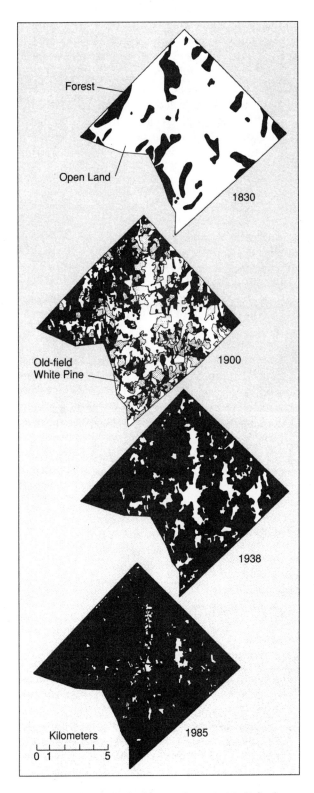

Figure 2 Modern spread of forest (shown in black) in the township of Petersham, Massachusetts, 1830 through 1985. White area is that considered suitable for agriculture; shaded portions in 1900 map indicate agricultural land abandoned between 1870 and 1900 that had developed forest of white pine in this period. From "Land-use History and Forest Transformations," by David R. Foster, in *Humans as Components of Ecosystems,* edited by M. J. McDonnell and S.T.A. Pickett, Springer-Verlag, New York, pp. 91–110, 1993.

dition a half-century into the future, to the year 2040. The results indicated that only slow changes were expected nationally in the major categories of land use and land cover: a loss in forest area of some 5% (a slower rate of loss than was experienced in the same period before); a similarly modest decline in cropland; and an increase in rangeland of about 5% through 2040. Projections are not certainties, however: they may either incorrectly identify the consequences of the factors they consider or fail to consider important factors that could alter the picture. Because of the significant impacts of policy, its role—notoriously difficult to forecast and assess— demands increased attention, in both its deliberate and its inadvertent effects.

Trends in the United States stand in some contrast to those in other parts of the developed world. While America's forest area continues to decline somewhat, that of many comparable countries has increased in modest degree, while the developing world has seen significant clearance in the postwar era. There has been substantial stability, with slow but fluctuating decline, in cropland area in the United States. In contrast, cropland and pasture have declined modestly in the past several decades in Western Europe and are likely to decline sharply there in the future as longstanding national and European Community agricultural policies subsidizing production are revised; as a result, the European countryside faces the prospect of radical change in land use and cover and considerable dislocation of rural life.

WHY DOES IT MATTER?

Land use and land cover changes, besides affecting the current and future supply of land resources, are important sources of many other forms of environmental change. They are also linked to them through synergistic connections that can amplify their overall effect.

Loss of plant and animal biodiversity is principally traceable to land transformation, primarily through the fragmentation of natural habitat. Worldwide trends in land use and land cover change are an important source of the so-called greenhouse gases, whose accumulation in the atmosphere may bring about global climate change. As much as 35% of the increase in atmospheric CO_2 in the last 100 years can be attributed to land use change, principally through deforestation. The major known sources of increased methane—rice paddies, landfills, biomass burning, and cattle—are all related to land use. Much of the increase in nitrous oxide is now thought due to a collection of sources that also depend upon the use of the land, including biomass burning, livestock raising, fertilizer application and contaminated aquifers.

> *In most of the world, both fossil fuel combustion and land transformation result in a net release of carbon dioxide to the atmosphere.*

Land use practices at the local and regional levels can dramatically affect soil condition as well as water quality and water supply. And finally, vulnerability or sensitivity to existing climate hazards and possible climate change is very much affected by changes in land use and cover. Several of these connections are illustrated below by examples.

Carbon emissions

In most of the world, both fossil fuel combustion and land transformation result in a net release of carbon dioxide to the atmosphere. In the United States, by contrast, present land use and land cover changes are thought to absorb rather than release CO_2 through such processes as the rapid growth of relatively youthful forests. In balance, however, these land-use-related changes reduce U.S. contributions from fossil fuel combustion by only about 10%. The use of carbon-absorbing tree plantations to help diminish global climate forcing has been widely discussed, although many studies have cast doubt on the feasibility of the scheme. Not only is it a temporary fix (the trees sequester carbon only until the wood is consumed, decays, or ceases to accumulate) and requires vast areas to make much of a difference, but strategies for using the land and its products to offset some of the costs of the project might have large and damaging economic impacts on other land use sectors of the economy.

Effects on arable land

The loss of cropland to development aroused considerable concern during the 1970s and early 1980s in connection with the 1981 National Agricultural Lands Study, which estimated high and sharply rising rates of conversion. Lower figures published in the 1982 National Resource Inventory, and a number of associated studies, have led most experts to regard the conversion of cropland to other land use categories as representing something short of a genuine crisis, likely moreover to continue at slower rather than accelerating rates into the future. The land taken from food and fiber production and converted to developed land has been readily made up for by conversion of land from grassland and forest. The new lands are not necessarily of the same quality as those lost, however, and some measures for the protection of prime farmland are widely considered justified on grounds of economics as well as sociology and amenities preservation.

Vulnerability to climate change

Finally, patterns and trends in land use and land cover significantly affect the degree to which countries and regions are vulnerable to climate change—or to some degree, can profit from it. The sectors of the economy to which land use and land cover are most critical—agriculture, livestock, and forest products—are, along with fisheries, among those most sensitive to climate variation and change. How vulnerable countries and regions are to climate impacts is thus in part a function of the importance of these activities in their economies, although differences in ability to cope and adapt must also be taken into account.

These three climate-sensitive activities have steadily declined in importance in recent times in the U.S. economy. In the decade following the Civil War, agriculture still accounted for more than a third of the U.S. gross domestic product, or GDP. In 1929, the agriculture-forest-fisheries sector represented just under ten percent of national income. By 1950, it had fallen to seven percent of GDP, and it currently represents only about two percent. Wood in 1850 accounted for 90 percent of America's total energy consumption; today it represents but a few percent. These trends suggest a lessened macroeconomic vulnerability in the U.S. to climate change, though they may also represent a lessened ability to profit from it to the extent that change proves beneficial. They say nothing, however, about primary or secondary impacts of climate change on other sectors, about ecological, health, and amenity losses, or about vulnerability in absolute rather than relative terms, and particularly the potentially serious national and global consequences of a decline in U.S. food production.

The same trend of lessening vulnerability to climate changes is apparent even in regions projected to be the most exposed to the more harmful of them, such as reduced rainfall. A recent study examined agro-economic impacts on the Missouri-Iowa-Nebraska-Kansas

Shifting patterns of land use in the U.S. and throughout the world are a proximate cause of many of today's environmental concerns.

area of the Great Plains, were the "Dust Bowl" drought and heat of the 1930s to recur today or under projected conditions of the year 2030. It found that although agricultural production would be substantially reduced, the consequences would not be severe for the regional economy overall: partly because of technological and institutional adaptation and partly because of the declining importance of the affected sectors, as noted above. The 1930s drought itself had less severe and dramatic effects on the population and economy of the Plains than did earlier droughts in the 1890s and 1910s because of land use, technological, and institutional changes that had taken place in the intervening period.

Shifting patterns in human settlement are another form of land use and land cover change that can alter a region's vulnerability to changing climate. As is the case in most other countries of the world, a disproportionate number of Americans live within a few miles of the sea. In the postwar period, the coastal states and counties have consistently grown faster than the country as a whole in population and in property development. The consequence is an increased exposure to hazards of hurricanes and other coastal storms, which are expected by some to increase in number and severity with global warming, and to the probable sea-level rise that would also accompany an increase in global surface temperature. It is unclear to what extent the increased exposure to such hazards might be balanced by improvements in the ability to cope, through better forecasts, better construction, and insurance and relief programs. Hurricane fatalities have tended to decline, but property losses per hurricane have steadily increased in the U.S., and the consensus of experts is that they will continue to do so for the foreseeable future.

CONCLUSIONS

How much need we be concerned about changes in land use and land cover in their own right? How much in the context of other anticipated environmental changes?

As noted above, shifting patterns of land use in the U.S. and throughout the world are a proximate cause of many of today's environmental concerns. How land is used is also among the human activities most likely to feel the effects of possible climate change. Thus if we are to understand and respond to the challenges of global environmental change we need to understand the dynamics of land transformation. Yet those dynamics are notoriously difficult to predict, shaped as they are by patterns of individual decisions and collective human behavior, by history and geography, and by tangled economic and political considerations. We should have a more exact science of how these forces operate and how to balance them for the greatest good, and a more detailed and coherent picture of how land in the U.S. and the rest of the world is used.

The adjustments that are made in land use and land cover in coming years, driven by worldwide changes in population, income, and technology, will in some way alter the life of nearly every living thing on Earth. We need to understand them and to do all that we can to ensure that policy decisions that affect the use of land are made in the light of a much clearer picture of their ultimate effects.

FOR FURTHER READING

Americans and Their Forests: A Historical Geography, by Michael Williams. Cambridge University Press, 599 pp, 1989.

An Analysis of the Land Situation in the United States: 1989–2040. USDA Forest Service General Technical Report RM-181. U.S. Government Printing Office, Washington, D.C., 1989.

Changes in Land Use and Land Cover: A Global Perspective. W. B. Meyer and B. L. Turner II, editors. Cambridge University Press, 537 pp, 1994.

"Forests in the Long Sweep of American History," by Marion Clawson. *Science*, vol. 204, pp 1168–1174, 1979.

China Shoulders the Cost of Environmental Change

By Vaclav Smil

Ecologists today have at their disposal a great deal of evidence about the processes that degrade our environment, ranging from coastal eutrophication to tropical deforestation. They have come to understand the intricacies of human-induced changes such as acid deposition and heavy metal accumulation. But they do not have realistic assessments of the economic losses that result from environmental pollution and ecosystem degradation. Possession of these kinds of numbers would strengthen their arguments for environmental management considerably. Persuasive figures could help reorient public policy, environmental law, and investors' thinking in favor of effective preventive action.

So, why haven't these figures been developed? First, there are (as yet) no generally accepted procedures for conducting such evaluations.[1] These methodological uncertainties leave individual researchers with no choice but to use subjective judgments when deciding which variables to include and how to treat them.

Suppose for example, researchers wanted to quantify the health effects of chronically high levels of urban air pollution. They could take the more minimalist approach, limiting the analysis to the value of the labor time lost due to higher upper-respiratory morbidity. Or they could attempt a more all-encompassing evaluation that would put a price on every individual discomfort and include the cost of premature death. As these researchers would quickly discover, the former approach is much easier than the latter. While a rich and fascinating literature on the price of personal suffering and the value of life does exist, objective criteria for putting a monetary value on respiratory discomfort, physical limitations, and anxiety induced by asthma attacks do not. As far as the value of life is concerned, actuarial practices, economic considerations, and moral imperatives offer estimates that may differ widely—sometimes by up to an order of magnitude.[2]

Having a standard set of procedures would not really simplify the task, however. The basic problem is that the kinds of specific figures required as basic inputs in such calculations are typically unavailable even in affluent countries, which have long had good statistical services.[3] This makes it impossible for researchers to avoid making subjective choices and simplifying assumptions, even though this weakens the persuasiveness of the eventual bottom line. As it is, the cumulative effect of even small departures from reality can easily halve or double a final figure.[4]

The second and perhaps most significant obstacle is the impossibility of putting a meaningful price on lost or reduced environmental services. For example, suppose a peasant living on a treeless plain takes straw from a field to light a fire to cook a meal. How can the loss of that straw be valued? The loss of the plant nutrients in the straw could be expressed rather easily by equating it with the cost of the synthetic fertilizers needed to replace those nutrients. But the straw also improved the soil's capacity to retain water and provided food for bacteria, fungi, and numerous invertebrates. Without these life forms, the soil will not be as productive and may be unable to sustain farming. How can these losses be valued?

Should these obstacles, daunting as they are, keep us from trying to evaluate economic cost? Absolutely not. It is simply necessary to always keep the limitations of the process in mind when interpreting the results. The evaluations will provide useful ranges of approximations—not correct single-figure answers. They will all be incomplete, and even the most comprehensive ones will almost certainly undervalue the real impact human actions have on the long-term integrity of environmental quality. However, despite these limitations, these assessments can be valuable, particularly when looking at the costs of environmental change in countries experiencing rapid growth, such as China.

A nation of more than 1.2 billion people, China's population grows by at least 13 million each year. This is the equivalent of adding the population of France in less than five years.[5] During the 1980s, China's gross domestic product (GDP) grew faster than that of all other nations except South Korea. In the early 1990s, China posted annual growth rates of up to 14 percent in GDP, the highest in the world.[6] But this rapid growth has exacted a price. irrational industrial and agricultural practices have left scars, and the ill effects of pollution already mar the air, water, and land of this nation endowed with absolutely large but on a per capita basis relatively limited resources.[7]

14. China Shoulders the Cost

Attempts to assess the economic cost of these changes have frequently met with difficulties. China's official government agencies are filled with masses of uncooperative bureaucrats prone to treat any unflattering figure as a state secret. Dubious statistics and unverifiable claims also abound. It seems to be one thing for the Dutch or Germans to assign a cost to environmental degradation in their countries and another thing entirely for the Chinese to do so.[8]

That said, the Chinese actually did some of the earliest studies. One, begun in 1984 and published in 1990, estimated the cost of pollution to be about 6.75 percent of the nation's annual GDP in 1983.[9] In response to this work, and with the goal of quantifying at least some of the major consequences of the degradation of China's ecosystems, I began work on a more comprehensive evaluation in 1993.[10] As this assessment was being completed, a group of Chinese researchers began a similarly comprehensive independent evaluation.[11]

Having both of these studies available creates an unparalleled opportunity to appraise the economic impact of environmental change in China. The following discussion will review some of the country's most important environmental concerns, highlight key conclusions from both studies, and explain the reasons for some of the major differences in their results. Readers interested in the studies' detailed assumptions are urged to refer to them specifically.

Air and Water Pollution

A small pioneering opinion survey published in China in 1994 uncovered some interesting facts. First, the public ranked air and water pollution second only to earthquakes and floods when asked to list environmental hazards. Respondents with science or engineering degrees, however, put the two pollution risks ahead of natural disasters.[12] China's severe air pollution problem stems from two sources: a traditionally high dependence on coal (whose combustion produces classic smog composed of suspended particulates and sulfur dioxide) and recent rapid increases in vehicular traffic (whose emissions of volatile organic compounds, nitrogen oxides, and carbon monoxide play key roles in the complex reactions that create photo-chemical smog).

While Chinese coal is of fairly good quality, only about one-fifth of it is cleaned and sorted before combustion. Typical conversion efficiencies remain very low in tens of millions of household stoves and thousands of small industrial and commercial boilers throughout the country, resulting in extraordinarily high emission factors per unit of delivered energy.[13] The high density of urban areas and their tendency to be both residential and industrial, improperly vented household stoves, and the burning of smoky biomass fuels in rural areas aggravate the situation.

Fossil fuel combustion in China now produces close to 20 million metric tons of sulfur dioxide and about 15 million metric tons of particulates a year. Monitoring shows the long-term averages of both pollutants to be multiples of the maximum limits recommended by the World Health Organization (WHO).[14] For example, WHO has established a limit of no more than 40–60 micrograms per cubic meter of sulfur dioxide as the annual mean. But even the cleanest suburbs of Beijing average 80 micrograms per cubic meter a year. In the city's most polluted neighborhoods, the annual mean is double that value. Still, these levels are low when compared to annual means of more than 400 micrograms per cubic meter in Taiyuan and Lanzhou and more than 300 micrograms in Linfeng, Chongqing (Sichuan), and Guiyang (Guizhou).

Particulates and sulfur dioxide cause relatively little damage to crops. Most of the yield losses are seen in suburban vegetable farms. They do, however, inflict considerably greater damage on materials, above all to buildings and corrodible surfaces. As acid deposition intensifies in the rainy South, the amount of this kind of damage will only increase.[15] A number of worrisome human health risks are also associated with this kind of pollution. At least 200 million Chinese are exposed to annual particulate concentrations of more than 300 micrograms per cubic meter; at least 20 million are exposed to twice that level. With the spread of new industries across the country, between 100 million and 200 million rural inhabitants may already be breathing air nearly as polluted as that in the cities.

Such high exposure levels call to mind conditions in the cities of western Europe and North America two to four generations ago. As was the case then, these high levels of exposure contribute to higher incidence of respiratory diseases, ranging from upper-respiratory infections to lung cancer. But assessing the share of these diseases attributable to outdoor air pollution alone is particularly difficult in China for two reasons. First, most people are also exposed to very high levels of indoor air pollution from inefficient stoves. Second, the nation abounds with smoking addicts.[16]

Polluted water is even more ubiquitous than polluted air in China. Several years ago, a survey of nearly 900 of the country's major rivers found that more than four-fifths of them were polluted to some degree, more than 20 percent so badly that their water could not be used for irrigation.[17] Drinking water meets official quality standards in only 6 out of the 27 largest cities that draw on surface sources and in just 4 out of those 27 that use underground sources.

It is common practice to release untreated municipal wastes into rivers, even in large cities. Half of Shanghai's waste is discharged into the Yangtze River and Hangzhou Bay. The Songhua River and Ji Canal, which flow through both the provinces of Jilin and Heilongjiang, still contain tens of metric tons of mercury, the legacy of uncontrolled releases prior to 1977 that resulted in mercury concentrations higher than those recorded in Japan's Minamata Bay.[18]

Enormous numbers of small to medium-sized rural and township enterprises have proliferated outside China's large cities in recent years. In Jiangsu Province, which surrounds Shanghai, roughly one such enterprise can be found per square kilometer. But along with jobs these plants have brought a variety of water pollutants into the Chinese countryside. Unknown volumes of untreated waste from these plants get dumped into streams and networks of canals, contaminating the water used for drinking (about half of the population draws its drinking water from surface sources), irrigating fields, and watering animals.

In addition to such common industrial pollutants as industrial oils, phenols, and heavy metals, high levels of nitrates leached from heavy fertilizer applications befoul the water. As the world's largest consumer of synthetic nitrogen fertilizers, China annually applies an average of around 140 kilograms of nitrogen per hectare. In the most intensively cultivated provinces, this figure surpasses 300 kilograms per hectare. Combined with animal and human wastes, these nitrogen loadings will eventually lead to serious nitrate contamination.[19]

The pollution of its water has created a number of negative economic impacts for China. Fish catches in streams and reservoirs have declined. Red tides now occur with greater frequency, jeopardizing the domestic shrimp aquaculture industry. Incidence rates of tumors among livestock have increased, as have those of cancers of the liver, stomach, and esophagus in humans. Studies of the localities suffering from the worst water pollution have found the incidence rate of cancers of the digestive system to be 3 to 10 times higher there than in unpolluted communities. These studies also uncovered instances of anemia, skin diseases, enlarged livers, premature hair loss, and a higher incidence of congenital deformities. Thanks in part to water pollution, viral hepatitis and dysentery top the list of infectious diseases in China.

In the countryside, waterborne pathogens and the eggs of parasites, which are concealed in organic wastes that are recycled to cropland, continue to be a major problem. The frequency of ascariasis, ancylostomiasis, and trichuriasis among China's vegetable farmers exceeds 90 percent in some regions.[20] The impossibility of using polluted water in industrial and agricultural production has also led to substantial economic losses even as the government has incurred the additional costs of tapping new resources.

My calculations put the total cost of China's air and water pollution at roughly 30 billion to 45 billion 1990 yuan. The Chinese study came up with a total close to 100 billion 1992 yuan. Even if the more conservative estimate were increased by about 20 percent to account for China's high rate of inflation during the early 1990s, the estimates of the Chinese study are still considerably higher. As Table 1 on this page shows, much of the explanation for these differences lies in the studies' treatments of the impact of air and water pollution on human health.

The two major factors are different assumptions about population totals exposed to particular pollution levels and different costs ascribed to typical treatments or lost labor hours. As far as the total of urban populations exposed to excessive air pollution is concerned, the two studies do not differ all that much. But, when it comes to averages of the costs of treating chronic bronchitis and lung cancer, they do. The Chinese study uses average treatment costs of 2,100 and 12,700 yuan, respectively. I estimated the same costs on the basis of a different set of Chinese figures as 800 and 5,000 yuan, respectively. Such disparities are typical among the many previous studies that also attempted to assess the toll pollution takes on human health.[21]

Changes in Land Use and Soil Degradation

This broad category embraces a wide range of phenomena, from the desertification of China's extensive interior grasslands to the disappearance of its coastal marshes. The most pressing problems, however, are the loss of farmland and the qualitative decline of arable soils. A combination of two factors explain these changes: China's relatively low per capita availability of agricultural land and the expectation that its population will swell by at least 300 million over the next 25 years.

China actually has more arable land area than the total of 95 million hectares claimed by the State Statistical Bureau. The best estimates (based on sample surveys and remote sensing) lie somewhere in the range of 120 million to 140 million hectares.[22] This means that in per capita terms China still has more than twice as much farmland as South Korea and Japan, its truly land-poor East Asian neighbors.

At the same time, however, official totals of recent losses of arable land may err on the low side. Cumulative losses over the past 40 years total to more than all of Germany's farmland. Since 1980, China has lost about half a million hectares annually. Much of this land was of good quality: Alluvial soils in the coastal provinces have been eaten up at the fastest rate by urban and industrial expansion.

Declines in soil quality resulting from increased soil erosion and a decrease in the extent and intensity of traditional recycling of organic waste have also adversely affected agricultural production. Improper irrigation and more intensive cropping, which relies on increased applications of agrochemicals, are the other two key culprits lurking behind the declines in soil quality.

A nationwide survey conducted on almost half of China's farmland identified various degrees of excessive soil erosion on 31 percent of the land. The phenomenon does not occur only in regions naturally prone to high erosion rates, such as China's Loess Plateau. In Sichuan, the country's most populous province with a population of more than 110 million, 44 percent of fields were eroding beyond a sustainable level during the late 1980s. This figure represented a fourfold increase in erosion compared to the early 1950s. Annual erosion losses on two million hectares of cultivated slopeland (nearly one-third of Sichuan's total) averaged 110 tons per hectare.[23] To get a sense of scale, consider that in the United States annual water erosion rates recently averaged just slightly more than 9 tons per hectare and wind erosion rates amounted to a little more than 7 tons per hectare. This combines for a total annual average loss of about 16 tons per hectare.

To quantify the economic costs of farmland losses and deteriorating soil quality, I calculated the value of lost harvests, decreased yields (or lower livestock production), and lost nutrients from eroded soils. I also estimated the costs (implicit as well as explicit) of siltation in reservoirs and canals (used both for irrigation and as sources of urban water supplies), higher damage from flooding,

Table 1. Economic costs attributable to air and water pollution and solid waste disposal

Category	Estimated costs in billions[a]	
	Smil (1990 yuan)	Xia Guang (1992 yuan)
Air pollution[b]	11.0-19.2	57.9
Damage to human health	3.8-6.5	20.2
Damage to plants	2.9-6.0	7.2
Damage to materials	2.0-4.0	16.5
Acid rain[c]		14.0
Water pollution[b]	9.7-14.0	35.6
Damage to human health	4.1-6.7	19.3
Losses of food production	2.6-3.4	2.5
Industrial losses	2.5-3.3	13.8
Solid waste disposal	9.0-10.5	5.1
Total	29.7-43.7	98.6

[a]1992 values are approximately 20 percent higher than 1990 values due to inflation. Thus, to convert 1990 values to 1992 values, increase them by 20 percent.

[b]Because some minor subcategories are not shown, the listed values may not add up to category totals.

[c]Smil's estimates for acid rain damage are included in the other air pollution categories.

SOURCE: V. Smil.

and the loss of key ecosystem services provided by lost paddy fields and wetlands and degraded grasslands.

As it turned out, the Chinese study's assessment of detrimental effects arising from changes in land use and soil degradation was remarkably similar. Given the nature of this accounting exercise, the two sets of calculations agree quite closely: My median value, (38.6) adjusted for inflation, differs from the Chinese total by less than 10 percent (see Table 2).

Deforestation

When it came to the analysis of deforestation in China, the two studies did not arrive at such similar conclusions. In fact, some of the most significant differences between them crop up in this analysis. By far the greatest disparity centers on estimates of the economic consequences of deforestation. This huge difference does not result from irreconcilable assumptions concerning soil erosion, stream silting, or the loss of water retention capacity. Indeed, a close look reveals greatly similar assumptions on all these accounts. The reason for the disparity lies in fundamentally different approaches to the problem.

The Chinese study adopted what could perhaps be best described as a deep ecological perspective. It estimated that China has lost approximately 290 million hectares of forest since the beginning of the country's history. Roughly 50 percent of this loss can be attributed to the conversion of forestland to farmland, settlements, and transportation networks, the unavoidable consequences of an ancient civilization's growth. The other half of the loss represents the costs of excessive logging. This land could have remained forested if properly managed and could now be harvested on a sustainable basis. Proceeding from these assumptions, the Chinese study calculated the impact of this excessive deforestation on the desiccation of the northern and northwestern regions of the country and on accelerated erosion, which causes streams and reservoirs to fill with silt, thereby leading to greater damage during floods.

In contrast, my study estimated the environmental cost of deforestation on the basis of the current extent of excessive cutting of mature growth. Curiously, if one were to take China's official statistics at face value, it would be impossible to take this approach. These statistics claim that China's total wood increment has surpassed the amount cut annually during most of the early 1990s. If this were true, there would be no net deforestation. But even if one accepted this obviously exaggerated statement, it appears in a different light once combined with an awareness of the changing composition of China's forests.

During the early 1990s, no less than three-fourths of China's timberlands were young or middle-aged stands. The growing stock ready for harvesting in mature forests amounted to less than one-fifth of all standing timber. This total could be cut in just seven to eight years. The amount of stock in forests approaching maturity will decline from almost one-third of all timber in the late 1980s to less than one-seventh by the year 2000.

According to figures compiled by the Chinese Ministry of Forestry, 25 out of the 131 state forestry bureaus in the most important timber production zones had exhausted their reserves by 1990. Only 40 of these bureaus post figures showing that they can harvest up to the year 2000. By 2000, almost 70 percent of China's state forestry bureaus will basically have no mature trees left to fell.[24] Furthermore, the official figure for the average amount of wood growing in forest plantings—28.27 cubic meters per hectare—makes it quite clear that the new plantings (whose growing stock may be yielding the statistical wood surplus) cannot hope to replace the felled mature forests, which had a growing stock of at least 70 to 80 cubic meters per hectare, for many decades. Consequently, even if it is real, the recent quantitative growth of Chinese forests hides a major qualitative decline.

In any case, estimates of the economic cost of deforestation in China should be based on figures that realistically represent long-term trends rather than capture short-term aberrations. Careful appraisal of the available evidence indicates that the over-cutting of mature forests—in other words, harvests above the average annual increment of wood in stands that store the largest volume of phytomass, harbor the greatest biodiversity, and are able to better provide various ecosystem services than recent plantings—has been proceeding recently at a rate of at least 50 and up to 100 million cubic meters per year.

With the volume of harvestable wood averaging around 90 cubic meters per hectare, this loss translates into the disappearance of half a million to one million hectares of mature forest per year. In terms of the loss of future timber supplies alone, this over-cutting would cost between 13 billion and 26 billion yuan a year. With the disappearance of these trees, however, comes the weakening or outright loss of some of the crucial functions they performed. The soil's water storage capacity diminishes. The land becomes even more vulnerable to wind and water erosion (whose rates are likely to increase by two orders of magnitude). And the local and regional climate may change. While the effects of such changes are difficult to quantify, they could contribute to changes in the biospheric carbon cycle and possibly global warming, which will have dire consequences for China's and Earth's biodiversity.

Putting a price on these effects remains a highly uncertain process. Multipliers range from 1.5 to more than 20 times the value of the cut timber.[25] Chinese foresters have estimated the combined ecosystem benefits of mature forests to be between 8 and 25 times the profit from harvested timber sales. For example, a detailed study of the Changbaishan natural reserve in Jilin concluded that to replace the forest's water storage capacity with a reservoir, control soil erosion by terracing the slopes, and use pesticides rather than forest-sheltered birds to control insects the cost would be about 49,000 yuan (in 1990 monies) per hectare, more than 20 times the value of sustainably harvested timber from the same area.[26]

Naturally, this ratio will rise if the forest's contributions to local and regional climate control and the value of its preservation of biodiversity are factored into the equation. Considerable value could also be imputed to the forest's future recreational worth and, in the long term, to the value of forests as carbon sinks. But even using 1.5 as the minimum multiplier value, the cost of lost ecosystem services due to excessive cutting ranges between 20 billion and 39 billion yuan. Including the value of timber lost due to unsustainable harvest methods and forest fires brings the grand total cost for forest mismanagement up to 40 billion to 70 billion yuan.

While I based my calculations on annual losses of half a million to one million hectares, the Chinese study based their estimates on a cumulative total loss of about 140 million hectares. Interestingly, the Chinese study's estimate of the cost—245 billion yuan—did not include any adjustments for lost ecosystem services, which represented the highest share of my estimates. If these costs were included, the unusual historical approach the Chinese took would have ended up with an even higher total. Their approach does have one advantage, however. It calls attention to the true extent of human impact on China's forests. On the other hand, some of the costs estimated in one part of the study were also considered in another (the effects of soil erosion in particular) so a simple addition of the two sets of estimates would involve some double-counting.

Wider Perspectives

Although both studies were intended to be as comprehensive as possible, in the end a number of critical effects had to be dropped from consideration. Major impacts that could not be quantified because of a lack of basic information included the effects of photo-chemical smog in and near China's large cities, damage attributable to China's nuclear weapons sector, declining fish catches in China's seas, and the foregone recreational value of lost forests, wetlands, and beaches.

Perhaps most importantly, neither set of calculations tried to attribute any monetary value to human discomfort and suffering. These reactions are provoked not only by excessive morbidity and premature mortality, but also from chronic exposure to high levels of noise in China's cities.[27] Finally, neither study could ascribe any definite value to China's loss of biodiversity and to the country's already huge and rising contribution to greenhouse gas emissions, a highly worrisome source of potential biospheric instability.

Both sets of calculations (the Chinese estimate of deforestation costs aside) are based on clearly conservative assumptions. Consequently, there can be no doubt that the economic burden of environmental pollution and ecosystem degradation in China was no less than 5 percent of the country's GDP in the early 1990s. The most likely conservative estimate falls somewhere in the range of 6 to 8 percent. Values around 10 percent would be consistent with a more comprehensive, although still far from all-inclusive, assessment. If prices were ever put on a number of the more elusive factors, the rate could conceivably rise to around 15 percent of annual GDP.

These burdens greatly surpass China's recent spending on environmental protection. During the 1980s and early 1990s, annual investment in this area equalled just 0.56 to 0.81 percent of GDP. Only in 1996 did officials promise to raise spending on environmental protection to just more than 1 percent by the year 2000. Even this raise, however, would be an order of magnitude lower than the most likely economic cost.

It is difficult to compare these burdens to those in other countries because converting yuan into a major currency such as the U.S. dollar is problematical. Using the official yuan-dollar exchange rate (the method favored until very recently by the World Bank) greatly underestimates the real values, putting China's GDP at less than $500 per capita. On the other hand, using a rate based on purchasing power parity (the approach favored by the International Monetary Fund) clearly exaggerates the real values, putting China's 1995 per capita GDP at nearly $3,000.

The World Bank's latest study argues that China's actual per capita gross domestic product in 1995 was about $2,000. This rate implies a purchasing power parity rate roughly four times the official exchange rate.[28] Using this conversion, the annual economic burden of air and water pollution would amount to about $50 billion (using the Chinese study's total). The annual price tag for land degradation and excessive deforestation comes to around $20 billion (using the Chinese study's total). My lowest estimate of the same cost is no less than $40 billion. Even the lowest likely grand total of about $90 billion U.S. (1992 value) is a huge sum, slightly larger than the value of all of China's exports in 1992.

In closing, the dual nature of these valuations must be stressed. These two studies were exploratory exercises based on a necessarily limited amount of information and requiring repeated assumptions. As such, they can make no claims to complete accuracy, can give no more than basic approximations, and are open to justifiable criticism. At the same time, all of their inherent weaknesses and uncertainties cannot negate the message carried by the bottom line.

First, there can be no doubt that recent environmental changes in China already carry economic costs roughly an order of magnitude higher than the country's annual spending on environmental protection. Even if the government tripled or quadrupled its outlays, they would still easily meet even the strictest benefit-cost criteria. Second, given the fact that the economic burden of environmental pollution and ecosystem degradation may already exceed one-tenth of China's annual gross domestic product, its recent aggressive quest for modernization must be a matter of serious national and international concern.

Vaclav Smil is a professor in the department of geography at the University of Manitoba, Winnipeg, Manitoba, Canada R3T 2N2 (tel: 204-474-9667).

Table 2. Economic costs attributable to changes in land use and soil

Category	Estimated costs in billions[a]	
	Smil (1990 yuan)	Xia Guang (1992 yuan)
Farmland loss and degradation	5.8-12.1	12.5
Soil erosion	11.0-26.4	17.2
Reduced crop yields	0.3-0.8	0.4
Loss of nutrients	5.0-15.0	16.2
Reservoir sitting	1.3-1.8	0.6
Clearing clogged canals	1.4-1.8	
Flooding damage	1.0-3.0	
Deterioration of grasslands	3.7-5.4	3.3
Total	20.5-43.9	35.8

NOTE: Because some minor subcategories are not shown, the listed values may not in fact add up to category totals.
[a]See Table 1 for conversion from 1990 yuan to 1992 yuan.
SOURCE: V. Smil.

Notes

1. For a detailed survey of possible valuation techniques, see A. J. Dixon et al., *Economic Analysis of Environmental Impacts* (London: Earthscan, 1994).
2. A. M. Freeman, *The Measurement of Environmental and Resource Values* (Washington, D.C.: Resources for the Future, 1993); and S. E. Rhoads, ed., *Valuing Life: Public Policy Dilemmas* (Boulder, Colo.: Westview Press, 1980).
3. This problem is obvious when one looks at the analytical framework recommended by the United Nations for national assessments of environmental

Environmental and Economic Accounting (New York, 1993).

4. For example, estimating that 80 percent of the people in a region are exposed to excessive concentrations of a pollutant whose effects cause a 40 percent rise in the incidence of upper-respiratory morbidity, and that a typical illness event is associated with a 30 percent increase in absence from work, there will be roughly a 10 percent rise in lost labor hours. Changing the fractions marginally to, respectively, 70, 30, and 20 cuts lost hours by more than half.

5. Thanks to a generation of fairly strict birth controls, China's relative population growth, recently at just around 1.1 percent a year, is much lower than in any other populous modernizing nation (India's rate has been about 1.9 percent, Brazil's 1.6 percent). However, the huge base makes the absolute additions still highly taxing.

6. China's inflation-adjusted gross domestic product averaged 9.4 percent a year between 1980 and 1991, compared to South Korea's 9.6 and India's 5.4 percent. Since 1991, China's growth rate of just above 10 percent has been unmatched worldwide.

7. For a comprehensive survey of China's current environmental ills, see V. Smil, *China's Environmental Crisis* (Armonk, N.Y.: M. E. Sharpe, 1993); and R. L. Edmonds, *Patterns of China's Lost Harmony* (London: Routledge, 1994).

8. Two studies—a Dutch one calculating pollution costs in 1985 and a West German account for 1983–85—ended up with very different conclusions. The Dutch study put the annual cost of air and water pollution and noise at just 0.5–0.9 percent of the country's gross domestic product, while the German total was 6 percent, an order of magnitude higher. See J. Nicolaisen et al., "Economics and the Environment: A Survey of Issues and Policy Options," *OECD Economic Studies* (Spring 1991): 7–43. The main reason for the higher German value lay in their accounting for the physical impact of air pollution now and for the impact of noise on property values.

9. National Environmental Protection

10. Agency, *Environment Forecast and Countermeasure Research in China in the Year 2000* (Beijing: Qinghua University Publishing House, 1990).

11. V. Smil, *Environmental Problems in China: Estimates of Economic Costs* (Honolulu, Hawaii: East-West Center, 1996).

12. M. Yushi et al., *Economic Costs of China's Environmental Change* (Report prepared for the Project on Environmental Change and Security, American Association of Arts and Sciences, Cambridge, Mass., 1997). Even before I completed my work, I thought it would be interesting if a small group of Chinese researchers undertook a similarly comprehensive evaluation independently. A comparison of the two studies could show both the usefulness and the limitations of these valuations. With the support of the Project on Environment, Population and Security directed by Thomas Homer-Dixon at the University of Toronto, I was able to ask Professor Mao Yushi, a noted Chinese economist, a member of the Chinese Academy of Social Sciences (CASS), and now the director of the Unirule Institute of Economics in Beijing to commission studies of the economic costs of environmental pollution, deforestation, and land degradation in China. Yushi chose professor Wang Hongchang of CASS to prepare a paper on deforestation and Professor Ning Datong of the Beijing Normal University to write about land-use changes. He selected Xia Guang of the National Environmental Protection Agency to work up an evaluation of the costs of air and water pollution.

13. Z. Jianguang, "Environmental Hazards in the Chinese Public's Eyes," *Risk Analysis* 14, no. 2 (1994): 163–67.

14. Conversion efficiencies range from just around 5 percent for steam locomotives and 10 to 15 percent for poorly designed traditional stoves to 30 to 40 percent for better urban stoves and 50 percent for small boilers. In contrast, the best household natural gas furnaces have efficiencies in excess of 90 percent, as do the largest industrial boilers.

15. For comparison of recent air pollution levels in the world's largest cities, see Earthwatch, *Urban Air Pollution in Megacities of the World* (London: Blackwell Publishers and the World Health Organization, 1992).

16. Z. Dianwu and H. M. Seip, "Assessing Effects of Acid Deposition in Southwestern China Using the MAGIC Model," *Water, Air, and Soil Pollution* 60, no. 1 (1991): 83–97.

17. On indoor air pollution, see K. Smith and Y. Liu, "Indoor Air Pollution in Developing Countries," in J. M. Samet, ed., *Epidemiology of Lung Cancer* (New York: Marcel Dekker, 1994), 151–84. China is now the world's largest producer of cigarettes, and its total of 350 million smokers is growing by 2 percent a year. The average number of cigarettes smoked per day rose from 10 in 1994 to 14 in 1996. See *China News Digest,* Internet Files, 25 November 1996.

18. W. Jusi, "Water Pollution and Water Shortage Problems in China," *Journal of Applied Ecology* 26, no. 3 (1989): 851–57.

19. G. Dazhi et al., "Mercury Pollution and Control in China," *Journal of Environmental Sciences* 3 (1991): 105–11.

20. For details on nitrogen enrichment of the biosphere, see V. Smil, *Cycles of Life* (New York: Scientific American Library, 1997).

21. L. Bo et al., "Use of Night Soil in Agriculture and Fish Farming," *World Health Forum* 14 (1993): 67–70.

22. Recent benefit/cost studies of controlling air pollution in the Los Angeles basin are a perfect example. Total annual health benefits from reduced morbidity were found to be as low as $1.2 billion (1990 U.S. dollars), or as high as $20 billion. See A. J. Krupnick and P. R. Portney, "Controlling Urban Air Pollution: A Benefit-Cost Assessment," *Science* 252 (26 April 1991): 522–28; and J. V. Hall et al., "Valuing the Benefits of Clean Air," *Science* 255 (14 February 1992): 812–17. Given that it may take $13 billion (1990 dollars) to clean up the basin's air, morbidity costs alone can either easily justify the effort or make it economically quite unappealing.

23. For more on China's changing farmland, see F. W. Crook, "Underreporting of China's Cultivated Land Area: Implications for World Agricultural Trade," *China Agriculture and Trade Report* RS-93 (1993): 33–39; and V. Smil, "Who Will Feed China?," *The China Quarterly* 143 (1995): 801–13.

24. H. Chunru, "Recent Changes in the Rural Environment in China," *Journal of Applied Ecology* 26, no. 3 (1989): 803–12.

25. L. Yongzeng, "Chinese Forestry: Crisis and Options," *Liaowang* (Outlook) 12, no. 12 (December 1989): 9–10.

26. See, among many others, D. Heinsdijk, *Forest Assessment* (Wageningen: Center for Agricultural Publishing and Documentation, 1975); and R. Repetto et al., *Accounts Overdue: Natural Resource Depletion in Costa Rica* (Washington, D.C.: World Resources Institute, 1991).

27. Q. Geping and L. Jinchang, *Population & the Environment in China* (Boulder, Colo.: Lynne Rienner Publishers, 1994).

28. For more on noise in China's cities, see V. Smil, *Environmental Change as a Source of Conflict and Economic Losses in China* (Cambridge, Mass.: American Academy of Arts and Sciences, 1992).

29. World Bank, *Poverty Reduction and the World Bank: Progress and Challenges in the 1990s* (Washington, D.C., 1996).

Unit 3

Unit Selections

15. **The Importance of Places, or, a Sense of Where You Are,** Paul F. Starrs
16. **The Rise of the Region State,** Kenichi Ohmae
17. **Metropolis Unbound: The Sprawling American City and the Search for Alternatives,** Robert Geddes
18. **Greenville: From Back Country to Forefront,** Eugene A. Kennedy
19. **Does It Matter Where You Are?** *The Economist*
20. **Low Water in the American High Plains,** David E. Kromm
21. **Water Resource Conflicts in the Middle East,** Christine Drake
22. **Boomtown Baku,** Richard C. Longworth
23. **Demographic Clouds on China's Horizon,** Nicholas Eberstadt

Key Points to Consider

❖ To what regions do you belong?

❖ Why are maps and atlases so important in discussing and studying regions?

❖ What major regions in the world are experiencing change? Which ones seem not to change at all? What are some reasons for the differences?

❖ What regions in the world are experiencing tensions? What are the reasons behind these tensions? How can the tensions be eased?

❖ Why are regions in Africa suffering so greatly?

❖ Discuss whether or not the nation-state system is an anachronism.

❖ Why is regional study important?

 **Links**   www.dushkin.com/online/

14. **AS at UVA Yellow Pages: Regional Studies**
    http://xroads.virginia.edu/~YP/regional.html
15. **Can Cities Save the Future?**
    http://pan.cedar.univie.ac.at/habitat/press/press7.html
16. **IISDnet**
    http://iisd1.iisd.ca/
17. **NewsPage**
    http://pnp1.individual.com/
18. **Telecommuting as an Investment: The Big Picture—John Wolf**
    http://www.svi.org/telework/forums/messages5/48.html
19. **The Urban Environment**
    http://www.geocities.com/RainForest/Vines/6723/urb/index.html
20. **Virtual Seminar in Global Political Economy//Global Cities & Social Movements**
    http://csf.colorado.edu/gpe/gpe95b/resources.html
21. **WWW-LARCH-LK Archive: Sustainability**
    http://www.clr.toronto.edu/ARCHIVES/HMAIL/larchl/0737.html

These sites are annotated on pages 6 and 7.

The Region

The region is one of the most important concepts in geography. The term has special significance for the geographer, and it has been used as a kind of area classification system in the discipline.

Two of the regional types most used in geography are "uniform" and "nodal." A uniform region is one in which a distinct set of features is present. The distinctiveness of the combination of features marks the region as being different from others. These features include climate type, soil type, prominent languages, resource deposits, and virtually any other identifiable phenomenon having a spatial dimension.

The nodal region reflects the zone of influence of a city or other nodal place. Imagine a rural town in which a farm-implement service center is located. Now imagine lines drawn on a map linking this service center with every farm within the area that uses it. Finally, imagine a single line enclosing the entire area in which the individual farms are located. The enclosed area is defined as a nodal region. The nodal region implies interaction. Regions of this type are defined on the basis of banking linkages, newspaper circulation, and telephone traffic, among other things.

This unit presents examples of a number of regional themes. These selections can provide only a hint of the scope and diversity of the region in geography. There is no limit to the number of regions; there are as many as the researcher sets out to define.

Paul Starrs's thought-provoking essay on the importance of place and the concept of region leads this unit. Then, "The Rise of the Region State" suggests that the nation-state is an unnatural and even dysfunctional unit for organizing human activity.

Cities are becoming city-regions, according to Robert Geddes in this article from *The American Prospect* series on urban America. The continuing rise of Greenville, South Carolina, is documented in the next article. Then, "Does It Matter Where You Are?" considers aspects of geographical location principles in the context of the new global economic systems.

The next two articles address the potential for problems in areas where fresh water is scarce. "Low Water in the American High Plains" focuses on mismanagement of a valuable water resource in the United States, while "Water Resource Conflicts in the Middle East" explains why future conflicts between countries may occur over access to fresh water in that region. "Boomtown Baku" discusses potential changes in this coastal city as the race for Caspian Sea petroleum heats up. Finally, China's changing demographic picture is discussed as the average age of its people increases.

The Importance of Places, or, a Sense of Where You Are[1]

In this mobile society where every shopping mall has the same stores, can we really say there are differences among the country's regions?

Paul F. Starrs

Paul F. Starrs is professor of geography at the University of Nevada.

"A time and a place for everything" goes a well-weathered adage. The meaning of time in this spare phrasing has never warranted a second thought. Unvarying since the sundial, time is a comfortable given governed by planetary rotation and counted through the cycle of seconds, hours, days and months. If the exact interval captured by a throwaway term like "just a moment" fluctuates depending on whether you're Hopi, Huron, Hasidic, a homeboy, 3 years old or 102, a clock nonetheless ticks through 24 inviolable and orderly hours in a day.

But consider again that initial sentence: While time is known, it is without doubt "a place" that is less defined. Can the region—a place broadly construed—really have any universal meaning, or is it just a vaguely named sacrament in the church of location? What are regions, and how should we receive them in an ultra-modern era of mass communication?

A thoughtful mention of any "region" immediately takes on a burden of complicated assumptions. Europe is a vexatious and fitting example. That it exists is incontestable. And yet, what exactly is Europe? A separate continent surely Europe is not; a uniform economic organization also assuredly no. True, Europe counts as its own a motley collection of highly assorted, if generally Caucasian, peoples (Basque, Catalan, Friesian, Welsh, Walloon), with a presumptive (if hardly universal) Christian religious heritage, and claims can be made to a very grossly familiar linguistic heritage. Beyond that, Europe is an eclectic assemblage whose political sovereignty is in essence a status deigned by United Nations recognition—Norway, Ireland, Portugal, the Netherlands, San Marino, Bosnia, Macedonia, Greece. These disparate parts amount to a decidedly unseemly whole. So if Europe is genuinely a single region, then what is the binder for its constituent parts? Obvious answer: "Europe" is, fact, a convention, a useful ploy hearkening to a generally common history that includes the World Wars, the Holy Roman Empire, Jenghis Khan's depredations, the Neanderthal and the European Economic Community (except for the countries that have not been allowed to join or elect not to). With such a messy match, Europe as a region is more organizing thought than any demonstrable fact.

While the "new regionalist" geographers, the breed of scholar whose domain generally is said to include places and regions, have of late been self-eviscerating like a sea cucumber over just what constitutes "a region," most essayists and geographical scholars agree that the region as a unit of analysis and description is basic and not to be rubbished. Remaining as a category fluid, elusive and mutable, regions are entrenched in common thought and vernacular speech.[2]

Reprinted with permissions from *Spectrum: The Journal of State Government,* Summer 1994, pp. 5–17. © 1994 by The Council of State Government.

As with fashions for couture and academe, regions erratically fall in and out of favor. They are today's hot ticket, especially as renewed in the guise of "regionalism," which last peaked in the United States during the 1940s heyday of the Tennessee Valley Authority, when economic needs were thought best explored in a great restructuring of the country into socioeconomic regions that had their own distinctive personalities (Odum & Jocher 1973; Campbell 1968; Dorman 1993; Archer 1993).[3]

Global forces both dismember and contribute to regional identity. In the 1990s, a major point of contention in academic geography is how to parse, map and understand the connections of regions to booming inter-regional phenomena: The globalization of finance (capital), information (cyberspace), communications (cellular, FAX, satellite), transportation (frequent flier flights and high-speed rail), language (the hegemony of American English), or culture otherwise construed (MTV, grunge, CNN, Rupert Murdoch's newspapers, music).[4] An accurate rendering would show the globe slashed by vast arrows of movement that reshape and deform, describing essential and ongoing patterns of geographical change. While post-modernists like Doreen Massey are able to write intelligently of "a new burst of time/space compression," a skeptic would reply that this "burst" is merely an acceleration of the same foreshortening of the world's borders, a death of certainty and control, which started in 1492.

For all the glitz of global systems and parlance about the world economy, regions remain vital. Teasing out the singularities and generalities of places are among the foremost skills that anyone, from politician to market researcher to planner, can trot out to make sense of the planet. If the Earth is not simply made up of nugget-like places that embody perfectly consistent traits, the region amounts to the most resourceful, utilitarian and creative of lies.

WHEN YOU'RE THERE, YOU'LL KNOW

Regions are, in general, more useful than real. What often are taken to be time-honored physical wholes—like the Southwest, the short-grass prairie, the Mid-west and the Middle East—turn out to have sloppy edges (Meinig 1971, Said 1979, Shortridge 1985). If a mental map, an image of shared traits and cultural stereotypes, is firm, when it comes actually to mapping boundaries and attributes the cartography of any region loses exactness. An old-time regional geographer could counter by saying that ephemera like planning districts, culture areas and homelands are hardly regions at all. By conservative reckoning, the only region is something with a self-evident physical presence. While for some of the tire-kicking school there remains a real pleasure toiling to map the Basin and Range physiographic province, imprecision is hardly a fatal flaw. Mathematicians working with fractals, physicists with sub-atomic particles and philosophers charting the ideological canyons of the mind are quick to admit that fuzziness can be more useful than the obsession of an accountant or a scientist with calculator exactitude. As with the edges of forests, deserts, tundra or other ecotones, the margins of a region frequently have the highest diversity and interest, and all borderlines say much of what lies inside.

As organization structures rather than geographical certainties, regions embody events, emotions, physical similarities, human activity, or history and economy. Find one unity or several, and the geographic elaboration of a region begins. Many of the "new regionalists" argue regions are social constructions, revealing economic or class practices. Regions, which are interesting primarily as a physical setting for social interactions, suffice, in essence, as a game board (Goldfield 1984, Thrift 1990, Pudup 1988). Others are content if a regional description stockpiles plenty of room for change, which regions do (Gilbert 1988). All regions are effective manipulations of nature, casting and organizing human practice and gathering up bundles of traits to make sense of our presence on earth. If regions did not exist, they would have to be invented. And because we need to know where and what the manifold human creations are, we have regions. There is poignancy to this metaphysics, as with any discussion of scholarship's fastidious fashions.[5] The reality is straightforward: Regions exist because we want, need and relish them.

The inherent problem with regions is as simple as it is true. Geographers, like the sociologists, conservation biologists, anthropologists and historians who have joined in the traffic of places, worry about criteria used to identify a region (Hough 1990). Are the keys solely economic? Religious? Ethnic? Territorial? Do regions originate in history or are they entirely contemporary? If an area where overlapping attributes meet forms a region, then the edges are a pronounced problem. Are places defined by watersheds and biological life—bioregions—a valid means of defining a region? Do the world's nations, discrete groupings of peoples that exist within and among the larger political entities that convention calls "countries" (think of Navajo, Provencal, Scots or Zulu territory), constitute regions? (Nietschmann 1993).

Predictably, questions raised are more persistent than easy. How should we contend with larger social and structural forces defining a region? Each place is linked to others. From the Internet to the global economy, from the tracing of transnational money flows to the Hollywood movies that bully the French cinematic self-identify, even the smallest Appalachian hollow, Mormon village and South Central Los Angeles gang are connected to additional parts of the world. The most erudite sort of formulations discussing the relationship between the local and the global admit that regions are indispensable (Lipietz 1993).

There is space aplenty for flouishing skepticism. Whenever a region is identified it is easy to carp about the criteria used to single it out.[6] Regions are ultimately a state of mind, a convention. They exist in untold numbers, interwoven and overlapping. And while scholars squirm about the squishy boundaries, if you're there, you generally know. Having a sense of where you are is not just street smart, it is a survival trait through human history and geography.[7]

Debate over regional structure turns around a problem that social theorists call reification, or, in clinical terms, setting a region in stone (Jameson 1981). Places are no more uniform than the people within them. Talking about the West or the Bible Belt or the Colorado Plateau or the Chicken Fried Steak Line is no more defensible than prattling on about what "American Catholics believe," or "Cuban Americans argue," or "environmentalists claim." Catholics, Cubans and environmentalists are many and diverse. By the same token, people who fall into a designated region do not all hold to whatever is being attributed to them. But regional characterizations allow each and every person to make sense of the world, and the categories developed often are remarkably acute in their capturing what is important about people who are, after all, geographically located on a discrete part of the earth.

Regions literally hold together. But each region is also part of larger processes and interactions. A secondary lesson is, then, simplicity itself: Always examine the meaning and assumptions behind any region. Almost certainly the region is real. How have those come to be, and what do they mean? Much elegant prose is devoted to the description of regions. In fact, some of the best writing, going from Joel Garreau to Jan Morris to John McPhee to Gretel Ehrlich, delves deeply into regions and their sumptuous character. But analyze always, for a region can be developed as a scientific fact as easily as an emotional and poetic necessity.

THE USES OF REGIONAL IDENTITY

Modern politics is replete with examples of regions created in an attempt to control problems too large for the traditional political structures of towns, cities and countries. Thrown together of opportunism and necessity, these utilitarian creations have become essential political and economic facts. There are regions, however, that are literary and spiritual so much as real. Places do have life, and regions have identity—as, for example, has the American South (Cash 1941; Odum 1947; Wilson & Ferris 1989). The range of forms and purposes behind singling out regions illustrates problems of the age. Regions meet an impressive assortment of ends. The past uses, boundaries and creators of regions prove as illuminating as where and how regions today are constituted. A region always accommodates its particular time and is build within the limitations of available space. For all that, there are six regional categories that are characteristically contemporary, and a few words on each will suggest why regions are likely to stick around. What distinguishes them is that each has to be mapped to have any meaning: Regions are pure and simple geographic creations, and however their charter may be construed, for each, the map is the territory, and the accuracy of the map directly reflects on the capacity and vigilance of the creator.

THE ECOSYSTEM

After a century of trying to manage natural resources according to boundaries that have far more to do with political accident, land division history and convenience than with biological necessities, government resource managers and scientists are attempting to piece together large coherent natural bodies. Unsurprisingly, this generally occurs in areas that are in some degree of crisis. Designated "ecosystems" have obvious value. While two—the Greater Yellowstone Ecosystem and the Everglades Ecosystem—are built around national parks, there are other prominent attempts, including the Forest Service-developed "Sierra Nevada Ecosystem Project" (SNEP, more commonly), which is struggling to bound and understand a swarming human presence in the Sierra Nevada foothills of California and Nevada. Ecosystem studies try to transcend politics, seeking biological and planning alternatives to the threats of unbridled growth in environments that are subject to an unusual degree of hazard, and which cannot be planned within the older city or country boundaries.

These vaunted "ecosystems" are by and large the creations of scientists-cum-managers, and they therefore fulfill and reinforce their own public credibility. Ironically, what the scientific land managers keep rediscovering is an old Alfred Korzybski line, "the map is not the territory, and the name is not the thing named" (Bateson 1979). Sadly, naming an ecosystem and designating experts to study it goes only a teeny distance toward solving crises often political, social and economic in nature. In being so vexedly human, problems are rarely readily accessible to the gimlet eye of control by scientific edict. Other recognizable ecosystems have not fared well—Amazonia, the Sahara, the Aral or the Caspian sea—in part because they extend into multinational space. Considering that earnest efforts like the "Biosphere Preserves" program of UNESCO are well-intentioned, it is obvious that conservation efforts within countries are far easier to develop and enforce than protections offered under international treaties or covenants.

THE REGIONAL AUTHORITY

Among the most successful regions are those embraced by regional authorities, which are created to handle questions ranging from transit problems, sewage, and hazard abatement to conservation districts, water systems, and comprehensive area planning. These umbrella agencies, as they are often called, have on occasion met with surprisingly good results. An experiment extending back decades, successful regional authorities accommodate themselves to political reality, but also can weather storms that might dislodge more local agencies.

Like corporations after passage of the 14th Amendment to the U.S. Constitution, regional authorities have come to be as real and authoritative as elected officials. They became, in fact, nearly human; physically real through legislative midwifery. In general, regional authorities are created when a job is too large (read costly) for a single political entity to handle, or when what is sought is a planning window directed far enough in the future that there is no apparent harm in many different political entities banding together for discussions. Politics and planning creates the region.

There are wonderful examples of regional authorities. The Tennessee Valley Authority and the Bonneville Power Authority are two historic cases, but others are less clouded with time. The Metropolitan Water District of greater Los Angeles is prime territory; born of the successes of the Los Angeles Department of Water and Power; which preceded the Metropolitan Water District. Together, these regional entities have done the impossible, bringing reliable and unchallengeable water to 12 million in the middle of a desert. Public utilities often are blended into regional authorities, thanks in part to their being unreachable under antitrust statutes. Their boundaries are political creations: Census tract lines, county divides, the foreseeable edges of planning areas or unincorporated city limits. Other examples of these sorts of authorities include regional park districts, regional planning authorities and the occasional super-agency, like the California Coastal Commission and its allied Coastal Conservancy. They have charters buried in time, but have assumed as charge ruling on the environmental and development future of a thousand miles of California coastline. The authority of these regional entities ebbs and flows with the urgency of constituency concerns, but can be vast.

THE VERNACULAR REGION

For all the unnaturalness of some planning and political regions, there also are valiant and ongoing attempts to recognize and name places that are plainly geographically independent and coherent. The clearest attempt comes with analyzing common speech and asking the residents of different areas where they see themselves living. This leads to the bounding of "vernacular regions," areas where much of the population has little doubt about where they live and who they are. When cast in large terms, this can establish carefully documented boundaries around common places like "The Midwest," of "The Southwest," or it can lead to the recognition of sub-regions like the Panhandle, the "Oil Patch," or the German Hill Country of Texas (Jordan 1979, Shortridge 1987).

A vernacular approach is the simplest regional take—people in and around a large area are asked to toss forward names, and the answers mapped. The resulting lines, or isonyms, designate areas where there is common acceptance of a regional name. The subtleties of use and meaning in this method can be telling—while residents of El Paso, Presidio, Pecos or Wink are given to consider themselves Southwesterners, the same can hardly be said of Texans who hail from Texarkana, Tyler, Pineland or Port Arthur, who hearken to the South. In matters of outlook on race or ethnicity or religion, in where people go for major shopping, or how area residents self-identify with a broad range of political, economic and social issues, these things matter.

BIOREGIONS AND WATERSHEDS

Offshoots of the physiographic regions of the 19th century, bioregions and watersheds have found a great and growing constituency among a number of contemporary essayists and poets like Gary Snyder; who press for a return to closer forms of community than anything favored by large cities (Parsons 1985). The bioregionalists hold that the good fight is best begun at home.[8] At its best, the reasoning has echoes of Pestalozzi and the reformist educators of the 1800s, who argued that learning was best conducted using the terrain, plants and animals—a local habitat—as ledger and classroom (Pestalozzi & Green 1912). Speaking (and especially writing) with passion and great literacy for attending to geographically immediate needs and understanding the physical whole of the watershed or bioregion ahead of vainly attempting to control the world's whole surface, bioregionalists pledge that community comes first. Although James Malin issued similar plaints in decades past, and Ray Dasmann placed the bioregionalist and "ecosystem" traditions in able contrast nearly 20 years ago, the rhetoric is hard to counter (Dasmann 1976; Opie 1983, Bogue 1981.)

In essence, bioregionalists, many of whom also preach for a "watershed consciousness," note that too much of humanity's destructive exploitation of the Earth is driven by quick movements between dissimilar areas, which feeds growth by acquiring and exploiting colonized places, emphasizing profit over knowledge. Know a place, and you are less likely to abuse it, goes the ethos. To understand environmental history is to grapple with local needs. Citing examples from traditional peoples who have managed lands around them for centuries with but modest deterioration, bioregional advocates voice a limited respect for the ambitious architects who preach for global environmental harmony. Their resolute business, however, is to get on with saving their own nest before instructing others in how to clean up the Earth. The intelligence and influence of bioregionalists is far in excess of their numbers.[9]

CULTURAL AREAS, ETHNIC REGIONS & HOMELANDS

Interest in coherent bodies of cultural attributes goes back thousands of years; thematic mapping thrives on locating collections of traits, and early thematic maps often plotted fairly singular behaviors. Geographers have reveled in tight areas where distinctive traits are uniquely preserved, and have mapped these for generations (Gastil 1975; Rooney et al 1982, Garreau 1981, Hart 1972). The Mormon Culture Region, Amish or Cajun country, the Bible Belt, the Sun Belt, are typical tips of a vast iceberg. The traits can be religious, political, economic, racial or linguistic. That this form of region is durable goes almost without saying. In effect, residents within the region are singled out for preserving traits that can be singular or archaic. The degree of uniformity varies hugely, place to place, region to region. By and large, the areas mapped in such regions are so distinctive as to be almost beyond argument (Arreola 1993; Conzen 1993).

Mapping regions is far more than an ideal scholarly exercise. Where the overlaps between regions are pronounced, or where there is quick shrinking, conflict or the loss of distinguishing traits can occur. Where the boundaries alter little with time, the area is unlikely soon to disappear to dilution or disturbance. And yet there are oddities. Some culture regions persist despite migration and a loss of language and economy; the Basque, if anything, have grown stronger in the last two decades, largely thanks to political liberalization. What nudges change is never entirely easy to say, but these regions can and will persist. To a point, they are self-selected and self-sustaining.

NODES IN THE GLOBAL EXCHANGE

Finally, regions are important parts of the global system. Whether the connections are those of politics, empire, economics, culture, or religion, there is no doubt that any region, in this era of mass communication, is both beholden and at times hapless before global forces, the powers that Fernand Braudel, Immanuel Wallerstein, William Appleman Williams and Donald W. Meinig have traced with rare wisdom. This bears comment, but also requires the remark. "So what?" In many a sense, this is nothing new; the links are just more obvious and better drawn than 100 years ago, and it requires more will to insist on self-reliance, to put down the cable remote, than before the World Wars. But the articulations between places are old hat, in either a historic or a geographic sense.

The existence of larger connections in no way eliminates the need or sensibility of regions. If anything, the distractions of a beckoning world act to strengthen regions. The emphasis on learning traditional languages (Welsh, Catalan, Hopi) and relearning ways of life once on the verge of extinction likely have never been so strong. Differences have become precious, and to be distinct and have a separate identity is more than faddish. It is to know and have a sense of yourself, even if the adopted identity is more fancied than honestly come by. This is the oddest part of the region as one small segment of the known world—as the boundaries really should be falling, and homogenization growing rampant, instead the ultimate luxury of 1990s society is to "discover your place."

CONCLUSION

... time is absorbed into place, and place into mind. The land becomes history, and history becomes thought as people cross space in awareness.
—Henry Glassie
Passing the Time in Ballymenone

The precision of time, whether dispensed by the clockwork grace of gnomic Swiss watchmakers or driven by quartz crystal Swatches, is not the stuff of which regions and places are made. Time is a certainty, and for all that, a subject of hate and dread. "Saving time" is no small act of desperation. But "saving a place" is geography, history and environmental preservation writ large. We live in time, but for places; they are our communities, where horses are trotted or buses colorfully "tagged" by graffiti territorialists. Here historic preservation committees solemnize over the sanctity of the past, arguing about whether "that stone house over by Fleming's" should be torn down; planning commissioners debate regional futures and decide whether a community garden ought to be fostered in the projects. Regions are among the most intelligent acts that we can work with as humans. As Buckaroo Banzai reminds us: "Remember: No matter where you go, there you are."[10] Of many stripes, places matter. And that, most of all, is why regions are relevant.

NOTES

1. Prepared with support from the S. V. Ciriacy-Wantrup Postdoctoral Fellowship in Natural Resource Economics at the University of California at Berkeley, 1993–94.

2. As good a measure as any flows from Current Contents, a vast data base for contemporary journal articles, book reviews and commentaries. Although "bioregion" appears in only one title, "culture region" is found in 44, "watershed" in 321 and "ecosystem" appears in 784 citations. Alas, a voguish term like "post-modern" is referred to in a paltry 166 items. On the other hand, "place" is cited 1,981 times, the word "region" is in 14,890 references, and when "regional" and "regionalism" are added, the count rises to 29,811 articles. Computers offer a false precision, but the bottom line is evident: Places matter.

3. The discipline of sociology had never before, and has never since, ridden so high. That the regionalist project grew moribund with economic recovery after World War II is history. However, the importance of regions (James Madison called them "sections" in his famous Federalist Number 10) is axiomatic through American history, from early days of the Republic to the Civil War. Joel Garreau's *Nine Nations of North America* suggests some of the current interest, but there is more. In California, for example, proposals regularly float through the Legislature and increasingly onto ballot measure referenda, asking for the state to be split into more "rational" divisions. So far, no joy; check in next year.

4. The globalization of information and mass communications is quixotic—as *The Economist* has noted, "And a network is more than links between places, it is itself a place" (Editors, *The Economist*, 1993). Often the changes imposed by such technology are not so direct or self-evident as the technologically-addicted might argue. William Gibson has put it nicely: "She was a courier in the city. . . . Was it significant that Skinner shared his dwelling with one who earned her living at the archaic intersection of information and technology? The offices the girl rode between were electronically counterminous—in effect, a single desktop, the map of distances obliterated by the seamless and instantaneous nature of communication. Yet this very seamlessness, which had rendered physical mail an expensive novelty, might as easily be viewed as porosity, and as such created the need for the service the girl provided." *(Virtual Light*, Bantam Books, New York, 1993): p. 93.

5. Academic debate over the intersections of place, time and space at their worst can toss even a hardened stomach. Perhaps fortunately, Patricia Nelson Limerick has dealt an elegant swat, if sadly unlikely to carry the force of a death blow, to the gamboling semi-literacy of academic fashion in her *New York Times Book Review* essay, "Dancing with Professors: The Trouble with Academic Prose," 31 October 1993, pp. 3, 23–24.

6. Take, for example, the sumptuous category of "the Southland" that has been the name for Southern California for at least five decades. How can Southern California be bounded by any single term—it now reaches into the Mojave Desert, nearly to the Nevada and Arizona boundaries, and is virtually across the Techachapi Ranges, into the San Joaquin Valley, and runs south to San Diego and Tijuana. Yet the term "Southland" is a sufficiently flexible regional label to encompass it all. See Starrs 1988, Davis 1990, and *The Economist* 1994.

7. I will apologize here to John McPhee for borrowing the great phrase that he used to describe the proxemics of Bill Bradley, who was a Princeton basketball player when McPhee first used the phrase. I hope the original author will take no exception to the usurpation of titles.

8. "It is not the ecologists, engineers, economists or earth scientists who will save spaceship earth, but the poets, priests, artists and philosophers," is how Lawrence Hamilton has put it in the Introduction to a volume he edited, *Ethics, Religion, and Biodiversity: Relations between Conservation and Cultural Values.* (Knapwell, Cambridge: The White Horse Press, 1993).

9. The bioregionalists are assuredly NOT to be confused with the "biospherians" whose "space capsule" existence in Biosphere 2 near Oracle, Arizona, shows far more en-

thusiasm for the ecosystem model than a modest biosphere consciousness.

10. The line is Peter Weller's in "The Adventures of Buckaroo Banzai Across the 8th Dimension," Twentieth Century Fox, 1984.

SOURCES

Archer, Kevin, 1993. "Regions As Social Organism: The Lamarckian Characteristics of Vidal de la Blache's Regional Geography," *Annals of the Association of American Geographers*, September; 83(3): pp. 498–513.

Arreola, Daniel D., 1993. "The Texas-Mexican Homeland," *Journal of Cultural Geography*, Spring-Summer; 13(2): pp. 61–74.

Bogue, Allan G., 1981. "The Heirs of James C. Malin: A Grassland Historiography," *Great Plains Quarterly*, Spring, 1(2): pp. 105–131.

Campbell, Robert D., 1968. "Personality as an Element of Regional Geography," *Annals of the Association of American Geographers*, December; 58(4): pp. 748–759.

Cash, W. J. [Wilbur Joseph], 1941. *The Mind of the South*; (New York: Alfred A. Knopf); 429 pages.

Conzen, Michael. 1993. "Culture Regions, Homelands, and Ethnic Archipelagos in the United States: Methodological Considerations," *Journal of Cultural Geography*, Spring-Summer; 13(2): pp. 13–30.

Dasmann, Raymond, 1976. "Future Primitive: Ecosystem People versus Biosphere People," *The CoEvolution Quarterly*, Fall, 11: pp. 26–31.

Davis, Mike. 1990. *City of Quartz: Excavating the Future in Los Angeles* (New York & London, Verso).

Dorman, Robert L., 1993. *Revolt of the Provinces: The Regionalist Movement in America, 1920–1945*; (Chapel Hill: University of North Carolina Press); 366 pages.

The Economist, 1994. "The Point of Los Angeles," [Editorial], *The Economist* [London], 22 Jan., p. 14.

The Economist, 1993. "Make Way for Multimedia," [Lead Editorial] *The Economist* [London], 16 October: pp. 15–16.

Entrikin, J. Nicholas, 1991. *The Betweenness of Place: Towards a Geography of Modernity*; [Critical Human Geography]; (London: Macmillan).

Garreau, Joel, 1981. *The Nine Nations of North America*. (New York: Avon Books).

Gastil, Raymond, 1975. *Cultural Regions of the United States* (Seattle, University of Washington), 366 pages.

Gilbert, Anne. 1988. "The New Regional Geography in English and French-speaking Countries," *Progress in Human Geography* 12:2, June, pp. 208–228.

Goldfield, David R., 1984. "The New Regionalism [Review Essay]," *Journal of Urban History*, February, 10(2): pp. 171–186.

Hart, John Fraser, 1991. "The Perimetropolitan Bow Wave," *Geographical Review*, January, 81(1): pp. 35–51.

Hart, John Fraser [editor], 1972. *Regions of the United States*; (New York: Harper & Row).

Hough, Michael, 1990. *Out of Place: Restoring Identity to the Regional Landscape*; (New Haven, Connecticut: Yale University Press); 230 pages.

Jameson, Frederic, 1981. *The Political Unconscious: Narrative as a Socially Symbolic Work* (London, Methuen).

Jordan, Terry G., 1978. "Perceptual Regions in Texas," *Geographical Review*, July, 68(3): pp. 293–307.

Lewis, Peirce, 1979. "Defining a Sense of Place," *The Southern Quarterly: A Journal of the Arts in the South*, Spring-Summer; 27(3 & 4): pp. 24–46.

Lipietz, Alain. 1993. "The Local and the Global: Regional Individuality or Interregionalism?" *Transactions of the Institute of British Geographers, New Series*; Vol. 18, pp. 8–18.

Massey, Doreen, 1992. "A Place Called Home?" *New Formations* Number 17, pp. 3–15.

Meinig, D. W. [Donald William], 1971. *Southwest: Three Peoples in Geographical Change, 1600–1970*; (New York: Oxford University).

Nietschmann, Bernard, 1993. "Authentic, State, and Virtual Geography in Film," *Wide Angle: A Quarterly Journal of Film History, Theory, Criticism, & Practice*, October, 15(4): pp. 5–12.

Odum, Howard W., 1947. *The Way of the South; Toward the Regional Balance of America*; (New York: Macmillan); 350 pages.

Odum, Howard W. and Katharine C. Jocher; [editors], 1973. *In Search of The Regional Balance of America*; [The University of North Carolina Sesquicentennial Publications]; (Westport, Conn.: Greenwood Press; orig. copyright 1945); 162 pages.

Opie, John, 1983. "Environmental History: Pitfalls and Opportunities," *Environmental Review*, 7(1): pp. 8–16.

Parsons, James J., 1985. "On "Bioregionalism" and "Watershed Consciousness"" *The Professional Geographer*, February, 37(1): pp. 1–6.

Pestalozzi, Johann Heinrich and John Alfred Green, 1912. *Pestalozzi's Educational Writings*. (New York, London: Longmans, Green & Co.; E. Arnold.) 328 pages.

Pudup, Mary Beth, 1988. "Arguments Within Regional Geography," *Progress in Human Geography*, September, 12(3): pp. 369–390.

Rooney, John F., Jr., Wilbur Zelinsky, and Dean R. Louder, [General editors], 1982. *This Remarkable Continent: An Atlas of United States and Canadian Society and Cultures*; Cartographic editor John D. Viteck; (College Station, Texas: Texas A & M University Press for The Society for the North American Cultural Survey); 321 pages.

Said, Edward W., 1979. *Orientalism*; (New York: Vintage Books, Random House); 368 pages.

Shortridge, James R., 1985. "The Vernacular Middle West," *Annals of the Association of American Geographers*, March, 75(1): pp. 48–57.

Shortridge, James R., 1987. "Changing Usage of Four American Regional Labels," *Annals of the Association of American Geographers*, September 1987, 77(3): pp. 325–336.

Starrs, Paul F., 1988. "The Navel of California and Other Oranges: Images of California and the Orange Crate," *The California Geographer*, Vol. 28, pp. 1–42.

Thrift, Nigel, 1990. "For a New Regional Geography 1," in *Progress in Human Geography*, June, 14(2): pp. 272–279.

Wilson, Charles Reagan, and William Ferris, [co-editors], 1989. *The Encyclopedia of Southern Culture*; (Chapel Hill: University of North Carolina Press for the Center for the Study of Southern Culture at the University of Mississippi); 1634 pages.

THE RISE OF THE REGION STATE

Kenichi Ohmae

Kenichi Ohmae is Chairman of the offices of McKinsey & Company in Japan.

The Nation State Is Dysfunctional

THE NATION STATE has become an unnatural, even dysfunctional, unit for organizing human activity and managing economic endeavor in a borderless world. It represents no genuine, shared community of economic interests; it defines no meaningful flows of economic activity. In fact, it overlooks the true linkages and synergies that exist among often disparate populations by combining important measures of human activity at the wrong level of analysis.

For example, to think of Italy as a single economic entity ignores the reality of an industrial north and a rural south, each vastly different in its ability to contribute and in its need to receive. Treating Italy as a single economic unit forces one—as a private sector manager or a public sector official—to operate on the basis of false, implausible and nonexistent averages. Italy is a country with great disparities in industry and income across regions.

On the global economic map the lines that now matter are those defining what may be called "region states." The boundaries of the region state are not imposed by political fiat. They are drawn by the deft but invisible hand of the global market for goods and services. They follow, rather than precede, real flows of human activity, creating nothing new but ratifying existing patterns manifest in countless individual decisions. They represent no threat to the political borders of any nation, and they have no call on any taxpayer's money to finance military forces to defend such borders.

Region states are natural economic zones. They may or may not fall within the geographic limits of a particular nation—whether they do is an accident of history. Sometimes these distinct economic units are formed by parts of states, such as those in northern Italy, Wales, Catalonia, Alsace-Lorraine or Baden-Württemberg. At other times they may be formed by economic patterns that overlap existing national boundaries, such as those between San Diego and Tijuana, Hong Kong and southern China, or the "growth triangle" of Singapore and its neighboring Indonesian islands. In today's borderless world these are natural economic zones and what matters is that each possesses, in one or another combination, the key ingredients for successful participation in the global economy.

Look, for example, at what is happening in Southeast Asia. The Hong Kong economy has gradually extended its influence throughout the Pearl River Delta. The radiating effect of these linkages has made Hong Kong, where GNP per capita is $12,000, the driving force of economic life in Shenzhen, boosting the per capital GNP of that city's residents to $5,695, as compared to $317 for China as a whole. These links extend to Zhuhai, Amoy and Guangzhou as well. By the year 2000 this cross-border region state will have raised the living standard of more than 11 million people over the $5,000 level. Meanwhile, Guangdong province, with a population of more than 65 million and its capital at Hong Kong, will emerge as a newly industrialized economy in its own right, even though China's per capita GNP may still hover at about $1,000. Unlike in Eastern Europe, where nations try to convert entire socialist economies over to the market, the Asian model is first to convert limited economic zones—the region states—into free enterprise havens. So far the results have been reassuring.

These developments and others like them are coming just in time for Asia. As Europe perfects its single market and as the United States, Canada and Mexico begin to explore the benefits of the North American Free Trade Agreement (NAFTA), the combined economies of Asia and Japan lag behind those of the other parts of the globe's economic triad by about $2 trillion—roughly the aggregate size of some 20 additional region states. In other words, for Asia to keep pace existing regions must continue to grow at current rates throughout the next decade, giving birth to 20 additional Singapores.

Many of these new region states are already beginning to emerge. China has expanded to 14 other areas—many of them inland—the special economic zones that have worked so well for Shenzhen and Shanghai. One such project at Yunnan will become a cross-border economic zone encompassing parts of Laos and Vietnam. In Vietnam itself Ho Chi Minh City (Saigon) has launched a similar "sepzone" to attract foreign capital. Inspired in part by Singapore's "growth triangle," the governments of Indonesia, Malaysia and Thailand in 1992 unveiled a larger triangle across the Strait of Malacca to link Medan, Penang and Phuket. These developments are not, of course, limited to the developing economies in Asia. In economic terms the United States has never been a single nation. It is a collection of region states: northern and southern California, the "power corridor" along the East Coast between Boston and Washington, the Northeast, the Midwest, the Sun Belt, and so on.

What Makes a Region State

THE PRIMARY linkages of region states tend to be with the global economy and not with their host nations. Region states make such effective points of entry into the global economy because the very characteristics that define them are shaped by the demands of that economy. Region states tend to have between five million and 20 million people. The range is broad, but the extremes are clear: not half a million, not 50 or 100 million. A region state must be small enough for its citizens to share certain economic and consumer interests but of adequate size to justify the infrastructure—communication and transportation links and quality professional services—necessary to participate economically on a global scale.

It must, for example, have at least one international airport and, more than likely, one good harbor with international-class freight-handling facilities. A region state must also be large enough to provide an attractive market for the brand development of leading consumer products. In other words, region states are not defined by their economies of scale in production (which, after all, can be leveraged from a base of any size through exports to the rest of the world) but rather by their having reached efficient economies of scale in their consumption, infrastructure and professional services.

For example, as the reach of television networks expands, advertising becomes more efficient. Although trying to introduce a consumer brand throughout all of Japan or Indonesia may still prove prohibitively expensive, establishing it firmly in the Osaka or Jakarta region is far more affordable—and far more likely to generate handsome returns. Much the same is true with sales and service networks, customer satisfaction programs, market surveys and management information systems: efficient scale is at the regional, not national, level. This fact matters because, on balance, modern marketing techniques and technologies shape the economies of region states.

Where true economies of service exist, religious, ethnic and racial distinctions are not important—or, at least, only as important as human nature requires. Singapore is 70 percent ethnic Chinese, but its 30 percent minority is not much of a problem because commercial prosperity creates sufficient affluence for all. Nor are ethnic differences a source of concern for potential investors looking for consumers.

Indonesia—an archipelago with 500 or so different tribal groups, 18,000 islands and 170 million people—would logically seem to defy effective organization within a single mode of political government. Yet Jakarta has traditionally attempted to impose just such a central control by applying fictional averages to the entire nation. They do not work. If, however, economies of service allowed two or three Singapore-sized region states to be created within Indonesia, they could be managed. And they would ameliorate, rather than exacerbate, the country's internal social divisions. This holds as well for India and Brazil.

The New Multinational Corporation

WHEN VIEWING the globe through the lens of the region state, senior corporate managers think differently about the geographical expansion of their businesses. In the past the primary aspiration of multinational corporations was to create, in effect, clones of the parent organization in each of the dozens of countries in which they operated. The goal of this system was to stick yet another pin in the global map to mark an increasing number of subsidiaries around the world.

More recently, however, when Nestlé and Procter & Gamble wanted to expand their business in Japan from an already strong position, they did not view the effort as just another pin-sticking exercise. Nor did they treat the country as a single coherent market to be gained at once, or try as most Western companies do to establish a foothold first in the Tokyo area, Japan's most tumultuous and overcrowded market. Instead, they wisely focused on the Kansai region around Osaka and Kobe, whose 22 million residents are nearly as affluent as those in Tokyo but where competition is far less intense. Once they had on-the-ground experience on how best to reach the Japanese consumer, they branched out into other regions of the country.

Much of the difficulty Western companies face in trying to enter Japan stems directly from trying to shoulder their way in through Tokyo. This instinct often proves difficult and costly. Even if it works, it may also prove a trap; it is hard to "see" Japan once one is bottled up in the particular dynamics of the Tokyo marketplace. Moreover, entering the country through a different regional doorway has great economic appeal. Measured by aggregate GNP the Kansai region is the seventh-largest economy in the world, just behind the United Kingdom.

Given the variations among local markets and the value of learning through real-world experimentation, an incremental region-based approach to market entry makes excellent sense. And not just in Japan. Building an effective presence across a landmass the size of China is of course a daunting prospect. Serving the people in and around Nagoya City, however, is not.

If one wants a presence in Thailand, why start by building a network over the entire extended landmass? Instead focus, at least initially, on the region around Bangkok, which represents the lion's share of the total potential market. The same strategy applies to the United States. To introduce a new top-of-the-line car into the U.S. market, why replicate up front an exhaustive coast-to-coast dealership network? Of the country's 3,000 statistical metropolitan areas, 80 percent of luxury car buyers can be reached by establishing a presence in only 125 of these.

The Challenges for Government

TRADITIONAL ISSUES of foreign policy, security and defense remain the province of nation states. So, too, are macroeconomic and monetary policies—the taxation and public investment needed to provide the necessary infrastructure and incentives for region-based activities. The government will also remain responsible for the broad requirements of educating and training citizens so that they can participate fully in the global economy.

Governments are likely to resist giving up the power to intervene in the economic realm or to relinquish their impulses for protectionism. The illusion of control is soothing. Yet hard evidence proves the contrary. No manipulation of exchange rates by central bankers or political appointees has ever "corrected" the trade imbalances between the United States and Japan. Nor has any trade talk between the two governments. Whatever cosmetic actions these negotiations may have prompted, they rescued no industry and revived no economic sector. Textiles, semiconductors, autos, consumer electronics—the competitive situation in these industries did not develop according to the whims of policymakers but only in response to the deeper logic of the competitive marketplace. If U.S. market share has dwindled, it is not because government policy failed but because individual consumers decided to buy elsewhere. If U.S. capacity has migrated to Mexico or Asia, it is only because individual managers made decisions about cost and efficiency.

The implications of region states are not welcome news to established seats of political power, be they politicians or lobbyists. Nation states by definition require a domestic political focus, while region states are ensconced in the global economy. Region states that sit within the frontiers of a particular nation share its political goals and aspirations. However, region states welcome foreign investment and ownership—whatever allows them to employ people productively or to improve the quality of life. They want their people to have access to the best and cheapest products. And they want whatever surplus accrues from these activities to ratchet up the local quality of life still further and not to support distant regions or to prop up distressed industries elsewhere in the name of national interest or sovereignty.

When a region prospers, that prosperity spills over into the adjacent regions within the same political confederation. Industry in the area immediately in and around Bangkok has prompted investors to explore options elsewhere in Thailand. Much the same is true of Kuala Lumpur in Malaysia, Jakarta in Indonesia, or Singapore, which is rapidly becoming the unofficial capital of the Association of Southeast Asian Nations. São Paulo, too, could well emerge as a genuine region state, someday entering the ranks of the Organization of Economic Cooperation and Development. Yet if Brazil's central government does not allow the São Paulo region state finally to enter the global economy, the country as a whole may soon fall off the roster of the newly industrialized economies.

Unlike those at the political center, the leaders of region states—interested chief executive officers, heads of local unions, politicians at city and state levels—often welcome and encourage foreign capital investment. They do not go abroad to attract new plants and factories only to appear back home on television vowing to protect local companies at any cost. These leaders tend to possess an international outlook that can help defuse many of the usual kinds of social tensions arising over issues of "foreign" versus "domestic" inputs to production.

In the United States, for example, the Japanese have already established about 120 "transplant" auto factories throughout the Mississippi Valley. More are on the way. As their share of the U.S. auto industry's production grows, people in that region who look to these plants for their livelihoods and for the tax revenues needed to support local communities will stop caring whether the plants belong to U.S.- or Japanese-based companies. All they will care about are the regional economic benefits of having them there. In effect, as members of the Mississippi Valley region state, they will have leveraged the contribution of these plants to help their region become an active participant in the global economy.

Region states need not be the enemies of central governments. Handled gently, region states can provide the opportunity for eventual prosperity for all areas within a nation's traditional political control. When political and industrial leaders accept and act on these realities, they help build prosperity. When they do not—falling back under the spell of the nationalist economic illusion—they may actually destroy it.

Consider the fate of Silicon Valley, that great early engine of much of America's microelectronics industry. In the beginning it was an extremely open and entrepreneurial environment. Of late, however, it has become notably protectionist—creating industry associations, establishing a polished lobbying presence in Washington and turning to "competitiveness" studies as a way to get more federal funding for research and development. It has also begun to discourage, and even to bar, foreign investment, let alone foreign takeovers. The result is that Boise and Denver now prosper in electronics; Japan is developing a Silicon Island on Kyushu; Taiwan is trying to create a Silicon Island of its own; and Korea is nurturing a Silicon Peninsula. This is the worst of all possible worlds: no new money in California and a host of newly energized and well-funded competitors.

Elsewhere in California, not far from Silicon Valley, the story is quite different. When Hollywood recognized that it faced a severe capital shortage, it did not throw up protectionist barriers against foreign money. Instead, it invited Rupert Murdoch into 20th Century Fox, C. Itoh and Toshiba into Time-Warner, Sony into Columbia, and Matsushita into MCA. The result: a $10 billion infusion of new capital and, equally important, $10 billion less for Japan or anyone else to set up a new Hollywood of their own.

Political leaders, however reluctantly, must adjust to the reality of economic regional entities if they are to nurture real economic flows. Resistant governments will be left to reign over traditional political territories as all meaningful participation in the global economy migrates beyond their well-preserved frontiers.

Canada, as an example, is wrongly focusing on Quebec and national language tensions as its core economic and even political issue. It does so to the point of still wrestling with the teaching of French and English in British Columbia, when that province's economic future is tied to Asia. Furthermore, as NAFTA takes shape the "vertical" relationships between Canadian and U.S. regions—Vancouver and Seattle (the Pacific Northwest region state); Toronto, Detroit and Cleveland (the Great Lakes region state)—will become increasingly important. How Canadian leaders deal with these new entities will be critical to the continuance of Canada as a political nation.

In developing economies, history suggests that when GNP per capita reaches about $5,000, discretionary income crosses an invisible threshold. Above that level people begin wondering whether they have reasonable access to the best and cheapest available products and whether they have an adequate quality of life. More troubling for those in political control, citizens also begin to consider whether their government is doing as well by them as it might.

Such a performance review is likely to be unpleasant. When governments control information—and in large measure because they do—it is all too easy for them to believe that they "own" their people. Governments begin restricting access to certain kinds of goods or services or pricing them far higher than pure economic logic would dictate. If market-driven levels of consumption conflict with a government's pet policy or general desire for control, the obvious response is to restrict consumption. So what if the people would choose otherwise if given the opportunity? Not only does the government withhold that opportunity but it also does not even let the people know that it is being withheld.

Regimes that exercise strong central control either fall on hard times or begin to decompose. In a borderless world the deck is stacked against them. The irony, of course, is that in the name of safeguarding the integrity and identity of the center, they often prove unwilling or unable to give up the illusion of power in order to seek a better quality of life for their people. There is at the center an understandable fear of letting go and losing control. As a result, the center often ends up protecting weak and unproductive industries and then passing along the high costs to its people—precisely the opposite of what a government should do.

The Goal is to Raise Living Standards

THE CLINTON administration faces a stark choice as it organizes itself to address the country's economic issues. It can develop policy within the framework of the badly dated assumption that success in the global economy means pitting one nation's industries against another's. Or it can define policy with the awareness that the economic dynamics of a borderless world do not flow from such contrived head-to-head confrontations, but rather from the participation of specific regions in a global nexus of information, skill, trade and investment.

If the goal is to raise living standards by promoting regional participation in the borderless economy, then the less Washington constrains these regions, the better off they will be. By contrast, the more Washington intervenes, the more citizens will pay for automobiles, steel, semiconductors, white wine, textiles or consumer electronics—all in the name of "protecting" America. Aggregating economic policy at the national level—or worse, at the continent-wide level as in Europe—inevitably results in special interest groups and vote-conscious governments putting their own interests first.

The less Washington interacts with specific regions, however, the less it perceives itself as "representing" them. It does not feel right. When learning to ski, one of the toughest and most counterintuitive principles to accept is that one gains better control by leaning down toward the valley, not back against the hill. Letting go is difficult. For governments region-based participation in the borderless economy is fine, except where it threatens current jobs, industries or interests. In Japan, a nation with plenty of farmers, food is far more expensive than in Hong Kong or Singapore, where there are no farmers. That is because Hong Kong and Singapore are open to what Australia and China can produce far more cheaply than they could themselves. They have opened themselves to the global economy, thrown their weight forward, as it were, and their people have reaped the benefits.

For the Clinton administration, the irony is that Washington today finds itself in the same relation to those region states that lie entirely or partially within its borders as was London with its North American colonies centuries ago. Neither central power could genuinely understand the shape or magnitude of the new flows of information, people and economic activity in the regions nominally under its control. Nor could it understand how counterproductive it would be to try to arrest or distort these flows in the service of nation-defined interests. Now as then, only relaxed central control can allow the flexibility needed to maintain the links to regions gripped by an inexorable drive for prosperity.

The Sprawling American City and the Search for Alternatives
Metropolis Unbound

by Robert Geddes

A new form of human settlement has emerged in the twentieth century, radically different from the cities of the past. The city has become a city-region. American city-regions' population growth is now dramatically outpaced by their geographic growth. In the two decades from 1970 to 1990, the New York region had a modest population increase of 8 percent, but it had an explosive growth of 65 percent in its built-up urbanized land. While Chicago grew 4 percent in population, its urbanized land increased 46 percent. Even places that were declining in their population were simultaneously growing in their urban area; Cleveland, for example, had a population decline of 8 percent, while it expanded geographically by 33 percent. This urban growth cycle is similar across America. City-regions are exploding into their surrounding countryside at growth rates that are eight to ten times greater than their population increases.

What is new is not the size of cities, but a change in their form. New York City, for example, used to have a concentric form surrounding Manhattan that resembled the growth rings of a tree. That was how it appeared when New York's Regional Plan Association, a civic organization, published its first plan 60 years ago. The Third Regional Plan published in 1996, however, describes a city-region with a population of 20 million people, extending 150 miles across and covering 13,000 square miles; its form now resembles a flower with petals radiating into five subregions in three states.

Ominously titled *A Region at Risk,* the regional plan warns of the dangers from the vast sprawl for New York's economy, environment, social fabric, and quality of everyday life. "Far more suddenly than people realize," write the authors, Robert Yaro and Tony Hiss, "super-sized metropolitan regions—areas hundreds of miles wide crowded with a dense mixture of aging cities, expanding suburbs, newer edge cities, and older farmlands and wildernesses—are emerging not just as a recognizable place but as humanity's new home base."

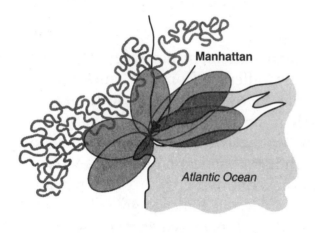

ANNIE BISSET

The everyday consequences for suburb and city alike are familiar enough: traffic congestion and inefficient transportation, unavailable and unaffordable housing, water and air pollution, social segregation and lack of community. In the decades after World War II, millions of Americans fled the cities to live in the suburbs, but in a sense the city has come after them. Nonetheless, the persistence of old political boundaries prevents the problems they face from being addressed together or even discussed coherently. The problems of transportation, housing, jobs, the environment, and social equity get scattered attention in public policy, but there is hardly any notice of the urban dynamic that lies behind them: the new form that American cities have taken. Nor is there much debate about the alternative paths of development that a few city-regions have taken in North America that could be the basis of a new paradigm for city-regions and neighborhoods in the next century.

THE EVOLVING CITY

The emergence of a new form of human settlement is relatively rare in human history. For thousands of years, human settlements grew slowly and predictably. Generally, they grew outward in concentric rings, each expansion being larger but still recognizably the same as its earlier form. For example, while the European city of Bruges was growing over a period of 500 years, its boundary walls were periodically moved outward, but it kept the same kind of shape. As a pre-industrial city, Bruges faced economic and technological limits on its size, such as everyday walking distances.

The relationships between how things grow and the shapes they take fascinated the biologist D'Arcy Thompson. In his 1917 book, *On Growth and Form,* Thompson analyzed both natural and manmade objects, from marine shells, teeth, fleas, and dinosaurs to soap bubbles and bridges, observing how and when their form accommodated and changed during growth.

If this pattern, as Thompson argued, applies to mechanical constructions like boilers and biological constructions like the marine shell Forminifera, it also applies to social constructions like cities. Before the industrial revolution, the size of towns and cities was constrained by natural limits, such as the capacity of the surrounding countryside to supply foodstuffs and the ability of people to move about by foot or on animals without mechanized vehicles. Railroads changed city form in two ways. Long-distance rail lines connecting to other cities and distant agricultural areas meant that a city's population size was no longer constrained by the food from its surrounding countryside. And short-distance rail lines extending into the country meant that the city's geographic size was no longer limited by walking distances. The city's form evolved into a star pattern, with new settlements—"railroad suburbs"—concentrated around rail stations, spaced a few miles apart. The legacy of America's dependence on rail lines and depots remains with us: The New York region, for example, has a rail network that is aging and somewhat disconnected but still includes 900 railroad stations.

The railroad suburb was a nineteenth-century invention, but it is also an alternative spatial model for the twenty-first century that retains some notable advantages compared to the sprawl of the more recent automobile suburb. The advantages of the star pattern come from its physical and social compactness, its preservation of the surrounding countryside, and its economy and efficiency of transport.

The automobile radically changed city form. The private car provided extraordinary flexibility, adaptability, and choice. Space and time were reconfigured. The city's edges—so clear in the old pre-industrial city and still evident along the finger-like corridors of the industrial city—melted away. Urban centers struggled to accommodate their new inhabitants—moving and parked vehicles. Centers kept their appeal—shopping centers, research centers, sports centers, health centers, to name a few—but each became a separate center. The city became a city-region of disjointed centers. Today, at its best, it is a galaxy; at its worst, it is chaos.

THE LOS ANGELES PARADIGM

Historically, two massive shifts of population have formed American city-regions. The farm-to-city shift after the Civil War is comparable to the massive city-to-suburb shift after World War II. Now more than half the nation's population lives in the suburbs. Although still separate legal jurisdictions, it no longer makes sense to talk of suburbs and cities as if they were separate; they are economically and ecologically joined in a new kind of human settlement, the city-region.

Periodically, a city seems to be the embodiment and image of the new. Historians call it the "shock city" of its time. Los Angeles has been the "shock city" of our time, as Manchester, England, was in the nineteenth century and New York was in this century's first half. Los Angeles is now seen as the first American city to remove itself from the European models of growth and form. Architect and urbanist Richard Weinstein argues that "the structure of the built-environment as it exists in Los Angeles now represents a paradigm of growth that already houses more than half of the [United States] population and is, with variations, the pattern of growth for most new settlements in the developed world."

The Los Angeles paradigm is an extended, open, unbounded matrix laced with linear corridors, from boulevards to commercial strips, and overlaid by freeways. Its keywords are *fragmented, incomplete, ad hoc, uncentered*. Concerning the Los Angeles environment, Weinstein argues that the open extended matrix, with all its in-between spaces, is more supportive of environmental health than denser, more continuous urban structures. There is more green, in-between.

But the Los Angeles urban form has had inequitable social consequences. Ethnic colonies have become isolated, the city fragmented. If the goal is to balance the economy, the environment, and social equity, is the open extended matrix of Los Angeles the inevitable model for American cities?

NORTHERN LIGHTS

On the North American continent, Toronto represents an alternative model of urban growth and form. In contrast to Los Angeles, Toronto generates vitality in its centers. Toronto's downtown is vibrant and pedestrian-friendly, and its neighborhoods retain their strength as places of sociability. By developing mass transit, Toronto succeeded, at least until the mid-1970s, in linking its centers and retarding the land-consuming and smog-producing dependence on the automobile. A key element in this achievement was that Toronto

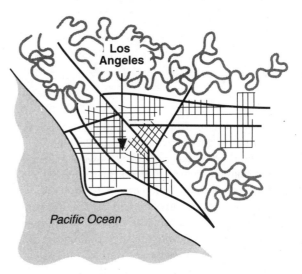

ANNIE BISSET

managed its postwar boom with a system of governance called Metro-Toronto that integrated urban and suburban decision-making. Metro-Toronto had jurisdiction over planning not only for five municipalities in the core metropolitan area, but also for the surrounding communities. Among its achievements was a light-rail transportation network financed by the core city.

Toronto has thus become a more equitable city than Los Angeles not only because of Canada's generous social programs, but also because the city has not isolated its less affluent residents. Ethnic minorities, the poor, and the elderly—thanks to public policy—are less segregated in Toronto than in other North American city-regions. Not only did Toronto build the transportation connections;

it has also created the continent's largest stock of dispersed mixed-income social housing.

In recent decades, however, the Toronto pattern of development has drifted away from this tradition. In 1972, the Ontario provincial government

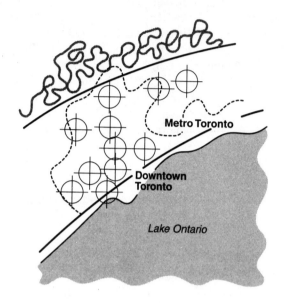

ANNIE BISSET

combined the surrounding communities into four mini-metro governments (Halton, Peel, York, and Durham), each having strong powers over their own region. According to Gardner Church, a political scientist at York University, the province failed to create any comprehensive planning authority or to sustain the earlier commitments to contain growth and coordinate transportation. Sprawl set in and the region stood in danger of becoming, as observers put it at the time, "Vienna surrounded by Phoenix." But recently, in an effort to reverse this backsliding, the province has made Metro-Toronto the unified government of the core metropolitan area and created a new super-regional authority, called Greater-Toronto, for transportation, social services, and economic development. The surrounding areas will share the costs of social services with Toronto. Church believes this new system "offers the potential for a return to comprehensive, progressive planning."

Another model for the future comes from the Pacific Northwest, where a chain of cities—including Portland, Seattle, and Vancouver—form a city-region now often called "Cascadia" (from the Cascade Mountains that parallel the Pacific coastline). Although this new city-region crosses state and international boundaries, the emerging idea of Cascadia provides an economically integrated vision of the settlements along a regional corridor, a "Main Street" called Interstate Highway 5. What is especially notable is that it also includes an ecologically integrated vision of the geology, vegetation, natural species, climate, and movement of water throughout the region.

Cascadia shows that an equilibrium of nature, society, and culture can still be the basis of city building. Think of Cascadia as a candidate for the historians' next "shock city." Its predecessors, Manchester, New York, and Los Angeles, all drew their image from their built landscape. Cascadia draws its power as a new paradigm from its natural landscape.

Portland, Seattle, and Vancouver have each pioneered in planning for environmental protection and the provision of greenspace (parks, riparian corridors, natural habitats) as parts of the urban fabric. Today, however, greenspace is at risk. The greatest challenge comes from rapid population growth and a pattern of human settlement that, like other American city-regions, is consuming land at an even faster rate. Sprawl development has led to inefficient use of land, energy, and other resources and has had profound impacts on air quality, the hydrology of watersheds, and the environmental health of the inhabitants. The question is whether Cascadia will go the way of Los Angeles. Or as Cascadian urbanists Ethan Seltzer, Ann Vernez Moudon, and Alan Artibise put it, "Will the legacy of our times result in the stewardship of the environment, or the destructive consumption of one of the most striking and abundant landscapes on the continent?"

Cascadia has also tried to meet the needs of socially diverse residents by regulating the form of urban development. Unlike most other city-regions, it has tried to define "urban growth boundaries" to promote compact development and "urban villages" with a mix of living, working, and leisure activities. Portland, for example, has set a growth boundary that is the most concrete commitment in North America to reversing trends toward racial and class segregation and the flight from inner cities. But Portland would never have been able to undertake this process if it had not been for action by its state.

LEADERSHIP IN THE STATES

In the American political system, cities have little autonomy. The authority to enact policies and programs that might effectively shape the development of cities lies with their state governments. Two states, Oregon and New Jersey, stand out as leaders.

Since 1973, Oregon has required each city to draw a growth boundary based on its assessment of economic development and community needs in the next 20 years. In turn, the city develops a comprehensive plan, including the steps it will take to create needed infrastructure for water and sewers, roads and transit, and other public facilities within the growth boundary. The growth boundary also influences state expenditures for highways and other roads. By 1986, to meet the state standards, all communities in Oregon had drawn up growth plans to limit their expansion.

Ethan Seltzer, who runs the Institute of Portland Metropolitan Studies at Portland State University, explains that the state expects land inside urban growth boundaries to be developed at urban densities and, in fact, allows developers to go to court for immediate approval if local jurisdictions fail to process permit applications for approved purposes within 120 days. "This means that multifamily development occurs by right and according to plan even in the suburbs!" Seltzer says. But outside the boundaries, he continues, "you cannot develop at urban densities, cannot get urban services, and face strict restrictions on what can be built in farm and forest zones. Even road widening for nonfarm uses is closely regulated outside of urban growth boundaries."

Seltzer notes,

> Creativity comes into play because, especially in recent years, the state is committed to accommodating growth through infill and redevelopment, and not just on vacant land at the edge. Today, the market is responding. In the last six months, 30 percent of our residential growth has been infill development in the region, 15 percent has been in attached housing/townhouses.... There is active development of housing in downtown Portland, and we will probably see a new public elementary school in downtown in the next few years.
>
> The Oregon program directs cities and investors to steward land committed to urban use much the way a farmer stewards his or her fields. Rather than [allowing] disinvestment, we pursue reinvestment. It comes at a cost. Currently we are struggling with our popularity, and what it means to live not in a cheap region but a desirable, valuable one.
>
> I guess what we've proven is that pursuing an end to sprawl is possible and desirable, but it won't by itself solve the problems of poverty or provide needed affordable housing.

He adds that while urban growth boundaries are not a "silver bullet," they "are great at what they do: stopping sprawl on farmland, directing attention back onto lands already committed to urban use, and in the metropolitan region here, suggesting to local elected officials that their future is a shared one best approached through a partnership with their brother and sister jurisdictions living within the same economy."

The growth and form of cities are critical issues for New Jersey, the only state to be entirely occupied by "metropolitan areas," according to the U.S. Census. In 1992, New Jersey produced its first state plan to "coordinate public and private actions to guide future growth into compact forms of development and redevelopment." Its

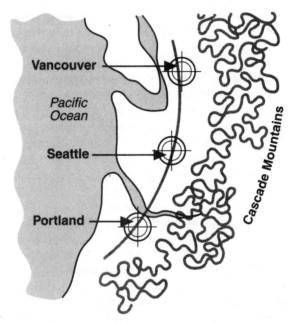

ANNIE BISSET

policies are like Oregon's: "encourage development, redevelopment, and economic growth in locations that are well situated with respect to present or anticipated public services and facilities, and to discourage development where it may impair or destroy natural resources or environmental qualities."

In New Jersey's search for a new model of urban growth and form, the keyword is *compact*. Comparing the traditional trends with the new policies proposed by the state plan, James Hughes and his colleagues at Rutgers University found that compact development would generate more jobs in accessible centers throughout the region, thereby reducing the jobless rates in inner cities. There would also be less destruction of the natural environment because forests, watersheds, and farmlands would be preserved. Local and state governments would save money because there would be less need for new infrastructure. For example, to accommodate growth until the year 2010, the traditional pattern would need 5,500 lane-miles of new local roads. For the same population and economy, the state plan would require only 1,600 new lane-miles. But the greatest benefit would be in the revitalization of neighborhoods.

Here Comes the Neighborhood

For revitalizing our cities, the "neighborhood" is almost always cited as the basic building block. Today in America there are two different concepts. The first is the idea of a "neighborhood" with a core and boundary. Spatially and socially, this "neighborhood" focuses on its core: local shops, a neighborhood school, perhaps a library and other community facilities for education, health care, and recreation. The neighborhood's population size and density, its network of roads and paths, even its image and character are linked to the neighborhood's core. At its boundary, the neighborhood's edges are marked by landscapes—generally, roads or parkways, or in cities, arterial streets. Neighborhoods, in this concept, are given names and generate loyalty; they are also inward-looking and intentionally static.

The city-building implications of this neighborhood concept are clear: Clusters of neighborhoods can create a district, and clusters of districts create the city. This "cluster" concept of the neighborhood, district, and city is the American vernacular. It is embodied in the postwar comprehensive plans for restructuring such old cities as Philadelphia and for the construction of such new cities as Columbia, Maryland. It is manifest in the power of "community boards" in large cities. And it is given lip-service by developers and their advertising agencies for suburban tracts.

The second concept, a "street-neighborhood," is radically different. It does not have the spatial and social clarity of the "core-and-boundary neighborhood." Instead, it idealizes the natural cohesion that comes from "neighboring" on the street and sidewalk. This sense of neighborhood is the consequence of face-to-face, casual, informal contacts in everyday city life. For the spatial setting of this concept of neighborhood, the gridiron street plan of such cities as Manhattan is especially useful. Paradoxically, the static, predictable, public structural form can support and stimulate the dynamic, small-scale, ad hoc, spontaneous life of everyone—residents and visitors, workers and walkers, insiders and outsiders.

The key to this concept of neighborhood is the street and sidewalk. The street is the armature, the skeleton, the structure of the street-neighborhood. To the streets are attached the social institutions that characterize a neighborhood: the schools, food stores, coffee shops, library and bookstores, movie theaters, local service stores, health clubs, parks and playgrounds, and of course, the workplaces and homes of the neighbors. The street-neighborhood is immensely popular. Throughout the United States, for example, old loft districts are being used for new living-working places; shopping malls are trying to simulate the life of a downtown street and sidewalk; and cities are recognizing that the key to the neighborhood is the street and its quality of life.

City Prospects

How can these concepts of neighborhoods serve an emerging new society profoundly affected by changes in communication and information technologies? They offer both positive and negative possibilities.

The core-and-boundary neighborhood can create a human-scale community and sense of place within a large city-region. Because it is a development unit that itself has edges, it can help establish an urban growth boundary. But the core-and-boundary neighborhood can turn pathological if the territorial boundary becomes hard-edged and gated, excluding outsiders from a segregated community

The street-neighborhood has the advantage that it does not intentionally create physical boundaries that exclude people. At its best, it is open, welcoming, and place-making. Diverse street-sidewalk places would be welcome insertions into conventional core-and-boundary neighborhoods, or even more, into the fabric of suburban sprawl. But the street-neighborhood also has pathological possibilities: The streets can be the territorial setting for intimidation and crime and, at their worst, these threats can destroy our cities.

Increasingly, "Main Street" is once again valued as a lively center of a surrounding neighborhood. In Toronto, for example, the ethnic diversity of the city-region is expressed by its many neighborhoods—Greektown, Chinatown, Portuguese Village—each with its own "Main Street." What had been St. Claire Avenue is now Corso Italia. Similarly, in northern Manhattan, Harlem's neighborhoods are anchored by their crosstown streets. The most famous is 125th Street, but others such as 116th and 135th Streets are each a string of lively places, central arteries for economic and cultural activity.

If, as Peter Drucker predicts, our future organization of work will be more akin to that in pre-industrial cities, with an intimate mixture rather than separation of living and working places, then the neighborhood street will once again be the vibrant setting for everyday life. More than ever, we will value places to meet, to see and be seen, to drink coffee together, and maybe, to bowl together.

But this will not happen automatically; the form of a city is a consequence of public policies. Four kinds of policies are needed: regional compacts to build and maintain infrastructure for transportation, water, and waste systems; community growth boundaries to contain the urban built-up land uses; regional compacts to preserve greenspaces and natural ecological systems; and public initiatives to support the centers of cities and neighborhoods.

Streets and sidewalks, buildings and plazas, gardens and parks profoundly affect our everyday lives and ought to be the subject of public debate. "By its form, as by the manner of its birth," wrote the French anthropologist Claude Levi-Strauss, "the city has elements at once of biological procreation, organic evolution and aesthetic creation. It is both a natural object and a thing to be cultivated; something lived and something dreamed. It is the human invention par excellence." We need the courage to create our cities again.

Robert Geddes, an architect and urban designer, is Dean Emeritus of the Princeton University School of Architecture and director of the Conference on Cities in North America, and editor of its book, *Cities in Our Future* (Island Press).

GREENVILLE FROM BACK COUNTRY to FOREFRONT

Eugene A. Kennedy

What factors are crucial in determining the success or failure of an area? This article explores the past and present and glimpses what may be the future of one area which is experiencing great success. The success story of Greenville County, S.C. is no longer a secret. This article seeks to find the factors which led to its success and whether they will provide a type of yardstick to measure the future.

The physical geography of this area is explored, as well as the economic factors, history, transportation, energy costs, labor costs and new incentive packages designed to lure new industries and company headquarters to the area.

Physical geography: advantageous

Greenville County is situated in the northwest corner of South Carolina on the upper edge of the Piedmont region. The land consists of a rolling landscape butted against the foothills of the Appalachian Mountains. Monadnocks, extremely hard rock structures which have resisted millions of years of erosion, rise above the surface in many places indicating that the surface level was once much higher than today. Rivers run across the Piedmont carving valleys between the plateaus. The cities, farms, highways and rail lines are located on the broad, flat tops of the rolling hills.

Climatologically, the area is in a transition zone between the humid coastal plains and the cooler temperatures of the mountains, resulting in a relatively mild climate with a long agricultural growing season. The average annual precipitation for Greenville County is 50.53 inches at an altitude of 1040 feet above sea level. The soil is classified as being a Utisoil. This type of soil has a high clay base and is usually found to be a reddish color due to the thousands of years of erosion which has leached many of the minerals out of the soil, leaving a reddish residue of iron oxide. This soil will produce good crops if lime and fertilizer containing the eroded minerals are added. Without fertilizers, these soils could sustain crops on freshly cleared areas for only two to three years before the nutrients were exhausted and new fields were needed. This kept large plantations from being created in the Greenville area. The climate and land are such that nearly anything could be cultivated with the proper soil modification. Physical potential, although a limiting factor, is not the only determining factor in the success of an area.

An open courtyard off Main Street, downtown Greenville, S.C.

18. Greenville

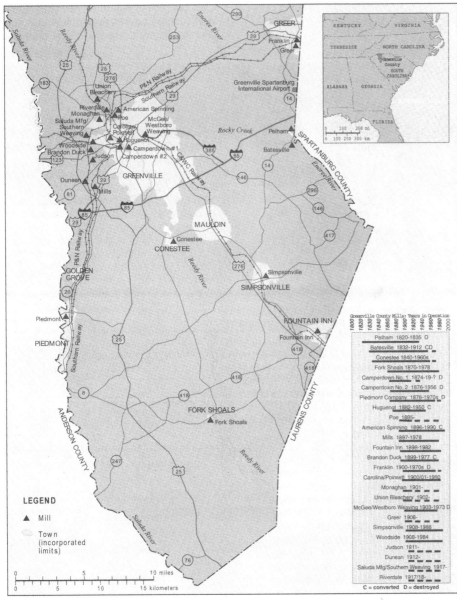

KAREN SEVERUD COOK, UNIVERSITY OF KANSAS MAP ASSOCIATES

As Preston James, one of the fathers of modern geography pointed out, the culture of the population which comes to inhabit the area greatly determines the response to that particular physical environment.

From European settlement through the textile era

An Englishman named Richard Pearis was the first to begin to recognize the potential of what was then known as the "Back Country." The area was off limits to white settlers through a treaty between the British and the Indians. In order to get around the law, Pearis married a native American and opened a trading post in 1768 at the falls of the Reedy River. He soon built a grist mill and used the waterfalls for power. Pearis prospered until the end of the Revolutionary War. He had remained loyal to the British, lost all his property when the new nation was established and left the country.

Others soon realized the potential of what was to become Greenville County. Isaac Green built a grist mill and became the area's most prominent citizen. In 1786, the area became a county in South Carolina and was named for Isaac Green. The Saluda, Reedy and Enoree Rivers along with several smaller streams had many waterfalls, making them excellent locations for mills during the era of water power. During the Antebellum period, Greenville County's 789 square miles was inhabited by immigrants with small farms and also served as a resort area for Low Country planters who sought to escape the intense summer heat and the disease carrying mosquitoes which flourished in the flooded rice fields and swampy low country of the coastal plains. Most of the permanent residents farmed and a few mills were built to process the grains grown in the area.

Although very little cotton was ever actually grown in Greenville County, cotton became the driving force behind its early industrialization. Beginning with William Bates in 1820, entrepreneurs saw this area's plentiful rivers and waterfalls as a potential source of energy to harness. William Bates built the first textile mill in the county sometime between 1830 and 1832 on Rocky Creek near the Enoree River. This was known as the Batesville Mill. Water power dictated the location of the early southern textile mills, patterned after the mills built in New England. Mill owners purchased the cotton from farms and hauled it to their mills, but lack of easy transportation severely limited their efforts until 1852. That year, the Columbia and Greenville Railroad finally reached Greenville County. Only the interruption of the Civil War kept the local textile industry from becoming a national force during the 1850s and 1860s.

The area missed most of the fighting of the Civil War and escaped relatively unscathed. This provided the area with an advantage over those whose mills and facilities had been destroyed during the war. The 1870 census reported a total $351,875 in textiles produced in the county. This success encouraged others to locate in Greenville. Ten years later, with numerous mills being added each year, the total reached $1,413,556.11. William Bates' son-in-law, Colonel H. P. Hammett, was owner of the Piedmont Company which was the county's largest producer. Shortly after the construction of the water powered Huguenot Mill in the downtown area of the City of Greenville County in 1882, the manufacture of cotton yarn would no longer be controlled by the geography of water power.

The development of the steam engine created a revolution in the textile industry. No longer was the location of the mill tied to a fast moving stream, to turn a wheel that moved machinery. Large amounts of water were still needed but the dependence upon the waterfalls was severed. Between 1890 and 1920, four textile plants were built in the county outside the current city limits of the City of Greenville. At least thirteen large mills were built near the city to take advantage of the rail system, as shown on the map, "Greenville County Mills." Thus, with cheaper and more efficient steam

109

3 ❖ THE REGION

F.W. Poe Manufacturing Co., Old Buncombe Road, Greenville, S.C. Built in 1895, purchased by Burlington Industries and closed in 1997. Palmetto State Dyeing and Finishing Co. opened in 1987. The company employs approximately 110 people.

power, transportation costs became a deciding factor. These mills built large boiler rooms adjacent to their plants and dug holding ponds for water.

Another drastic change took place in the textile industry around 1900. This change would provide even greater flexibility for the mill owners. A hydroelectric dam was constructed on the Saluda River, five miles west of the city of Greenville. It was completed in 1902 and would provide cheap electricity for the county. John Woodside, a local mill owner who foresaw electricity as the next step in the evolution of the industry, built what was then the largest textile mill in the world in the city of Greenville that same year. He located it further from a water source than previously thought acceptable. However, John had done some primitive locational analysis and chose the new site well. It was located just beyond the city boundary to limit his tax liability and directly between the lines of two competing rail companies—the Piedmont Railroad and the Norfolk and Western (now known as Norfolk and Southern). John Woodside's mill proved to be a tremendous success. With water no longer a key factor of location, the owners identified transportation as the key factor of location. Others began to build near rail lines.

The textile industry made Greenville County very prosperous. The mills needed workers and shortly outstripped the area's available labor supply. Also, many did not want to work in the hot, poorly ventilated, dangerous conditions found in the mills. When most of the mills were still built of wood, the cotton fibers floating in the air made fire a very real danger. Many businesses sprang up to service the needs of the workers and the textile mill owners. Farmers, sharecroppers, former slaves and children of former slaves were recruited to work in the mills. Housing soon became scarce and the infrastructure wasn't equipped to handle the influx of new workers. To alleviate the problem, the mill owners built housing for their workers. These were very similar to the coal camps of Appalachia and other factory owned housing in the north. They were very simple dwellings built close to the mill so the workers could easily walk to and from work. They also provided company-owned stores, doctors and organized recreational activities for their employees, creating mill communities. Many people who worked for the mills would have told you they lived at Poe Mill or Woodside, the names of their mill communities, rather than Greenville.

In the 1960s, rail transportation of textiles was a cost the owners wished to lower. They found a cheaper, more versatile form of transportation in the trucking industry. The interstate highway system was now well developed and provided a means of keeping costs down for the operators. In the 1970s, owners began to identify wages and benefits as a major factor in their cost of operation and many firms relocated in foreign countries, which offered workers at a fraction of the wages paid in the United States and requiring few if any benefits.

Meeting the challenge of economic diversification

Greenville County used its natural physical advantage to become the "Textile Capital of the World." Many of the other businesses were tied directly or indirectly to the textile industry. These ranged from engineering companies who designed and built textile machinery to companies which cleaned or repaired textile machines. Employment in the textile industry in Greenville County peaked in 1954 with 18,964 workers directly employed in the mills. As the industry began to decline, the leaders of the industry along with local and state leaders showed great foresight by combining their efforts into an aggressive move to transform Greenville County into a production and headquarters oriented economy. A state sponsored system of technical schools greatly facilitated this effort. Workers could get the training they needed to pursue almost any vocation at these centers. This system still is a factor in Greenville County's success.

Table 1
CORPORATE HEADQUARTERS IN GREENVILLE COUNTY

1. American Leprosy Mission International
2. American Federal Bank
3. Baby Superstores, Inc.
4. Bowater Inc.
5. Builder Marts of America Inc.
6. Carolina First Bank
7. Delta Woodside Industries Inc.
8. Ellcon National Inc.
9. First Savings Bank
10. Heckler Manufacturing and Investment Group
11. Henderson Advertising
12. Herbert-Yeargin, Inc.
13. JPS Textile Group Inc.
14. Kemet Electronics Corp.
15. Leslie Advertising
16. Liberty Corp.
17. Mount Vernon Mills Inc.
18. Multimedia Inc.
19. Ryan's Family Steakhouses
20. Span America
21. Steel Heddle Manufacturing Co.
22. Stone Manufacturing Co.
23. Stone International.
24. TNS Mills Inc.
25. Woven Electronics Corp.

18. Greenville

> What ultimately swayed the automaker to choose Greenville? One of the main reasons was physical location.

The group emphasized the ability to make a profit in Greenville County. The focus of their efforts was turned to creating a sound technical education network along with the flexibility to negotiate packages of incentives to lure large employers. Incentives included negotiable tax and utility rates, plus a strong record of worker reliability due to South Carolina's nonunion tradition, with very few work stoppages. The foresight of this group has paid off handsomely. The majority of the textile mills which provided the backbone of the economy of Greenville County are no longer in business. Many of the old buildings still stand. Ten of the mills built before 1920 now are used in other capacities. American Spinning was built in 1896 and now is used as a warehouse, office space and light manufacturing all under one roof. Most of the mills are used for warehouse space or light manufacturing such as the Brandon Duck Mills, which operated between 1899 and 1977 as a cotton mill. It now houses two small factories which assemble golf clubs and part of the mill is used as a distribution center. The low lease cost (from $1 to $15 per square foot) is an enticement for other businesses to locate in these old buildings.

The old Huguenot Mill, the last water powered mill built in the county, was recently gutted and has been rebuilt as offices for the new 35 million dollar Peace Center entertainment complex in downtown Greenville. The Batesville Mill, the first in the county, was built of wood. It burned and was rebuilt in brick in 1881. It closed its doors in 1912 because the water-powered mill was not competitive. The mill was purchased by a husband and wife in 1983, converted into a restaurant, and was the cornerstone and headquarters of a chain of FATZ Restaurants until it burned again in 1997. So, in considering diversification, one of the first steps was to look for other uses for the facilities which already existed.

Other efforts also met with great success. As businesses began to look south during the 1970s for relocation sites, Greenville began to use its natural advantages to gather some impressive companies into its list of residents. By 1992, the combination of these efforts made Greenville County the wealthiest county in the state of South Carolina. Twenty-five companies have their corporate headquarters in the county, as shown in Table 1.

Forty-nine others have divisional headquarters in the county, as shown in Table 2. This constitutes a sizable investment for the area, yet even this list does not include a 150 million dollar investment by G. E. Gas Turbines in 1992 for expansion of their facility. This was the largest recent investment until 1993.

Along with American companies, foreign investment was sought as well. Companies such as Lucas, Bosch, Michelin, Mita and Hitachi have made major investments in the county. Great effort has been put into reshaping the face of Main Street in Greenville as well. The city is trying to make a place where people want to live and shop. Many specialty stores have opened replacing empty buildings left by such long time mainstays as Woolworths. The Plaza Bergamo was created to encourage people to spend time downtown. The Peace Center Complex provides an array of entertainment choices not usually found in a city the size of Greenville. The Memorial Auditorium, which provided everything from basketball games, to rodeo, concerts, high school graduations and truck pulls has closed its doors and was demolished in 1997 to make way for a new 15,000 seat complex which will be named for its corporate sponsor. It will be called the Bi-Lo Center. Bi-Lo is a grocery store chain and a division of the Dutch Company, Ahold. A new parking garage is being built for this center and two other garages have recently been added to improve the infrastructure of the city. City leaders have traveled to cities such as Portland, Oregon to study how they have handled and managed growth and yet kept the city friendly to its inhabitants.

The largest gamble for Greenville County came in early 1989. The automaker BMW announced that it was considering building a factory in the United States. Greenville County and the state of South Carolina competed against several other sites in the midwest and southeast for nearly two years. On June 23, 1992, the German automaker chose to locate in the Greenville-Spartanburg area. Although the plant is located in Spartanburg County,

Panorama of downtown Greenville, S.C.

3 ❖ THE REGION

Table 2
DIVISIONAL HEADQUARTERS IN GREENVILLE COUNTY

1. Ahold, Bi-Lo.
2. BB&T, BB&T of South Carolina.
3. Bell Atlantic Mobile
4. Canal Insurance Company
5. Coats and Clark, Consumer Sewing Products Division.
6. Cryovac—Div. of W. R. Grace & Co.
7. Dana Corp., Mobile Fluid Products Division.
8. DataStream Systems, Inc.
9. Dodge Reliance Electric
10. Dunlop Slazenger International, Dunlop Slazenger Corp.
11. EuroKera North America, Inc.
12. Fluor Daniel Inc.
13. Fulfillment of America
14. Frank Nott Co.
15. Gates/Arrow Inc.
16. General Nutrition Inc., General Nutrition Products Corp.
17. Gerber Products Co., Gerber Childrenswear, Inc.
18. GMAC
19. Goddard Technology Corp.
20. Greenville Glass
21. Hitachi, Hitachi Electronic Devices (USA)
22. Holzstoff Holding, Fiberweb North America Inc.
23. IBANK Systems, Inc.
24. Insignia Financial Group, Inc.
25. Jacobs-Sirrine Engineering
26. Kaepa, Inc.
27. Kvaerner, John Brown Engineering Corp.
28. Lawrence and Allen
29. LCI Communications
30. Lockheed Martin Aircraft Logistics Center Inc.
31. Manhattan Bagel Co.
32. Mariplast North America, Inc.
33. Michelin Group, Michelin North America.
34. Mita South Carolina, Inc.
35. Moovies, Inc.
36. Munaco Packing and Rubber Company.
37. National Electrical Carbon Corp.
38. O'Neal Engineering
39. Personal Communication Services Dev.
40. Phillips and Goot
41. Pierburg
42. Rust Environment and Infrastructure
43. SC Teleco Federal Credit Union
44. Sodotel
45. South Trust Bank
46. Sterling Diagnostic Imaging, Inc.
47. Umbro, Inc.
48. United Parcel Service
49. Walter Alfmeier GmbH & Co.

the headquarters are in Greenville and both counties will profit greatly. When the announcement was made, the question was: what ultimately swayed the automaker to choose Greenville? One of the main reasons was physical location. The site is only a four hour drive from the deepwater harbor of Charleston, SC. Interstates 26 and 85 are close by for easy transportation of parts to the assembly plant. The Greenville-Spartanburg Airport is being upgraded so that BMW can send fully loaded Boeing 747 cargo planes and have them land within five miles of the factory. Plus, the airport is already designated as a U.S. Customs Port of Entry and the flights from Germany can fly directly to Greenville without having to stop at Customs when entering the country. Other incentives in the form of tax breaks, negotiated utility rates, worker training and state purchased land helped BMW choose the 900 acre site where it will build automobiles.

Huguenot Mill (lower left). Built in 1882, on Broad Street, Greenville, S.C., is the last waterpowered mill in Greenville County. It is being refurbished to become part of the Peace Center Complex at right.

The large incentive packages might appear self-defeating but BMW's initial investment was scheduled to be between 350 and 400 million dollars. The majority of the companies supplying parts for BMW also looked for sites close enough to satisfy BMW's just-in-time manufacturing needs. The fact that Michelin already made tires for their cars here, Bosch could supply brake and electrical parts from factories already here and J.P. Stevens and others could supply fabrics for automobile carpets and other needs readily from a few miles away also was a factor.

One major BMW supplier, Magna International, which makes body parts for the BMW Roadster and parts for other car manufacturers, located its stamping plant in Greenville County. Magna invested $50 million and will invest $35 million more as BMW expands. Magna needed 100 acres of flat land without any wetlands and large rock formations. This land needed to be close enough to provide delivery to BMW. After studying several sites, Magna chose South Donaldson Industrial Park, formerly an Air Force base, just south of the city of Greenville. The county and state will help prepare the location for their newest employer.

Road improvements, the addition of a rail spur and an updating of water and sewer facilities will all be provided to Magna in this agreement. Also, Magna will receive a reduced 20 year fixed tax rate along with other incentives for each worker hired. These incentive packages may seem unreasonable but they have proven to be necessary in the 1990s when large organizations are deciding where to locate.

The future: location, location and location

From the time of the earliest European settlers, the natural advantages of Greenville County helped bring it to prosperity. The cultural background of the settlers was one of industry and a propensity for changing the physical environment to maximize its industrial potential. Nature provided the swift running rivers and beautiful waterfalls. The cultural background of the settlers caused them to look at these natural resources and see economic potential.

The people worked together to create an environment which led Greenville County to be given the title of "Textile Center of the World" in the 1920s. Then, again taking advantage of transportation opportunities and economic advantages, the area retained its textile center longer than the majority of textile centers.

Today, after 30 years of diversification, economic factors now are normally the deciding factor in the location of a new business or industry. Greenville County with its availability of land, reasonable housing costs, low taxes, willingness to negotiate incentive packages, and positive history of labor relations helped make it a desirable location for business. Proximity to interstate transportation, rail and air transport availability help keep costs low. The county's physical location about half way between the mega-growth centers of Charlotte, North Carolina and Atlanta, Georgia places it in what many experts call the mega-growth center of the next two decades. Now, with BMW as a cornerstone industry for the 1990s and beyond, Greenville County looks to be one of the areas with tremendous growth potential.

Thus, a combination of physical, environmental and cultural factors greatly influence the location of businesses. Transportation costs, wage and benefit packages and technical education availability are all interconnected.

The newest variable involves incentive packages of tax, utility reduction, worker training, site leasing and state and local investment into improving the infrastructure for attracting employers. The equation grows more and more complex with no one factor outweighing another; however, economic costs of plant or office facilities, wages and benefits and transportation seem to be paramount. Greenville County is blessed with everything it needs for success. It will definitely be one the places "to be" in the coming years.

> A combination of physical, environmental and cultural factors greatly influence the location of businesses.

Eugene A. Kennedy is a native of West Virginia who attended Bluefield State College, Bluefield West Virginia, and received an M.A. in geography from Marshall University in Huntington. He is currently a public educator in the Greenville County School system, Greenville S.C. He was awarded a "Golden Apple" by Greenville television station WYFF in 1997, has been a presenter at the South Carolina Science Conference, and a consultant to the South Carolina State Department of Education. He can be reached at GEOGEAK@aol.com

References and further readings

DuPlessis, Jim. 1991. Many Mills Standing 60 Years After Textile Heydays. *Greenville News-Piedmont.* July 8. pp. 1c–2c.

Greater Greenville Chamber of Commerce, 1990. *1990 Guide to Greenville.*

Greenville News-Piedmont. 1991–1993. *Fact Book 1991; 1992; 1993.*

Patterson, J. H. 1989. *North America.* Oxford University Press: N.Y. Eighth edition.

Scott, Robert. 1993. Upstate Business. *Greenville-News Piedmont.* 15 August. pp. 2–3.

Shaw, Martha Angelette. 1964. *The Textile Industry in Greenville County.* University of Tennessee Master's Thesis.

Strahler, Arthur. 1989. *Elements of Physical Geography.* John Wiley and Sons: N.Y. Fourth edition.

Does it matter where you are?

The cliché of the information age is that instantaneous global telecommunications, television and computer networks will soon overthrow the ancient tyrannies of time and space. Companies will need no headquarters, workers will toil as effectively from home, car or beach as they could in the offices that need no longer exist, and events half a world away will be seen, heard and felt with the same immediacy as events across the street—if indeed streets still have any point.

There is something in this. Software for American companies is already written by Indians in Bangalore and transmitted to Silicon Valley by satellite. Foreign-exchange markets have long been running 24 hours a day. At least one California company literally has no headquarters: its officers live where they like, its salesmen are always on the road, and everybody keeps in touch via modems and e-mail.

Yet such developments have made hardly a dent in the way people think and feel about things. Look, for example, at newspapers or news broadcasts anywhere on earth, and you find them overwhelmingly dominated by stories about what is going on in the vicinity of their place of publication. Much has been made of the impact on western public opinion of televised scenes of suffering in such places as Ethiopia, Bosnia and Somalia. Impact, maybe, but a featherweight's worth.

World television graphically displayed first the slaughter of hundreds of thousands of people in Rwanda and then the flight of more than a million Rwandans to Zaire. Not until France belatedly, and for mixed motives, sent in a couple of thousand soldiers did anyone in the West lift so much as a finger to stop the killing; nor, once the refugees had suddenly poured out, did western governments do more than sluggishly bestir themselves to try to contain a catastrophe.

Rwanda, of course, is small (population maybe 8m before the killings began). More important, it is far away. Had it been Flemings killing Walloons in Belgium (population 10m) instead of Hutus slaying Tutsi in Rwanda, European news companies would have vastly increased their coverage, and European governments would have intervened in force. Likewise, the only reason the Clinton administration is even thinking about invading Haiti is that it lies a few hundred miles from American shores. What your neighbours (or your kith and kin) do affects you. The rest is voyeurism.

The conceit that advanced technology can erase the contingencies of place and time ranges widely. Many armchair strategists predicted during the Gulf war that ballistic missiles and smart weapons would make the task of capturing and holding territory irrelevant. They were as wrong as the earlier seers who predicted America could win the Vietnam war from the air.

In business, too, the efforts to break free of space and time have had qualified success at best. American multinationals going global have discovered that—for all their world products, world advertising, and world communications and control—an office in, say, New York cannot except in the most general sense manage the company's Asian operations. Global strengths must be matched by a local feel—and a jet-lagged visit of a few days every so often does not provide one.

Most telling of all, even the newest industries are obeying an old rule of geographical concentration. From the start of the industrial age, the companies in a fast-growing new field have tended to cluster in a small region. Thus, in examples given by Paul Krugman, an American economist, all but one of the top 20 American carpet-makers are located in or near the town of Dalton, Georgia; and, before 1930, the American tire industry consisted almost entirely of the 100 or so firms carrying on that business in Akron, Ohio. Modern technology has not changed the pattern. This is why the world got Silicon Valley in California in the 1960s. It is also why tradable services stay surprisingly concentrated—futures trading (in Chicago), insurance (Hartford, Connecticut), movies (Los Angeles) and currency trading (London).

History's Heavy Hand

This offends not just techno-enthusiasts but also neo-classical economics: for both, the world should tend towards a smooth dispersion of people, skills and economic competence, not towards their concentration. Save for transport costs, it should not matter where a tradable good or service is produced.

The reality is otherwise. Some economists have explained this by pointing to increasing returns to scale (in labour as well as capital markets), geographically uneven patterns of demand and transport costs. The main reason is that history counts: where you are depends very much on where you started from.

The new technologies will overturn some of this, but not much. The most advanced use so far of the Internet, the greatest of the world's computer networks, has not been to found a global village but to strengthen the local business and social ties among people and companies in the heart of Silicon Valley. As computer and communications power grows and its cost falls, people will create different sorts of space and communities from those that exist in nature. But these modern creations will supplement, not displace, the original creation; and they may even reinforce it. Companies that have gone furthest towards linking their global operations electronically report an increase, not a decline in the face-to-face contact needed to keep the firms running well: with old methods of command in ruins, the social glue of personal relations matters more than ever.

The reason lies in the same fact of life that makes it impossible really to understand from statistics alone how exciting, say, China's economic growth is unless you have physically been there to feel it. People are not thinking machines (they absorb at least as much information from sight, smell and emotion as they do from abstract symbols), and the world is not immaterial: "virtual" reality is no reality at all; cyberspace is a pretence at circumventing true space, not a genuine replacement for it. The weight on mankind of time and space, of physical surroundings and history—in short, of geography—is bigger than any earthbound technology is ever likely to lift.

Low Water in the American High Plains

David E. Kromm

David E. Kromm, professor of geography at Kansas State University, has authored and coauthored numerous articles and one book on water management issues.

Depletion of the Ogallala Aquifer once threatened to return the region to a Dust Bowl, but much is being done to conserve groundwater today.

Water defines many regions. It determines their character and sustains their well-being. Ironically, this is especially true of areas with an inherent water scarcity. What would Southern California and central Arizona be like without water imported from distant sources? The green fields and golf courses would give way to dry plains and hills, populated more by grazing animals than by people.

A huge dry area in the middle of the United States depends no less on water to transform a dusty outback into a productive garden. This is the High Plains overlying the Ogallala Aquifer.

There exists widespread concern that America's largest underground water reserve is drying up. The vast Ogallala Aquifer that underlies 174,000 square miles of the High Plains from west Texas northward into South Dakota has been partially depleted as more than 150,000 wells pump water for irrigation, municipal supply, and industry. In some areas the wells no longer yield enough water to make irrigation possible. In others there remains sufficient water, but it lies 300 or more feet below the surface. The cost of lifting water from such depths makes it uneconomical for many uses.

Touching base with some key words in the groundwater vocabulary helps tell the Ogallala story more precisely. One is *aquifer*. An aquifer is a zone of water-saturated sands and gravels beneath the earth's surface. It is not an underground lake or river. It is the porous rock structure that contains the water that we tap with wells. Some aquifers release water to wells readily, whereas others hold the water tightly. This affects the rate at which water can be withdrawn and is called the *specific yield* of a well. *Depth to water* describes the vertical distance from ground level to the aquifer. Together with the volume and quality of water, the depth and specific yield define the economic limit of water withdrawal. The fresh water in an aquifer is called *groundwater*, in contrast to surface waters such as rivers and lakes. Over half the nation's population depends on groundwater for its drinking water. Although there are several aquifers, the Ogallala ranks as the main formation in the High Plains aquifer system. Most people in the region refer to the entire system as the Ogallala.

The unconsolidated sand and gravel that form the Ogallala aquifer were laid down by fluvial deposition from the Rocky Mountains about 10 million years ago. More recent, near-surface deposits of the late Tertiary and Quaternary ages compose the High Plains aquifer.

3 ❖ THE REGION

The volume of available water varies from place to place. Nebraska has about two-thirds (65 percent) of the High Plains groundwater, followed by Texas (12 percent), and Kansas (10 percent). The total water in storage is enough to fill Lake Huron or to cover the state of Colorado to a depth of 45 feet. Physical characteristics of the aquifer result in one farmer in a county having little or no access to water while another has substantial reserves. Ogallala water moves gradually from west to east, following the general slope of the land surface. This movement does not affect use or depletion, as it takes nearly a century and a half to flow a mere 10 miles.

Changing views

When Stephen Long explored the region in the early nineteenth century, he discovered a sea of grass that contrasted with the trees that dominated landscapes to the east. Few lakes or rivers can be found. These signs of aridity led Long to name the area "The Great American Desert." Settlers on their way to rainy Oregon viewed the plains as worthless territory that had to be crossed in order to reach a land more promising. Even those few who attempted to break the prairie sod found it tough to plow. The native Americans and buffalo herds were initially displaced by cattle ranching, not by cultivation.

A whole new perception of the plains burst forth in the middle of the nineteenth century as railroads penetrated the region and touted its virtues. The promoters encouraged the development of farming communities that would depend on the rails for goods and the delivery of grain to eastern markets. New steel plows cracked the soil and many boosters came to believe that "the rain follows the plow." Once the ground was broken and crops planted, the aridity would end. Hopeful farmers took up homesteads from the government or bought lands directly from the railroad companies.

There ensued frequent dry years and just enough wet ones to allow many of the farmers to hang on. The rosy view of the High Plains as a garden was soon tarnished but never wholly abandoned. This changed with the 1930s. Everything went wrong. Depression weakened markets and reduced capital availability and drought destroyed the crops. A new name was coined for the High Plains—the Dust Bowl. Outmigration swelled and the popular image was that of skies darkened by blowing dirt, desolate farmsteads surrounded by drifting soil, and impoverished families attempting to leave with whatever they could

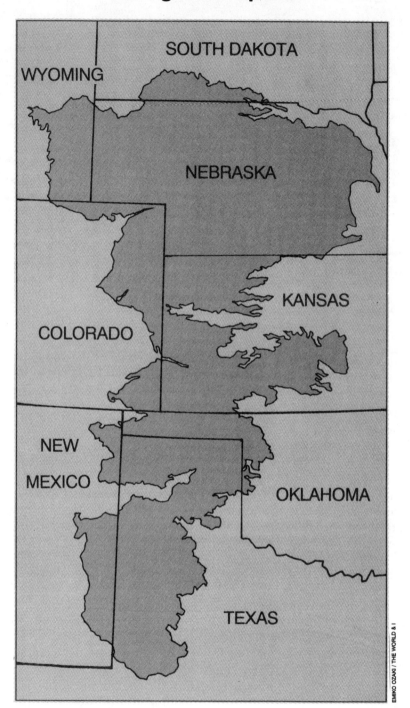

The Ogallala Aquifer

The promotional railroad poster, 1887: the advantages of settling the High Plains.

carry in their dilapidated vehicles. John Steinbeck chronicled the depth of misery in *The Grapes of Wrath*.

Emergence of a new Corn Belt

Even as the last settlers were taking up land in the 1890s, irrigation was beginning. It occupied small patches dependent on surface sources such as the Arkansas River or shallow wells. The Dust Bowl brought renewed interest in soil and water conservation and encouraged irrigation to create a stable moisture supply. Farmers suffered greatly from climatic variability in the High Plains. In a given year it might be a land of bountiful rain or a sun-drenched and parched desert.

Irrigation expanded in the 1930s and even more so in the dry years of the 1950s, as new technologies made possible large-scale tapping of the Ogallala aquifer. Pump engines came on the market that could lift large quantities of water from sources deep underground. New watering methods, such as center-pivot sprinklers, allowed irrigation of sandy and hilly areas. The green circles seen from aircraft flying more than 30,000 feet above the surface became the most prominent feature in the High Plains. The perception of the 1960s and 1970s was of a new Corn Belt flourishing in the land of the underground rain. Thanks to the Ogallala, stability had at last come to the High Plains.

Or had it? For the most part the Ogallala is a fossil water aquifer. It was formed millions of years ago and has been minimally recharged since. Only in a few areas such as the Nebraska Sandhills and the sand-sage prairie of southwestern Kansas was new water being added. Elsewhere, whatever was pumped out was forever gone. Wells sunk into zones with limited saturation thickness dried up early. Many were in west Texas, where large-scale irrigation began. By the late 1970s the whole High Plains region seemed threatened again, no less than it had been in the dirty thirties. Between 1978 and 1987, the irrigated area harvested in the High Plains sharply fell from nearly 13 million acres to 10.4 million acres.

Depletion ranks as such a serious problem that quality and streamflow issues are easily overlooked. In its natural state, the High Plains aquifer is generally of high quality. It is suitable for a wide array of uses without filtration. Where soils are sandy, however, precipitation can reach the water table, bringing with it nitrates from fertilizers and other surface contaminants. Abandoned water wells directly link pollutants and the aquifer. Where irrigation development is intense, decline in stream flows often follows. In Kansas, more than 700 miles of once permanently flowing rivers are now dry. The blue band on maps representing the Arkansas River in western Kansas hides the reality of a dry streambed.

An integrated agribusiness economy

Fears of impending disaster were real as the regional economy had expanded and grown relatively prosperous on the basis of irrigated agriculture. Except for the Platte River valley in Nebraska and Colorado, and minor surface sources in other places, groundwater sustained irrigation. In the southern reaches of the High Plains, irrigated cotton supports gins, oil mills, and a denim factory. Throughout most of the region farmers cultivate forage crops to feed cattle. Much of the cattle industry centers on huge feeder lots that collectively consume many tons of grain daily. The cattle in turn supply large meat processing plants, some of which are among the largest in the world.

Through the multiplier effect, irrigated agriculture affects all aspects of the regional economy. Where irrigation prevails, most rural towns seem vital, alive, and healthy. Where irrigation is absent, much less economic activity occurs, far fewer people are employed in commerce, and a decreasing volume and array of goods and services are available. Boarded-up storefronts provide visual evidence of the decline. One measure of civic progress in these communities is the tearing down of no longer needed buildings. Another is keeping the weeds under control in empty lots where structures once stood.

Observing this depopulation and decay, Deborah and Frank Popper of Rutgers University have recommended that the entire area be turned over to native grasses and animals. They call for a buffalo commons. The Poppers see a natural process of people moving away, with the population becoming so small and scattered that basic educational, medical, and commercial services could no longer be provided. To be fair, it should be noted that extensive cultivation and cattle ranching would continue to occupy the land, leaving little space for buffalo. Nonetheless, the reduction in the number of towns and farmsteads would approximate the appearance of the High Plains of more than a century ago. This general contraction of people and services occurs where dryland farming prevails. It could become the norm for irrigated areas, where scarcity results from high pumping costs as water is lifted from ever-greater depths, reduction in specific yield, or lack of water as wells are physically depleted.

The High Plains has exemplified what Wes Jackson, director of the Land Institute in Kansas, calls "the failure of success." The success of irrigation was sowing the seeds of its own destruction through depletion. Relying on a single water source also creates vulnerability. Take away the aquifer and the High Plains is without sufficient precipitation or surface water to sustain intensive agriculture or most other activities.

Responding to water scarcity

Few residents in the High Plains are standing idly by, merely watching a drama of decline being played out as irrigation becomes less and less a part of the regional economy. Water conservation has become more prevalent and is seen as the answer to sustaining irrigation and the integrated agribusiness economy it supports. Municipalities, factories, and other water users have lessened water use, but the major concern remains irrigation. More than 90 percent of the water consumed in the High Plains goes to irrigation. If irrigators are able to significantly reduce water consumption, for most there should be adequate and accessible water for many decades to come.

What are irrigators doing to conserve water? In 1988–90, with support from the Ford Foundation, the author and fellow geographer Stephen E. White looked into this question. A questionnaire was mailed to 1,750 irrigators in Texas, Oklahoma, Kansas, and Nebraska, asking them what water-saving practices they had

COURTESY OF DAVID KROMM

■ Center-pivot irrigation has permitted the short-term cultivation of sandy, sloping land. Such areas are suitable only for grazing in the long term.

adopted. Over 40 percent (709) returned a completed survey. Those who responded cited over 21 conservation practices now in wide use throughout the High Plains. Some of these techniques are: periodically checking pumping plant efficiency, planting drought-tolerant crops, replacing open ditches with underground pipes, and recovering runoff from fields. On the center-pivot sprinkler system, low-pressure heads are being installed on drop tubes to reduce overwatering and wind drift.

Several of the widely used practices are sophisticated, information-based techniques that can be highly effective. Scheduling irrigation based on moisture need serves as an example. The goal is to ensure that just enough water is in the root zone of a plant to accomplish the management objectives of the farmer. Soil water levels and plant stress are monitored so that the farmer can determine how much water to apply and when. The days of large amounts of water draining off the fields into ditches or leaching downward beyond the root zone are largely gone.

Institutions play a major role in how the Ogallala and other aquifers are used and who controls the decisions affecting water use. Water law provides an example. Two general forms of water rights exist in the United States, and both function in the High Plains. In Texas, *riparian water rights* prevail, wherein the owner of land overlying an aquifer has the right to use the groundwater beneath. In the remaining states in the region differing forms of the *appropriation doctrine* exist. Under appropriation, the first party to establish the right to use a specified amount of groundwater (or surface water such as a river) has priority of use in times of scarcity. Junior rights may be reduced or even curtailed when they endanger senior rights. The doctrine may be summarized as "first in time, first in right." It has served the western states well, in that insufficient water is available to provide enough for all landowners, as is necessary to apply the riparian doctrine. Most states have developed a priority in water use, with domestic and livestock consumption usually ranking first.

These water rights are administered by various levels of organizations and agencies. Water largely falls under state control, but substate or local districts are often permitted by law. These authorities are formed usually to ensure that existing rights, often for irrigation, are protected. The authority of the local district varies from largely educational, establishing programs to advocate water conservation, to requiring a land-use plan or metering of water flow for all irrigators with water rights. Local districts frequently have enforcement authority. As a rule, a main goal of both state and local groups is to protect existing water rights, most of which are held by irrigators. Increasingly, water markets are being introduced that allow rights to be sold to other users. A farmer, for example, may sell a water right to a municipality or to a developer converting land from agricultural to residential use.

Much remains to be done, but the institutions and technologies of the High Plains show that the region is responding effectively. Most of the states now have water or resource management districts that are responsible for ensuring wise use of the Ogallala. A variety of policies have emerged. In various areas, wells are spaced so as not to interfere with each other, flow meters are required to measure water use, and drilling of new irrigation wells is prohibited. Although planned and orderly depletion constitutes the primary goal, one groundwater management district in western Kansas is considering a zero-depletion policy.

Innovation in irrigation technology has been striking. Many of the water-saving practices and devices used by farmers in the region and elsewhere were developed or first applied in the High Plains. Most are manufactured in the area. Land-grant universities throughout the region lead the way in developing techniques to conserve water. Nonprofit groups such as the Nebraska Groundwater Foundation support public education efforts. The High Plains ranks as a global center of technology and institutions for improving efficiency in irrigation.

What next?

Although numerous writers continue to cite the High Plains as a region where significant groundwater depletion occurs and where irrigation is doomed by its own excesses, current practices and conditions no longer support these views. Because of institutional and technical innovation, educational campaigns, and the land and water stewardship of most farmers, sustainable irrigation appears likely in the High Plains. The Great American Desert will not reappear as the Buffalo Commons.

Article 21

Water Resource Conflicts in the Middle East

by Christine Drake

DEPARTMENT OF POLITICAL SCIENCE AND GEOGRAPHY, OLD DOMINION UNIVERSITY, NORFOLK, VIRGINIA

Water resource issues are critical in the Middle East. Not only is the region located in the arid zone, but it is experiencing increasing pressures on its scarce water resources, pressures that could erupt into serious conflict. This article analyzes the causes of these increasing pressures and the conditions and major sources of conflict in the three major international river basins of the Middle East—the Tigris-Euphrates, the Nile, and the Jordan-Yarmuk, and investigates possible solutions. **Key words:** *water resources, Middle East, conflict, Tigris-Euphrates, Nile, Jordan-Yarmuk.*

In the Middle East, water may be more important than either oil or politics. Whereas the proven oil reserves in the area are estimated to last at least 100 years, water supplies are already insufficient throughout the region, and competition for them is inevitably going to increase in the years ahead. Already there have been a number of clashes between countries over water, and several political leaders have suggested that future conflicts may well center on access to water, both surface and sub-surface. Water is, after all, the most basic of resources, critical to sustainable development in the Middle East and to the well-being of the area's population (Bakour and Kolars 1994, 139).

The purpose of this article is to provide a background for teachers at all levels on the water situation in the Middle East. It examines the causes of the increasing pressure on water resources, considers the conditions and major sources of conflict in the three major international river basins in the Middle East, and investigates possible solutions.

Geographical and Historical Background

At the very root of the problem of limited water resources is the physical geography of the Middle East, for this region is one of the most arid in the world (Figure 1). Descending air (which can hold more moisture as it sinks) and prevailing northeast trade winds that blow from a continental interior region to a warmer, more southerly location explain why almost all of the Middle East is dry. Only Turkey, Iran, and Lebanon have adequate rainfall for their needs because they are located farther north and have more mountainous topography, which intercept rain- and snowbearing westerly winds in winter. Every other country has at least part of its territory vulnerable to water shortages and/or is dependent on an exogenous source of water (a source of water originating outside its boundaries).

It has been estimated that about 35 percent of the annual renewable water resources of the Middle East is provided by exogenous riv-

21. Water Resource Conflicts

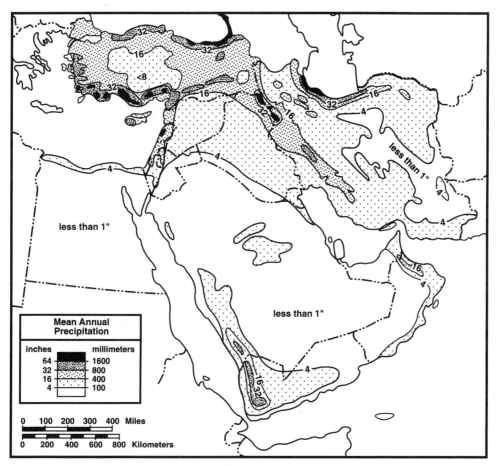

Figure 1. Adapted from Colbert C. Held's Middle East Patterns: Places, Peoples, and Politics. 1994.

ers (Sadik and Barghouti 1994, 3). Certainly, the two major river systems that bring water into the region, the Nile to the Sudan and Egypt, and the Tigris-Euphrates primarily to Syria and Iraq, have sources outside the arid zone—the Nile in the heart of East Africa (the White Nile) and especially in Ethiopia (the Blue Nile), and the Tigris and Euphrates in Turkey (and to a limited extent in Iran). Other rivers, such as the Jordan, Yarmuk, Orontes, Banias, etc., are too small to be of much significance, yet, in the case of the Jordan-Yarmuk, are large enough to quarrel over.

Droughts are common and a natural part of the climate. Yet the prolonged recent droughts suggest that even the natural aridity of the area may have been exacerbated by such human actions as the increase in greenhouse gases from burning more fossil fuels. Even without human intervention, rainfall is seasonal. The water resource problem thus concerns not only the total volume of water available but also its seasonality—the shortage of water in the dry, hot summers. In addition, most of the Middle East rainfall is very irregular, localized, and unpredictable. Furthermore, the region suffers from high evapotranspiration rates, a factor that diminishes the value of the water that is available.

In the past, people adapted to the seasonality of the rainfall and to the periodic droughts and were able to produce enough food to meet local demand. They devised a variety of ingenious ways to store water and to meet the needs of both rural and urban populations. But since the middle of the 20th century such measures have become inadequate, and a new balance must be found among competing needs for water.

Causes of Conflict

Escalation of conflict over water issues in the Middle East results from the confluence of a number of factors, especially rapidly growing populations, economic development and increasing standards of living, technological developments, political fragmentation, and poor water management. The inadequacy and relative ineffectiveness of international water laws as a means of settling and regulating freshwater

TABLE 1

Key Population Characteristics of the Major Countries in the Middle East Involved in Water Conflicts (1996 Estimates)

	Population (million)	Rate of Growth (%)	Doubling Time at present rate	Projected Population 2025 (million)	Percent Urban
Tigris-Euphrates Basin					
Turkey	63.9	1.6	43	78.3	63
Syria	15.6	3.7	19	31.7	51
Iraq	21.4	3.7	19	52.6	70
Nile Basin					
Ethiopia	52.7	3.1	23	129.7	15
Sudan	28.9	3.0	23	58.4	27
Egypt	63.7	2.2	31	97.6	65
Jordan-Yarmuk Basin					
Jordan	4.2	2.6	27	8.3	78
Syria	15.6	3.7	19	31.7	51
Israel	5.8	1.5	47	8.0	90
West Bank	1.7	3.2	22	3.4	-
Lebanon	3.8	2.0	34	6.1	86

Source: 1996 World Population Data Sheet. Washington, DC: Population Reference Bureau, 1996.

issues as well as the lack of any real enforcement mechanisms compound the problem.

Population Growth

Underlying and exacerbating the conflict over water resources in the Middle East is the enormous increase in population (Table 1). (The Middle East is defined here as the traditional southwest Asian countries, and including Turkey, Iran, and Egypt, but excluding the former Soviet republics and the other countries of North Africa.) From somewhere around 20 million in 1750, the population tripled to around 60 million by 1950, but then almost quintupled to 286 million by 1996 (Population Reference Bureau 1996). It is estimated that at present rates of growth, the population will double again in less than 30 years.

Immigration has also been a significant part of the problem, particularly in the Jordan-Yarmuk watershed area, as hundreds of thousands of Jews from all over the world have moved to Israel since its establishment in 1948, and hundreds of thousands of Palestinians have relocated to Jordan from Kuwait in the wake of the Gulf War. If an independent Palestinian state comes into being, water shortages could be compounded by up to 2.2 million Palestinians currently registered worldwide as refugees who could return and settle there.

As populations grow, per capita availability of water decreases. At present, around half of the population in the Middle East depends directly on water for its livelihood. Despite efforts to attain food self-sufficiency, the region imports more than 50 percent of its food requirements now, and that proportion is expected to grow considerably over the next decades (Sadik and Barghouti 1994, 8). Indeed, the Middle East now has the fastest-growing food deficit in the world, and Starr (1995) believes that food insecurity will increasingly pit nation against nation.

Rapid Economic Development

Rapid economic development and rising standards of living, spurred on at least in part by the oil boom, have raised the demand for water both by industry and domestic users. Urbanization also adds to the strain; over half the population of the Middle East now lives in urban areas where populations consume 10 to 12 times as much water per capita as village dwellers.

Technological Developments

Technological developments now enable people to alter their environments in unprecedented ways. The balance between people and their environments and their direct dependence on the natural seasons and cyclical availability of water has been changed by people's ability to build huge dams and create vast reservoirs where increased evaporation occurs, to construct large irrigation schemes where much water is wasted through inefficient watering methods, to extract large quantities of shallow groundwater resulting in lowered water tables, and to damage or destroy rivers and aquifers by polluting them, often irrevocably.

Political Fragmentation

Whereas in times past major empires covered the entire area and dampened conflict among the different peoples within them, the end of World War I saw the dissolution of the one major empire that had controlled much of the Middle East for over 500 years, the Ottoman Empire. Similarly, British colonialism and administration of most of the Nile basin reduced friction over water until the 1950s. With the creation of independent states and increasing ethnic consciousness, growing disparities and rivalries have developed within and among the very diverse populations in the region—Arabs, Turks, Iranians, Kurds, Ethiopians, and Israelis—to name just the main protagonists. All have become more competitive and nationalistic.

Overuse and pollution of rivers and shared aquifers (underground water-bearing formations) are a source of growing tension. Water was one of the underlying causes of the 1967 Arab-Israeli war and continues to be a stumbling block in the search for peace. Forty percent of Israel's water, for example, is obtained from aquifers beneath the West Bank and Gaza. Water was a major rallying cry both for the Palestinian Intifada in the Occupied Territories and for conservative parties in Israel. Whereas, in the past, the Cold War had a restraining effect on the likelihood of major conflict, that lid has now been removed.

Poor water management

Poor water management exacerbates the problem of both water quantity and quality. Great quantities of water are lost through inefficient irrigation systems such as flood irrigation of fields, unlined or uncovered canals, and evaporation from reservoirs behind dams. Pollution from agriculture, including fertilizer and pesticide runoff as well as increased salts, added to increasing amounts of industrial and toxic wastes and urban pollutants, combine to lower the quality of water for countries downstream (called downstream riparians), increase their costs, and provoke dissatisfaction and frustration, again creating irritations that can lead to conflict.

International Water Laws

Existing international water laws are underdeveloped and inadequate and in some respects do not seem geared to the problems of arid developing countries, having been developed in the temperate and better watered areas of Europe and North America. Various legal principles exist, but there are no legally binding international obligations for countries to share water resources (Morris 1992, 36). Agreements must depend upon the mutual goodwill of co-riparians (the countries bordering a specific river) in any particular drainage basin. The likelihood of conflict or cooperation depends very much upon a number of geopolitical factors. These include the relative positions of the co-riparians within the drainage basin, the degree of their national interest in the problem, and the power available to them both externally and internally to pursue their policies (Jones 1995).

Major Regions of Conflict

Although there are many international rivers and several important shared aquifers in the Middle East, and all have potential for water disputes, the potential for the greatest conflicts occurs in the three major international river basins: the Tigris-Euphrates, the Nile, and the Jordan-Yarmuk.

The Tigris-Euphrates Basin

In the case of the Tigris-Euphrates, there are several fundamental issues. Turkey is the source area for more than 70 percent of the united Tigris-Euphrates flow and owns large portions of the drainage basins of the two rivers (Figure 2). It also has the upstream position and so the opportunity to use the waters of the Tigris-Euphrates as it pleases. Indeed, the creation of Turkey's Southeast Anatolia Development Project (Turkish acronym GAP) is evidence of its felt rights. This very ambitious plan is to build 22 dams on the Euphrates in order to increase irrigation and electricity generation and to bring greater prosperity to a heretofore neglected Kurdish

3 ❖ THE REGION

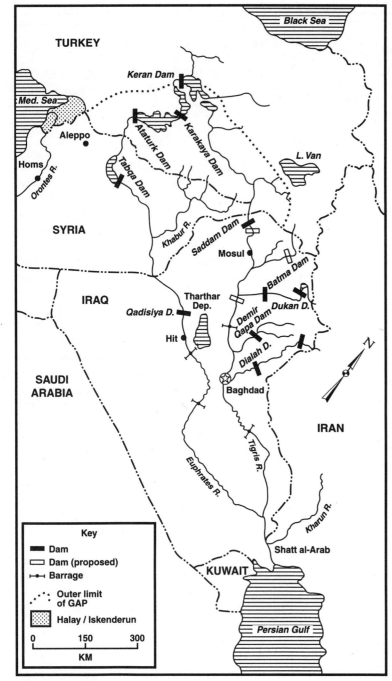

Figure 2.

region of the country. Turkey argues that the GAP project will actually benefit all three riparian countries, Syria and Iraq as well as Turkey, as it will reduce damage from floods and even out the river's flow, storing excess water from the wet season and snow melt so it can be used in the dry summer season and soften the impact of droughts.

Inevitably, however, as more of the Euphrates water is withdrawn and used in Turkey, less will be available for downstream riparians. Indeed, the GAP project, if completed, is expected to reduce the flow of the Euphrates by 30 to 50 percent within the next 50 years (Bakour and Kolars 1994, 134). Furthermore, the water will be of lower quality, as increased amounts of salts, fertilizers, pesticides, and other pollutants enter the river after having been used for irrigation.

Syria depends on the Euphrates for over half its water supply and has a population growing at 3.8 percent, with almost no effort being made to reduce that extraordinarily high rate of growth. It also has plans to expand its irrigation projects. Iraq is even more dependent than Syria on the waters of the Euphrates and Tigris and claims historic rights to the water, but its position as the lowest riparian state renders it vulnerable to decreased water supply from both Turkey and Syria. A 1987 protocol in which Turkey promised 500 cubic meters per second at its border with Syria has not been solidified into a firm agreement or treaty, nor has Syria's pledge to deliver 290 cubic meters to Iraq, an amount that is only about half of what Iraq claims (570 cubic meters per second) and is clearly far below its needs (Bakour and Kolars 1994, 139). Iraq's population is expected to grow from 21 million in 1996 to around 35 million in 2010. The only ameliorating fact here is that Iraq can compensate for lack of water in the Euphrates by taking water from the Tigris, which at present is underutilized.

A related problem is the current dispute over the actual size of the annual flow of the Euphrates. Data on average discharge vary enormously. Syria's claims that Turkey is deliberately reducing the flow of the Euphrates are countered by Turkey's claim that the region suffers from periodic droughts.

The Nile Basin

The Nile catchment is shared by 10 countries, but the main disputes over water so far involve only the three largest: Ethiopia, Sudan, and Egypt (Figure 3). Of the three, Egypt is

21. Water Resource Conflicts

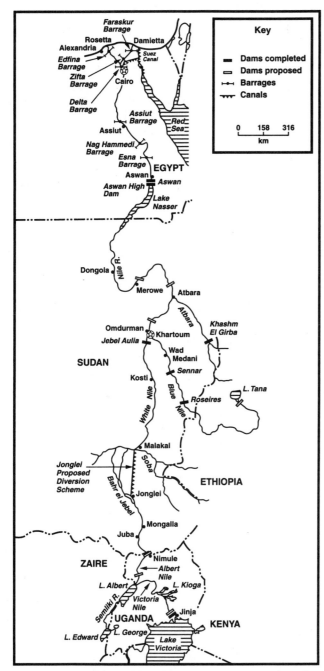

Figure 3. Adapted from Kliot 1994. *Water Resources and Conflict in the Middle East.* New York: Routledge.

the country with the most obvious water crisis, and the situation is becoming more severe each year. Egypt's population of about 64 million is growing annually by almost one- and one-half million. Egypt is almost totally dependent on the Nile (while contributing virtually nothing to it) and also claims that its prior usage entitles it to a disproportionate share of the Nile waters. Not only is over 95 percent of its agricultural production from irrigated land, but Egypt needs both to expand its agricultural land and to reduce the salt-water intrusion of the Mediterranean into the Nile delta, goals threatened by growing water shortages.

Since about 85 percent of the flow of the Nile into Egypt originates on the Ethiopian plateau, Egypt is most concerned about its relationship with Ethiopia. Egypt has repeatedly warned Ethiopia not to take any steps that would affect the Blue Nile discharges (Kliot 1994, 68). On several occasions, however, Ethiopia has claimed that it reserves its sovereign right to use Blue Nile (and Sobat) river water for the benefit of its own rapidly increasing population (at 3.1 percent a year). Indeed, it has extensive plans to develop about 50 irrigation projects and to expand its hydroelectric generation potential as well, plans which are more feasible now that its civil war has ended. Experts believe it is highly possible that Egypt may experience a modest reduction in the amount of Nile water available to it as Ethiopia claims a larger share of the Nile headwaters in the future (Whittington and McClelland 1992).

Egypt is also concerned about its immediate upstream neighbor, Sudan. Although Sudan is incapable of expanding its water use much at present, racked as it is by civil war, economic recession, and a shortage of foreign investment, that situation could well change. Sudan has the potential to become the bread basket of the Middle East, but that would be possible only with increased use of Nile water. Since 1929, the two countries have had an agreement allocating the Nile waters. The 1929 allocation was adjusted, however, in 1959, by an agreement which gave Sudan more water, reducing Egypt's relative share from 12:1 to 3:1.

What will happen when Ethiopia and Sudan begin demanding more of the Nile's water? Will Egypt accept the Helsinki and International Law Commission rules that irrefutably entitle Ethiopia and Sudan to a larger portion of Nile water (Kliot 1994, 98)? Will Egypt try to change those rules in order to give greater weight to the principle of prior use? Or will Egypt be tempted to use its position as the most powerful nation in the Nile basin to assure its present allocation, even if this means the use of military force and international conflict?

The Jordan-Yarmuk Waters

It is in the Jordan-Yarmuk basin that tensions have run the highest (Figure 4). All of the countries and territories in and around the Jordan River watershed—Israel, Syria, Jordan, and the West Bank—are currently using between 95 per-

3 ❖ THE REGION

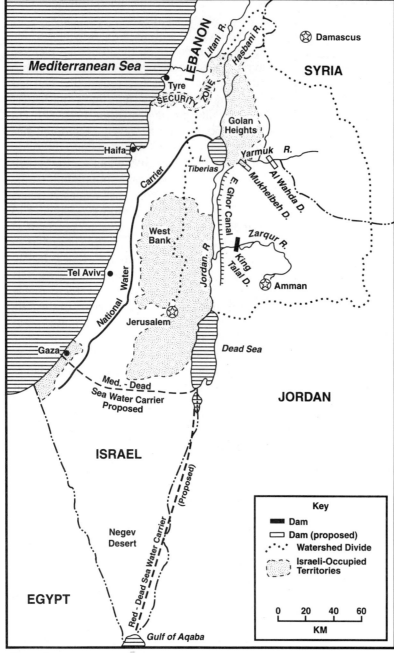

Figure 4.

65 percent of the country's total utilizable surface water (Sadik and Barghouti 1994, 16).

Scarce water resources have either precipitated or exacerbated much of the recent political conflict in the Jordan river basin (Wolf 1995). Indeed, the Jordan basin has been described as "having witnessed more severe international conflict over water than any other river system in the Middle East ... and ... remains by far the most likely flashpoint for the future" (Anderson 1988, 10).

For Syria, Jordan, and Israel, the proportion of water derived from international sources is very high. Over 90 percent of Syria's water resources are shared with her neighbors, Turkey, Iraq, Israel, Lebanon, and Jordan. Jordan gets more than 36 percent of its water from sources shared with Syria, West Bank, and Israel. And more than half of Israel's water resources are shared with Syria, Lebanon, Jordan, and the Palestinians. The economies and societies of the countries in the basin of the Jordan-Yarmuk are very vulnerable to any restrictions in their water supplies; hence, the situation is highly volatile.

Water conflicts among the protagonists have been longstanding, although the situation has worsened in recent years. Both overt and covert conflict has occurred over the division of the Jordan-Yarmuk waters, as both Israel and the Arab countries have tried to divert water: Israel through its National Water Carrier to expand agriculture in the Negev and the Arabs through their attempts to divert water from the Jordan basin to Lebanon, Syria, and Jordan. Disagreement over water was a major contributing cause of the 1967 Arab-Israeli war. Through its victory, Israel enhanced its water resources by capturing the Golan Heights and the West Bank aquifer. These captured sources supply as much as 25 percent of Israel's total water needs (Waterbury 1994) but have led to charges that Israel is stealing Arab water. On the West Bank, Israeli authorities have prevented the Pal-

cent and more than 100 percent of their annual renewable freshwater supply (Wolf 1995, 1). Shortfalls have been made up by overpumping limited groundwater. Water tables are being lowered at alarming rates. It is obvious that all the surface water and groundwater resources are over-stretched and over-utilized. Jordan's situation is perhaps the most serious. Only 5 percent of its land area receives sufficient rainfall to support cultivation and less than 10 percent of its agricultural land is irrigated at present, even though irrigation in the Jordan valley consumes about

estinians from digging new wells or even finding alternative sources of water to compensate for water lost as a result of withdrawals by Jewish settlements. The disparity between the water allocations to Jewish and Arab settlements on the West Bank is enormous: the average aggregate per capita consumption for the Jewish settlements ranges between 90 and 120 cubic meters, whereas for Arab settlements the consumption is only 25–35 cubic meters per capita (Kliot 1994).

No comprehensive agreements have been reached over an equitable allocation of the Jordan-Yarmuk waters, although water was a vital and sometimes overriding factor in the peace process of the early 1990s. The October 1994 peace treaty between Israel and Jordan, which was only formalized after the last and most contentious issue—shared water resources—was agreed to, included an agreement for a Joint Water Committee to develop additional water resources, including two new dams, one each on the Yarmuk and the Jordan (Wolf 1995). Also, the creation of a Palestinian Water Authority was an important aspect of the September 1993 Declaration of Principles that allowed Israelis and Palestinians to recognize one another as legitimate political entities. Yet further agreements are urgently needed to include not only surface water but also groundwater (which desperately needs to be replenished). But given the inherent hostility in the region, it is likely that water will remain a major source of conflict and instability for the foreseeable future.

It is in the Jordan River basin, perhaps more than any other, that water has become "a highly symbolic, contagious, aggregated, intense, salient, complicated zero-sum power-and-prestige-packed crisis issue, highly prone to conflict and extremely difficult to resolve" (Frey and Naff 1985, 77).

Possible Solutions

Gideon Fishelson (1989, 180) of the Armand Hammer Fund for Economic Cooperation in the Middle East has stated:

> The danger of war over water hangs over the heads of the Middle East countries, yet there is also the possibility of cooperation and harnessing new technologies and capital that would prevent such wars. Solving the water issue is one of the essential prerequisites to achieving a meaningful and lasting peace in the Middle East.

Solutions other than conflict clearly exist. Some of the measures that can help to mitigate the problems and relieve tension over scarce water resources follow.

Conservation

More water can be made available through reducing waste in irrigation, municipal, and industrial uses, in transport, and in distribution. Far more can be done to encourage the more frugal use of water, through education and media campaigns, but especially by more realistic pricing for water use. However, such changes will be hard to implement and will run into enormous opposition, since water is such a politically sensitive issue; traditionally, water has been regarded as a free resource available to all who need it.

Better Management

More water-efficient methods and technologies can be adopted. For example, in agriculture, it is essential to move away from inundation (flood) techniques, which currently account for about 95 percent of irrigation practices, to sprinkler techniques, trickle irrigation techniques (which take water directly to the plant roots from perforated plastic pipes), or subsurface irrigation techniques (which control the height of the water table and irrigate plants through capillary action, thus minimizing evaporation). Other management techniques include lining canals to prevent seepage losses, covering them to minimize evaporation, repairing pipes to reduce leakage, improving drainage to reduce soil salinization, minimizing evaporation by better field preparation and water application, and irrigating at night and early in the morning when evaporation rates are lower. In addition, more water can be recycled and treated sewage used at least on non-food and ornamental plants, while water quality can be preserved by preventing the contamination of water sources.

Prioritizing Uses

Administrative solutions include managing the demand by reallocating water from comparatively low-value uses, such as agriculture, which currently consumes 80 to 90 percent of the region's water supplies, to essential domestic use and higher-value, industrial uses. Light industry yields about 30 times more input to GNP than does agriculture per unit of water used. However, such a strategy would lead to increased reliance upon foreign sources for food supplies, an outcome that seriously worries many policy-makers because of food security vulnerability. Such prioritizing is realistic, though, because there is enormous capacity to save water in agriculture, not only through more efficient irrigation techniques, but also by moving from water-consumptive crops such as rice and sugar cane to more water-efficient crops and increasing the use of more salt-tolerant crops, such as certain oil-seeds.

Technological Solutions

There are many technological ways to help solve problems of both water quantity and water quality. The obvious and traditional response to water shortages has been to construct dams in order to retain runoff from floods, to store water for use throughout the year, and even to replenish shallow aquifers. But all dams have an impact on development downstream.

Interbasin water transfers are another technological solution, such as Turkey's proposed two "Middle East Peace Pipelines," designed to carry surplus Anatolian water from the Seyhan and Ceyhan rivers to Syria, Jordan, and Israel, and on through Iraq to Saudi Arabia and the Gulf states. Countries, however, are very reluctant to be dependent on others for their most basic need—water, which could be interrupted for political reasons. In addition, the costs for all of these projects would be enormous, probably higher than by obtaining water through desalination.

Developing cost-effective solar desalination technology is obviously a goal for the future. Already desalination of brackish water is competitive, although desalination of sea water is invariably expensive and largely influenced by petroleum prices. Another goal is to find a way to reduce evaporation on storage lakes behind dams.

Increasing Cooperation Among Co-riparians

In order to obtain the maximum use of water, it is essential to develop each river basin in its entirety, in a fully integrated way, with comprehensive approaches to the sustainable management of water resources. Better and more detailed data need to be obtained and shared among all riparians. Such an approach is now beginning to occur with the most recent movement toward peace in the Jordan River basin.

Developing Better, Enforceable, International Water Laws

Clearly it is necessary to develop water laws that encompass all surface water and groundwater both within and outside a river basin, involve all co-riparian states, provide for sustainability of water use, prioritize water uses, and offer more incentives for water preservation and conservation (Kliot 1994). Only then will countries have their allocation assured and be able to plan their own use of water, free from threats and uncertainty. Better mechanisms to enforce water laws and solve conflicts over water resources must also be created.

Reducing Population Growth Rates

Finally, but perhaps most significantly, it is crucial that all states in the region drastically reduce their population growth rates, for in a context of rapidly increasing population, all other efforts to ensure water availability over the long term may well be futile. Certainly population stability (and some would argue for not just reducing growth rates but also the absolute size of the population) is a long-term strategy, but it may well be the most critical strategy of all. Unfortunately, Middle Eastern traditions value large families, while in some countries with large ethnic or religious diversity promoting family planning is politically unacceptable.

Conclusion

Experts disagree on the likelihood for future conflict over water. Some argue that more water conflicts are inevitable because of the combination and synergistic effect of the causes already discussed: growing water scarcity, increasing populations, rising standards of living, and higher consumption levels. Many of the rivers and aquifers in the region are shared, and the lack of adequate treaties and international laws, added to the absence of adequate enforcement mechanisms, increases the likelihood of confrontation. There has also been a history of hostility among some of the countries and a growing nationalistic self-awareness of the differences among the varied peoples in the Middle East. One could also suggest, perhaps a bit cynically, that countries "need" enemies to deflect attention from internal divisions, political corruption, and economic hardships, and to help to unify or integrate the population. Some even contend that countries need also to justify their military forces and keep them busy!

Others argue, however, that future water conflicts are not likely for a number of substantial reasons. For one thing, cooperation is cheaper than conflict. As one person put it: "Why go to war over water, when for the price of one week's fighting you could build five desalination plants?" (Wolf 1995, 76). There is considerable international pressure to avoid war over water, partly because it could escalate into war over other, even more intractable issues. In addition, there are a number of external geopolitical forces that are indirectly exerting pressure for peaceful cooperation on water allocation issues, such as the end of the cold-war manipulation of states in the Middle East, the progress toward a peace between Israel and its neighbors, and Turkey's goal to enter the European Union. The fact that in each river basin there is one stronger military power (Turkey, Egypt, and Israel) further deters conflict. Lack of capital will probably delay and may even prohibit development of some water-using projects. Moreover, if the solutions suggested earlier are implemented, conflict will be less likely.

Furthermore, states have it in their power actually to decide to treat water use as a vehicle for cooperation. Throughout the years of hostilities in the Middle East, water issues have actually been the subject of occasional secret talks and even some negotiated agreements between the states in the region (Wolf 1995). In regional peace talks, cooperation on regional water planning or technology might actually help provide momentum toward negotiated political settlement. According to Frey and Naff (1985, 67),

> Precisely because it is essential to life and so highly charged, water can—perhaps even tends to—produce cooperation even in the absence of trust between concerned actors.

In any case, water is only one of many factors at work in the Middle East—certainly an important one, but only one. Israel's settlements on the West Bank and its occupation of the Golan Heights, radical groups within not only the Palestinian population and Israel, but also among the Kurds and other disadvantaged groups, irredentist pressures, economic pressures—these and many other pressures could produce bitter conflict that uses water either as a weapon or as an excuse for hostility.

Much depends on leadership in the region, including the ability of governments to control radical or conservative elements that want to exploit water issues, the ability to obtain capital for the development of industry (which will take some of the pressure off agriculture), the ability to obtain secure food sources from outside the area, the ability to reduce population growth, the ability to educate the public on water issues, develop an ethos of conservation, and change water pricing systems, and finally, the ability to promote cooperation and encourage the sharing of technology, data, and research. One has to hope that the benefits of cooperation in the development of river basins and the rule of law will be seen to outweigh the costs of conflict.

References

Anderson, E. 1988. Water: The next strategic resource. In *The Politics of scarcity: Water in the Middle East*, eds. Starr, Joyce, and Daniel Stoll, 1–22. Boulder, CO: Westview Press.

Bakour, Yahia, and John Kolars. 1994. The Arab Mashrek: Hydrologic history, problems and perspectives. In *Water in the Arab world: Perspectives and prognoses*, eds. P. Rogers and P. Lydon, 121–45. Cambridge, MA: Harvard University Press.

Fishelson, Gideon, ed. 1989. *Economic cooperation in the Middle East*. Boulder, CO: Westview Press.

Frey, Frederick, and Thomas Naff. 1985. Water: An emerging issue in the Middle East? *Annals of the American Academy of Political and Social Science* 482:65–84.

Held, Colbert C. 1994. *Middle East patterns: Places, peoples, and politics*. 2nd ed. Boulder, CO: Westview Press.

Jones, Mark C. 1995. Critical factors in transnational river disputes: An analytical framework for understanding the India-Bangladesh water scarcity dispute over the Ganges River. Unpublished Master's thesis. The Miami University at Oxford, Ohio, Department of Geography.

Kliot, Nurit. 1994. *Water resources and conflict in the Middle East*. London: Routledge.

Morris, Mary E. 1992. Poisoned wells: The politics of water in the Middle East. *Middle East Insight* 8:2, 35–39.

Population Reference Bureau. 1996. 1996 World population data sheet. Washington, DC: Population Reference Bureau, Inc.

Sadik, Abdul-Karim, and Shawki Barghouti. 1994. The water problems of the Arab World: Management of scarce resources. In *Water in the Arab world: Perspectives and prognoses*, eds. P. Rogers and P. Lydon, 1–38. Cambridge, MA: Harvard University Press.

Starr, Joyce S. 1995. The Middle East food alarm. *The Christian Science Monitor*, 26 October, p. 19.

Starr, Joyce, and Daniel Stoll, eds. 1988. *The politics of scarcity: Water in the Middle East*. Boulder, CO: Westview Press.

Waterbury, John. 1994. Transboundary water and the challenge of international cooperation in the Middle East. In *Water in the Arab world: Perspectives and prognoses*, eds. P. Rogers and P. Lydon, 39–64. Cambridge, MA: Harvard University Press.

Whittington, Dale, and Elizabeth McClelland. 1992. Opportunities for regional and international cooperation in the Nile basin. In *Country experiences with water resources management: Economic, technical, and environmental issues*, eds. G. LeMoigne, G. S. Barghouti, G. Feder, L. Garbus, and M. Xie. World Bank Technical Paper No. 175. Washington, DC: The World Bank.

Wolf, Aaron T. 1995. *Hydropolitics along the Jordan River: Scarce water and its impact on the Arab-Israeli conflict*. Tokyo: United Nations University Press.

Article 22

BOOMTOWN BAKU

By Richard C. Longworth

BAKU, THE CAPITAL OF THE NEWLY independent nation of Azerbaijan, was a gracious city in the past and no doubt will be again, when new oil begins to flow. The city's hills rise from the slightly salty waters of the Caspian Sea, and many streets are still graced by fine buildings erected a century ago, when Baku wells pumped more than half the world's oil.

But you must look close to see the grace. Seventy years of Soviet rule left Baku coated with grime and riddled with decay, saddled with an economy that only a Moscow planner could love. When the Soviet Union collapsed seven years ago, so did the economy, and it is only beginning to recover. The streets are more potholes than pavement. Old men and women stand on street corners selling pathetic handfuls of lemons or nuts. Doctors and teachers make about $10 or $20 a month, officially,

Richard C. Longworth, a senior writer and foreign correspondent at the Chicago Tribune, *is a member of the* Bulletin's *Board of Directors.*

and survive by extorting bribes from patients and pupils.

In the countryside, poverty is virtually total. Markets sell rotting cabbages and fatty lamb, and hospitals are seldom more than a room, a table, two chairs, and a screen for undressing. Most factories have closed and 800,000 refugees, the detritus of the war with Armenia over Nagorno Karabakh, slowly rot in prefab huts or in caverns dug from the salt-white earth, lined with carpets and covered by mud-and-reed roofs. They have no jobs, nor even the hope of jobs, and are kept alive by international relief organizations.

Bosnia with oil

Yet there is money in Azerbaijan—big money, the result of an oil boom that is just beginning. The big nations of the world are here, elbowing and jostling and conniving for a piece of the same nascent boom. In this dirt-poor land tucked unhappily between Russia and Iran, the United States has staked a strategic interest. Azerbaijan is, in fact, the focal point

of the next round in the Great Game of Nations, a dangerous, hot-headed place with a Klondike of wealth beneath it. It is Bosnia with oil.

The money glints incongruously from the grime, like a diamond on a beggar. Limestone mansions rise in walled compounds near impoverished villages or on the hills above Baku, overlooking the hovels on the flats below. Within the fifteenth-century walls of the old city, oil companies from the United States, France, Italy, Japan, Russia, and more have rehabbed old, honey-colored villas as headquarters for their assault on the Caspian's riches. The Turks and Chinese are here, and so are all the big names in the oil business—Amoco, Shell, Pennzoil, Elf, Statoil, Lukoil, Itochu, Agip—not just the Seven Sisters, but their nieces and nephews, too. Down by the harbor, British Petroleum occupies the old Nobel compound—much of the money that funded the original Nobel Prizes came from Baku oil 100 years ago.

The Americans have installed an embassy in a gracious old mansion and the Russians are building a handsome new legation—possibly the first tasteful public building erected under Moscow's aegis since 1917—just down the street from the Hyatt Regency Hotel, where the oil workers drink with British Airways stewardesses in the Lord Nelson Pub. Down on Neftchilar Prospekti (Oilmen's Boulevard), beside the sea and around the walls of the old city, new shops are opening—a Pierre Cardin salon here, an Yves Rocher boutique there, an American supermarket nearby, a Ragin' Cajun, where the oil workers can eat Tex-Mex. In gaudy restaurants like the Cidir, Turkish belly dancers gyrate and Belarussian chanteuses croon for tables of fat, dark-suited men too bored to applaud. The president's son and his cronies gamble in shiny new casinos. Lesser moguls shop for new Volvos and BMWS, just arrived from Dubai; there are no auto dealerships, but everyone knows the street corners along Nizami Street where such cars are sold.

Lurking beneath the Caspian are 100 billion barrels of oil, or maybe 200 billion.

Bechtel and other construction companies are here to build the pipelines that will carry oil and natural gas, not only from Azerbaijan but from the other two Caspian nations, Kazakhstan and Turkmenistan, on the eastern shore of the sea. The Turks, Iranians, Georgians, and Russians all hope the pipelines will cross their territory and are willing to pay whatever to whomever to get them; Azerbaijan ranks with Russia and Nigeria as one of the world's most corrupt nations, and a willing bribe seldom lacks a palm to cross.

In this oil-soaked cockpit, the prospects for both wealth and trouble are simply stupendous. The oil companies figure that the potential reserves of oil lurking beneath the Caspian or locked into the soil around it amount to no less than 100 billion barrels, maybe 200 billion barrels. At current prices, that's a bonanza of $4 trillion. It is the biggest pool of reserves on earth, after the Persian Gulf itself, and represents a major alternative source of supply if ever the Gulf nations turn off their tap.

Buying protection

The comparison with Bosnia is irresistible. The Caucasus is a land of ancient vendettas and warring tribes that makes the Balkans look straightforward by comparison, and the Caspian is where this bloody region meets Central Asia and the Middle East. It's where Orthodox Christianity meets both Sunni and Shi'ite Islam, where Iranians traveling north have met Russians coming south. It's an area once ruled by Iran, then by Russia, now contested by Turkey, whose language and civilization dominate the region.

In this macedoine of traditional foes and feudal paranoias, the new major player is the United States. On the surface, this is natural. The big American oil companies are fighting for a piece of the action and, where oil is concerned, the flag often follows trade. But in this part of the world, nothing is ever simple.

For one thing, Azerbaijan is a desperately poor country of seven million, just freed from the Soviet empire, seldom independent in its history and, with Russia to the north and Iran to the south, profoundly uncomfortable. It needs a protector and, with all that oil, it can pay for one. Enter Washington.

"This is our chance to strengthen our independence," said Vafa Gulizade, the chief foreign policy adviser to President Heydar Aliyev, when I interviewed him earlier this year. "Oil is policy. It's energy, and energy plays a very strategic role."

Hasan Hasanov, the foreign minister, agreed: "We see the United States as a guarantor, one of the guarantors, of our independence." (About two weeks after I interviewed him, Hasanov, a long-time Communist Party functionary under the Soviet regime, was fired by President Aliyev, then arrested for allegedly taking kickbacks on the construction of Baku's newest hotel, the Europa.)

Most Americans are not sure just where Azerbaijan is and would be surprised to know that they have become the designated bodyguard for an oil-rich nation in one of the world's scrappiest areas. Yet the American government has let Aliyev and his ministers assume that the Americans will protect Azerbaijan from its surly neighbors. Both the Azeris and the Americans deny this means that the United States would fight for Azerbaijan, as it did for Kuwait, but there doesn't seem to have been much discussion on this point and the actual parameters of U.S. involvement seem vague.

"The Azeris see themselves threatened potentially by Russia and Iran," a Western diplomat in Baku told me. "They would like to count on the United States in this case. This is part of their calculation in opening their resources to American Western firms. I'm not so sure they're wrong."

The Azeri payback so far has been substantial. American firms—Amoco, Exxon, Pennzoil, and Unocal—dominate the Azerbaijan International Operating Co. (AIOC), a consortium of 11 companies from eight nations that is drilling the first oil here since independence. Amoco and British Petroleum are the two biggest partners, with 17 percent each. The not-so-subtle Azeri hint: There are more good contracts to come for companies whose governments support Azerbaijan.

So Washington, like Baku, now finds itself squeezed between Iran and Russia. If Russia dislikes the expansion of NATO toward its western frontier, it is livid about the growing American influence on its southern flank. So far, Azerbaijan has bought off the Russians by giving Lukoil, the Russian oil firm, 10 percent of the AIOC, the drilling consortium. But the Russians, who also border the Caspian, feel cheated out of a larger share of the oil under the sea. Iran, meanwhile, is being shut out. No Iranian company belongs to AIOC, and the United States is pressuring the Caspian countries not to ship any oil or natural gas south through Iran, even though it is the cheapest route from the landlocked sea to the outside world.

A personal affair

The Aliyev government expects Washington to protect more than Azerbaijan. It also expects the Americans to protect Aliyev himself.

The president is a vigorous 74-year-old, a former KGB general and Communist Party member of the Politburo under the Soviets. He took power in 1993 in a coupe against his ineffective but democratically elected predecessor, Abullaz Elchibey. He immediately jailed his opponents, then won a rigged election. Since then he has run something less than a democracy but more than a dictatorship. The state-controlled television dutifully carries Aliyev's interminable speeches and the press avoids any criticism of him. But newspapers otherwise are relatively free and the president's opponents, now mostly out of jail, speak openly.

Azerbaijan is far from a police state. Elchibey plans to oppose Aliyev in the next presidential election later this year. Foreign observers say the president, who has brought some badly needed stability to his country, would probably win 70 percent of the vote in a fair election. Nevertheless, he is likely to rig it again, if only from force of habit.

Since 1993, there have been two attempted coups against Aliyev, one allegedly run from Russia, the other from Turkey. Aliyev scotched them both, but the coups indicated the fragility of the stability he has imposed on his country. Unable to trust any of the regional powers, Aliyev has looked to American backing for his personal political future. Once again, the payoff is oil. Lest any Westerner forget the president's personal control over his nation's reserves, he has made his son, Ilham, the vice president of Socar, the state oil company and a key partner in AIOC.

Twin foreign policies

This Azerbaijani dependence on the United States is remarkable, considering that Washington has two foreign policies toward the region, one pro-Azeri, the other anti-Azeri. The pro-Azeri policy belongs to the administration, which listens to the oil companies. The anti-Azeri policy belongs to Congress, which listens to the Armenian lobby.

Congress got there first, and the reason is Nagorno Karabakh, a mountainous enclave within Azerbaijan that has traditionally been inhabited by ethnic Armenians. (Stalin made it part of Azerbaijan and, at the same time, gave Nakhichevan, a similar enclave, to Azerbaijan, even though it is cut off from Azerbaijan by Armenian territory. The old dictator, a Caucasian himself from Georgia, was either playing divide-and-rule with his neighbors or indulging a malevolent joke.)

The Karabakh Armenians simmered for years within Azerbaijan. Then, as Moscow's rule began to weaken in 1988, they voted to join Armenia. This sparked an anti-Armenian pogrom in Sumgait, a grim industrial town near Baku. War broke out and raged for six years, killing at least 20,000. When a truce was called in 1994, Armenia controlled not only Karabakh and a corridor linking it to Armenia, but six other Azeri regions, amounting to about 20 percent of Azeri territory. Hundreds of thousands of Azeris fled to what remained of Azerbaijan and are there to this day. Azerbaijan slapped an embargo on Armenia, arguing that no sane nation would allow goods to flow freely to a neighbor who happened to be occupying one fifth of its territory.

Negotiations, called the Minsk Talks, were organized by the Organization for Security and Cooperation in Europe (OSCE), under the leadership of the United States, Russia, and France. Over the years, the two sides inched toward a compromise—that Nagorno Karabakh would remain part of Azerbaijan but would have virtually autonomous status. As part of the deal, the occupied regions would return to Azerbaijan and the refugees could go home. The Karabakh Armenians never agreed; they insist now on nothing short of full independence. But their demand is a nonstarter, because no other nation would agree to this change in national frontiers by force. Both Aliyev and Armenian President Levon Ter-Petrossian recognized this and bought the compromise. Both risked reprisals by nationalist hotheads, and Ter-Petrossian paid the price. He was forced from office, replaced by Robert Kocharian, a hard-line nationalist from Karabakh who rejected the Minsk compromise.

Congress makes policy

The Caucasus, like the Balkans, has too much history. Its peoples nurse mutually exclusive memories of glorious empires, ancient wrongs, historical grievances. Only the most foolish outsider would try to argue that all the atrocities in the war were on one side, or assign blame for the standoff, or take sides. But that's what Congress did. In 1992, it passed the Freedom Support Act providing aid for the 15 former Soviet republics, and then passed an amendment, Article 907, that said none of the aid could go to the Azeri government until it took steps to lift the embargo. Azerbaijan was, and is, the only former Soviet government treated this way.

Article 907 is the result of political persuasion, a patient's gratitude, and pillow talk. It was promoted by the Armenian Assembly of America, a highly professional lobby based on Capitol Hill that is able to tap the vocal, generally affluent and self-conscious Armenian-American communities across the nation on any issue important to Armenia. The group's arm in Congress is the Armenian Caucus, made up of 62 congressmen. One, a California Democrat named Anna Eshoo, is of Armenian heritage, and she played a relatively small role in the passage of 907. The real leadership was provided by Sen. Bob Dole, out of gratitude to an Armenian-born surgeon who helped repair his grievous World War II wounds, and by Cong. John Porter, an Illinois Republican.

Porter has never been to either Armenia or Azerbaijan, but he is powerfully influenced by his wife, Kathryn, a human rights activist who has said that "I would die for the rights of the people in Nagorno Karabakh." (Kathryn Porter has never been to Azerbaijan either—"I haven't done Baku yet," she says—but she has been to Karabakh and says the mountain peasants there "reminded me of Americans at an earlier time in our own country." Armenians call her "the angel of Karabakh." Azeris call her a menace.)

American aid does go to Azerbaijan, although much less than to Armenia, which receives the fourth-highest amount of U.S. aid per capita. By law, no U.S. aid can go through the Azeri government, although a 1997 amendment permits some humanitarian aid to do so. Instead, it goes directly from USAID to the Baku office of Save the Children, which then distributes it to CARE, Relief International, and other charities operating in the refugee camps.

Save the Children must insure that none of the aid touches government hands. Until recently, charities had to store medicine beneath tarpaulins in fields because all the warehouses were owned by the government. Similarly, the charities are not allowed to train doctors or nurses working in state-owned hospitals, which is to say, in any of the hospitals in Azerbaijan; or to help the refugees find jobs, because then they would pay taxes to the government.

Relief International, unable to work with the hospitals, has hired 40 of its own doctors and nurses and has trained them well. If it could have worked with hospitals, it could have trained hundreds, leaving a large and self-sustaining pool of expertise behind when it leaves. The fact is that, in a Third World country seven years removed from communism, it's hard to do anything that doesn't involve the government.

This policy was up and running before the administration became aware, about three years ago, that Azerbaijan might hold one of the keys to America's energy future and that Article 907 wasn't exactly making friends in Baku. Clinton has since feted Aliyev at the White House and the State Department is doing all it can, within the limits of 907, to improve relations with the Azeri government.

There is no Azeri lobby to speak of in the United States, but the Azeris have a de facto lobby in the oil companies, who want the Baku government to love them and have lobbied vigorously against 907. Their payroll includes such luminaries as James A. Baker III, Dick Cheney, Brent Scowcroft, Zbigniew Brzezinski, John Sununu, Lloyd Bentsen, and other Washington veterans who have been able to persuade Congress to modify 907, but not to scrub it.

Thomas Goltz, a journalist who lived in Baku and wrote a book called *Azerbaijan Diary*, has noted that U.S. aid to Azerbaijan effectively promotes revolution. If aid workers could work with the government, he says, they could help promote privatization, or small business programs, or press freedom, or education. As it is, "not one dime of American money has been invested in the process of reforming governmental institutions in Azerbaijan." Instead, aid can only go to non-governmental organizations. In a demi-democracy like Azerbaijan, this usually means helping groups that the government sees as the opposition at best and subversives at worst.

Goltz records the "ultimate irony" that should any of these groups "take power and actually be able to effect the sort of societal and institutional reforms they learned at the knee of their American mentors, at that very moment they, as the government, would be subject to the same restrictions on U.S. aid as their predecessor—that is, they would become U.S. aid pariahs." The deposed government would suddenly become eligible for American largesse.

Article 907, in short, is a perfect example of the kind of mindless damage that a big, sparsely informed nation like the United Sates can do when its foreign policy is driven by a mixture of ethnic politics, human rights idealism, and public ignorance. That the antidote to this policy in Azerbaijan does not seem to be an outraged electorate but pressure from oil companies, acting out of total self-interest, says more about American policy-making in these post–Cold War days than it does about the Caucasus and its issues.

In the snarls of Caucasian politics, alliances can take shape without planning, and the United States has become the leader of one. The administration's approach has angered Armenia and its Washington lobby, which do not accept that there may be two sides to the Karabakh problem. Russia, in the meantime, has sent arms to Armenia and helped it overcome the blockade. Turkey and Armenia, of course, are ancient enemies and the Turks have helped the Azeri blockade. Iran, the target of U.S. sanctions, needs friends wherever it can find them. The result is a new alignment, with the United States, Azerbaijan, and Turkey on one side, and Russia, Iran, and Armenia, on the other. It seems doubtful that anyone in Washington sat down and planned this, but it exists nonetheless.

The new Great Game

The Caspian drama is just beginning. The overthrow of Ter-Petrossian has put the Minsk Talks back to square one. Political stability in Azerbaijan, or any of the Caucasus, is not exactly a tradition. About the time Armenia's Ter-Petrossian was being overthrown and Azerbaijan's Foreign Minister Hasanov arrested, the president of Georgia, Edward Shevardnadze, barely escaped an assassination attempt. AIOC has just begun to pump its first oil. More contracts are to be let and much more oil is to come.

As important as the oil are the pipelines needed to get it out, and local politics have thoroughly tangled this issue. The first oil is going now through a rickety old pipeline to Novorossiysk, a Black Sea port in Russia, and a new line is being built to Supsa, in Georgia. But neither could begin to carry the 800,000 barrels a day that the Caspian is expected to eventually produce, so a big new line is needed. A decision where to lay this line must be made later this year. Going through Iran is cheapest, but the United States opposes that route and probably has the clout to keep it from happening. Instead, Washington wants a pipeline through Georgia and down to the Turkish Black Sea port of Ceyhan, because this would bypass both Russia and Armenia, and reward Turkey, America's ally.

This angers the Russians and has provoked Armenian threats of sabotage. It also has raised eyebrows among the oil companies because it would cost two or three times as much as a line through Iran. Who, they ask, is going to pay this extra cost? Some government subsidies, Turkish or American or both, may be needed to sweeten the deal.

At the turn of the last century, when the oil companies were carving up the Middle East, the American, British, and Dutch governments saw their role as the companies' diplomatic outriders. What was good for Standard Oil was good for America, so the thinking went, and the State Department did the oilmen's bidding.

This long-time alliance broke down in the 1970s, as the Nixon administration tilted ever more sharply toward Iran in an attempt to balance the power of the Soviets along the Soviet Union's southern flank. The Nixon/Kissinger policy ended badly. The increased power of oil producers, rather than oil companies, enabled the Shah—and then OPEC itself—to double and triple prices overnight. That led to the oil crisis of 1973 and 1979 and, indirectly, to the overthrow of the Shah and the collapse of American hegemony in the region.

We are at another turn of a century, and again State Department policy, at least in that part of the world, is being driven by oil companies. The consequences are unpredictable. A new Great Game, it seems, has just begun.

DEMOGRAPHIC CLOUDS ON CHINA'S HORIZON

by Nicholas Eberstadt

While there is little about China's position in the year 2025 that we can predict with confidence, one critical aspect of China's future can be described today with some accuracy: her population trends. Most of the Chinese who will be alive in 2025, after all, have already been born.

The most striking demographic condition in China today is the country's sparse birth rate. Though most of the population still subsists at Third World levels of income and education, fertility levels are remarkably low—below the level necessary for long-term population replacement, in fact. This circumstance of course relates to the notorious "One Child" policy of China's Communist government, applied with varying degrees of force for nearly two decades.

Ironically, by laboring so ferociously to avoid one set of "population problems"—namely, "overpopulation"—Beijing has helped to ensure that another, even more daunting set of problems will emerge in the decades ahead. Those population problems will be, for Beijing and for the world, utterly without precedent. While impossible to predict their impact with precision, they will impede economic growth, exacerbate social tensions, and complicate the Chinese government's quest to enhance its national power and security.

How can we know fairly well what China's demography will look like 25 years from now? Because according to the latest estimates by the U.S. Bureau of the Census, about a billion of the 1.2 billion Chinese living on the mainland today will still be alive in 2025—accounting for about seven out of every ten of the 1.4 billion Chinese then alive.

The main population wildcard in China's future is fertility. The Census Bureau suggests that the nation's total fertility rate (TFR) now averages a bit under 1.8 births per woman per lifetime (significantly below the 2.1 births necessary for long-term population stability). For broad portions of the Chinese populace, fertility appears to be even lower—as depressed as 1.3 lifetime children per woman in some cities. In Beijing and Shanghai, TFRs may actually have fallen under one by 1995!

The Census Bureau assumes Chinese fertility will average about 1.8 births per woman through 2025. But today's child-bearing takes place under the shadow of the country's severe—and coercive—anti-child campaign. Might not the birth rate leap up if that program were discarded or reversed? It's impossible to be sure, but bits of evidence suggest that a revolution in attitudes about family size has swept China since Mao's death—and that this would prevent fertility from surging back toward more traditional patterns, even if all governmental controls were relaxed. Consequently, the Bureau projects that China will be reaching zero population 25 years from now.

But China's population will look quite different than it does today, as the nearby chart reveals. China in 2025 will have fewer children: The population under 15 years of age is projected to be almost 25 percent smaller than today. The number of people in their late twenties may drop nearly 30 percent. But persons in their late fifties stand to swell in number by over 150 percent, and there will be more people between the ages of 55 and 59 than in any other five-year age span. Persons 65 years or older are likely to increase at almost 3.5 percent a year between now and 2025, accounting for over three-fifths of the country's population growth.

In short, if Census Bureau projections prove correct, China's age structure is about to shift radically from the "Christmas tree" shape so familiar among contemporary populations to something more like the inverted Christmas trees we see out for collection after the holidays. While in 1997 there were about 80 Chinese age 65 or older for every 100 children under age five, by 2025 China would have more than 250 elderly for every 100 preschoolers.

China's coming demographic transformation will bring three sets of serious social problems: rapid aging, declining manpower, and a protracted bride shortage.

China's "graying" will be as swift as any in history. In 1995, the median age in China was just over 27 years. By 2025 it will be about 40.

Although several European nations are already at China's 2025 median age, they got there much more slowly, and with much more societal wealth available to cushion the effects. A similar "graying" over the last four decades in Japan has emerged as an intense concern of Tokyo policymakers, who wonder how the nation is going to manage its growing burden of pensioners. Like the China of 2025, Japan's median age is currently around 40 years. But Japan is vastly richer today than China could hope to be by 2025. Even if its current brisk pace of economic progress continues, China will still be by far the poorest country ever to cope with the sort of old-age burden it will face.

For despite its recent progress, China remains a land of daunting economic disparities and crushing poverty. World Bank research has indicated that the proportion of Chinese suffering "absolute poverty" (defined by the Bank as living on less than $1 a day) was nearly 30 percent from 1981–95. Though that figure is undoubtedly lower today, under almost any plausible scenario for future income growth, China in 2025 will still have hundreds of millions of people with incomes not much different from today's average ones— but with a dramatically older population.

How will such a nation care for its elderly? Under current arrangements, the only social security system for most of the country's poor is their family. Scarcely any public or private pension funds operate in the remote rural areas where the overwhelming bulk of China's poor reside. In 2025, grandparents will by and large be the parents of the "One Child" era, so they will have few offspring to offer them shelter in old age. A small but significant group will have no surviving children.

A fourth or more could have no surviving sons—thus finding themselves, under Chinese culture, in the unenviable position of depending on the largesse of their son-in-law's household, or, worse, competing for family resources against their son-in-law's parents.

Problem two will be declining manpower. Over the past generation, China's brisk economic growth has been partly due to an extraordinary increase in the work force. From 1975 to 1995, China's "working-age population" grew by over 50 percent (or nearly 300 million persons). In a final burst due to population increase, it will grow by over 12 million persons a year at the beginning of the 2000's. Then the growth of potential workers will abruptly brake. By around 2015, China's working-age group will have peaked at just under 1 billion. In 2025, it is projected to be about 10 million persons smaller than a decade earlier. Thereafter, the decline may accelerate, with the workforce shrinking by as many 70 million people over 15 years.

Having fewer workers may complicate China's quest for economic growth. Younger people tend to be better educated than their elders, and as they stream into a work force this increases its average skill. But China's demographic trends will slow down the improvement in education and skill-levels among the working-age population. Today. the rising cohort of 10- to 14-year-olds represents roughly a seventh of the country's working-age population. By 2015 it will be only a twelfth. This presages a sharp slowdown of education-based improvements in labor productivity.

China will not be the first country in the world to wrestle with an aging populace, or a shrinking workforce. But China's third major demographic challenge is unprecedented: a coming imbalance between men and women of marriageable age.

CHINA'S AGE PYRAMID 1995 & 2025

Ages	1995 Males	1995 Females	2025 Males	2025 Females
80+	3	6	10	16
70-79	16	19	43	51
60-69	35	34	78	80
50-59	48	44	115	113
40-49	77	71	94	91
30-39	96	90	112	102
20-29	126	119	93	81
10-19	102	95	92	83
0-9	118	105	79	75

Source: U.S. Bureau of the Census

Beginning with the advent of the nation's "One Child" policy, Chinese sex ratios began a steady, and eerie, rise. By 1995, a Chinese sample census counted over 118 boys under the age of five for every 100 little girls. Part of this imbalance is a statistical artifact—the combination of strict government birth quotas and the strong Chinese preference for sons has caused some parents to hide or "undercount" their newborn daughters, so that they might try again for a boy. But the larger portion of the reported imbalance appears to be real—the consequence of sex-selective abortion and, to a lesser extent, female infanticide.

This tragic imbalance between boys and girls will mean a corresponding mismatch of prospective husbands and brides two decades hence. By Chinese tradition, virtually everyone able to marry does. But the arithmetic of these unnatural sex imbalances is unforgiving, implying that approximately one out of every six of the young men in this cohort must find a bride from outside of his age group—or fail to continue his family line.

In China's past, any problem of "excess" males was generally solved by the practice of marrying a younger bride, which worked well when each new generation was larger than the one before. But with today's low fertility, each new generation in China will typically be smaller than the one before. So if young men try to solve their marriage problem by pairing off with a younger woman, they will only intensify the "marriage crisis" facing men a few years their junior. Nor will searching abroad for a Chinese wife be very promising: By 2020, the surplus of China's twentysomething males will likely exceed the entire female population of Taiwan!

In early modern Europe, bachelorhood was an acceptable social role, and the incidence of never-married men was fairly high. In China, however, there is no such tradition. Unless it's swept by a truly radical change in cultural attitudes toward marriage over the next two decades, China is poised to experience an increasingly intense, perhaps desperate, competition among young men for the nation's limited supply of brides.

A 1997 essay in the journal *Beijing Luntan* predicted direly that "such sexual crimes as forced marriages, girls stolen for wives, bigamy, visiting prostitutes, rape, adultery... homosexuality... and weird sexual habits appear to be unavoidable." Though that sounds overly dramatic, the coming bride shortage is likely to create extraordinary social strains. A significant fraction of China's young men will have to be re-socialized to accept the idea of never marrying and forming their own family. That happens to be a condition in which men often exhibit elevated rates of crime and violence.

Many of China's young men may then be struck by a bitter irony: At a time when (in all likelihood) their country's wealth and power is greater than ever before, their own chances of establishing a family and comfortable future will look poor and worsening. Such a paradox could invite widespread disenchantment.

There is little any future Chinese government will be able to do to address this problem. China's involuntary bachelors will simply have to "handle punishment they have received as a result of... the mistakes of the previous generation," suggests the *Beijing Luntan*. How they will accept this remains to be seen—and it will bear directly on the character and behavior of the China that awaits us.

Nicholas Eberstadt is a visiting scholar at the American Enterprise Institute and a visiting fellow of the Harvard Center for Population and Development Studies.

Unit 4

Unit Selections

24. **Transportation and Urban Growth: The Shaping of the American Metropolis,** Peter O. Muller
25. **Bridge to the Past,** Scott Elias
26. **County Buying Power, 1987–97,** Brad Edmondson
27. **Puerto Rico, U.S.A,** Brad Edmondson
28. **Do We Still Need Skyscrapers?** William J. Mitchell
29. **Indian Gaming in the U.S.: Distribution, Significance and Trends,** Dick G. Winchell, John F. Lounsbury, and Lawrence W. Sommers
30. **For Poorest Indians, Casinos Aren't Enough,** Peter T. Kilborn

Key Points to Consider

❖ Describe the spatial form of the place in which you live. Do you live in a rural area, a town, or a city, and why was that particular location chosen?

❖ How does your hometown interact with its surrounding region? With other places in the state? With other states? With other places in the world?

❖ How are places "brought closer together" when transportation systems are improved?

❖ What problems occur when transportation systems are overloaded?

❖ How will public transportation be different in the future? Will there be more or fewer private autos in the next 25 years? Defend your answer.

❖ How good a map reader are you? Why are maps useful in studying a place?

 **Links**   www.dushkin.com/online/

22. **Edinburgh Geographical Information Systems**
    http://www.geo.ed.ac.uk/home/gishome.html
23. **GIS Frequently Asked Questions and General Information**
    http://www.census.gov/geo/gis/faq-index.html
24. **International Map Trade Association**
    http://www.maptrade.org/
25. **PSC Publications**
    http://www.psc.lsa.umich.edu/pubs/abs/abs94-319.html
26. **U.S. Geological Survey**
    http://www.usgs.gov/research/gis/title.html

These sites are annotated on pages 6 and 7.

Spatial Interaction and Mapping

Geography is the study not only of places in their own right but also of the ways in which places interact. Places are connected by highways, airline routes, telecommunication systems, and even thoughts. These forms of spatial interaction are an important part of the work of geographers.

In "Transportation and Urban Growth: The Shaping of the American Metropolis," Peter Muller considers transportation systems, analyzing their impact on the growth of American cities. "Bridge to the Past" recounts the work of scientists in clearing up the mystery of the age of the Bering Land Bridge. The next two articles from *American Demographics* use the power of the choropleth map to tell their stories. "Do We Still Need Skyscrapers?" questions the need for high density structures in the new era of extensive communications. The article on Indian gaming presents a series of maps to make its points. In the next article, lack of accessibility to population concentrations leaves the Pine Ridge Reservation poor despite its casino, according to Peter Kilborn.

It is essential that geographers be able to describe the detailed spatial patterns of the world. Neither photographs nor words could do the job adequately, because they literally capture too much of the detail of a place. There is no better way to present many of the topics analyzed in geography than with maps. Maps and geography go hand in hand. Although maps are used in other disciplines, their association with geography is the most highly developed.

A map is a graphic that presents a generalized and scaled-down view of particular occurrences or themes in an area. If a picture is worth a thousand words, then a map is worth a thousand (or more!) pictures. There is simply no better way to "view" a portion of Earth's surface or an associated pattern than with a map.

Article 24

Transportation and Urban Growth

The shaping of the American metropolis

Peter O. Muller

In his monumental new work on the historical geography of transportation, James Vance states that geographic mobility is crucial to the successful functioning of any population cluster, and that "shifts in the availability of mobility provide, in all likelihood, the most powerful single process at work in transforming and evolving the human half of geography." Any adult urbanite who has watched the American metropolis turn inside-out over the past quarter-century can readily appreciate the significance of that maxim. In truth, the nation's largest single urban concentration today is not represented by the seven-plus million who agglomerate in New York City but rather by the 14 million who have settled in Gotham's vast, curvilinear outer city—a 50-mile-wide suburban band that stretches across Long Island, southwestern Connecticut, the Hudson Valley as far north as West Point, and most of New Jersey north of a line drawn from Trenton to Asbury Park. This latest episode of intrametropolitan deconcentration was fueled by the modern automobile and the interstate expressway. It is, however, merely the most recent of a series of evolutionary stages dating back to colonial times, wherein breakthroughs in transport technology unleashed forces that produced significant restructuring of the urban spatial form.

The emerging form and structure of the American metropolis has been traced within a framework of four transportation-related eras. Each successive growth stage is dominated by a particular movement technology and transport-network expansion process that shaped a distinctive pattern of intraurban spatial organization. The stages are the Walking/Horsecar Era (pre-1800–1890), the Electric Streetcar Era (1890–1920), the Recreational Automobile Era (1920–1945), and the Freeway Era (1945–present). As with all generalized models of this kind, there is a risk of oversimplification because the building processes of several simultaneously developing cities do not always fall into neat time-space compartments. Chicago's growth over the past 150 years, for example, reveals numerous irregularities, suggesting that the overall metropolitan growth pattern is more complex than a simple, continuous outward thrust. Yet even after developmental ebb and flow, leapfrogging, backfilling, and other departures from the idealized scheme are considered, there still remains an acceptable correspondence between the model and reality.

Before 1850 the American city was a highly compact settlement in which the dominant means of getting about was on foot, requiring people and activities to tightly agglomerate in close proximity to one another. This usually meant less than a 30-minute walk from the center of town to any given urban point—an accessibility radius later extended to 45 minutes when the pressures of industrial growth intensified after 1830. Within this pedestrian city, recognizable activity concentrations materialized as well as the beginnings of income-based residential congregations. The latter was particularly characteristic of the wealthy, who not only walled themselves off in their large homes near the city center but also took to the privacy of horse-drawn carriages for moving about town. Those of means also sought to escape the city's noise and frequent epidemics resulting from the lack of sanitary conditions. Horse-and-carriage transportation en-

24. Transportation and Urban Growth

abled the wealthy to reside in the nearby countryside for the disease-prone summer months. The arrival of the railroad in the 1830s provided the opportunity for year-round daily commuting, and by 1840 hundreds of affluent businessmen in Boston, New York, and Philadelphia were making round trips from exclusive new trackside suburbs every weekday.

As industrialization and its teeming concentrations of working-class housing increasingly engulfed the mid-nineteenth century city, the deteriorating physical and social environment reinforced the desires of middle-income residents to suburbanize as well. They were unable, however, to afford the cost and time of commuting by steam train, and with the walking city now stretched to its morphological limit, their aspirations intensified the pressures to improve intraurban transport technology. Early attempts involving stagecoach-like omnibuses, cablecar systems, and steam railroads proved impractical, but by 1852 the first meaningful transit breakthrough was finally introduced in Manhattan in the form of the horse-drawn trolley. Light street rails were easy to install, overcame the problems of muddy, unpaved roadways, and enabled horsecars to be hauled along them at speeds slightly (about five mph) faster than those of pedestrians. This modest improvement in mobility permitted the opening of a narrow belt of land at the city's edge for new home construction. Middle-income urbanites flocked to these "horsecar suburbs," which multiplied rapidly after the Civil War. Radial routes were the first to spawn such peripheral development, but the relentless demand for housing necessitated the building of cross-town horsecar lines, thereby filling in the interstices and preserving the generally circular shape of the city.

The less affluent majority of the urban population, however, was confined to the old pedestrian city and its bleak, high-density industrial appendages. With the massive immigration of unskilled laborers, (mostly of European origin after 1870) huge blue-collar communities sprang up around the factories. Because these newcomers to the city settled in the order in which they arrived—thereby denying them the small luxury of living in the immediate company of their fellow ethnics—social stress and conflict were repeatedly generated. With the immigrant tide continuing to pour into the nearly bursting industrial city throughout the late nineteenth century, pressures redoubled to further improve intraurban transit and open up more of the adjacent countryside. By the late 1880s that urgently needed mobility revolution

(Library of the Boston Athenaeum)

Horse-drawn trolleys in downtown Boston, circa 1885.

was at last in the making, and when it came it swiftly transformed the compact city and its suburban periphery into the modern metropolis.

The key to this urban transport revolution was the invention by Frank Sprague of the electric traction motor, an often overlooked innovation that surely ranks among the most important in American history. The first electrified trolley line opened in Richmond in 1888, was adopted by two dozen

4 ❖ SPATIAL INTERACTION AND MAPPING

(Library of the Boston Athenaeum)

Electric streetcar lines radiated outward from central cities, giving rise to star-shaped metropolises. Boston, circa 1915.

other big cities within a year, and by the early 1890s swept across the nation to become the dominant mode of intraurban transit. The rapidity of this innovation's diffusion was enhanced by the immediate recognition of its ability to resolve the urban transportation problem of the day: motors could be attached to existing horsecars, converting them into self-propelled vehicles powered by easily constructed overhead wires. The tripling of average speeds (to over 15 mph) that resulted from this invention brought a large band of open land beyond the city's perimeter into trolley-commuting range.

The most dramatic geographic change of the Electric Streetcar Era was the swift residential development of those urban fringes, which transformed the emerging metropolis into a decidedly star-shaped spatial entity. This pattern was produced by radial streetcar corridors extending several miles beyond the compact city's limits. With so much new space available for homebuilding within walking distance of the trolley lines, there was no need to ex-

Before 1850 the American city was a highly compact settlement in which the dominant means of getting about was on foot, requiring people and activities to tightly agglomerate in close proximity to one another.

tend trackage laterally, and so the interstices remained undeveloped. The typical streetcar suburb of the turn of this century was a continuous axial corridor whose backbone was the road carrying the trolley line (usually lined with stores and other local commercial facilities), from which gridded residential streets fanned out for several blocks on both sides of the tracks. In general, the quality of housing and prosperity of streetcar subdivisions increased with distance from the edge of the central city. These suburban corridors were populated by the emerging, highly mobile middle class, which was already stratifying itself according to a plethora of minor income and status differences. With frequent upward (and local geographic) mobility the norm, community formation became an elusive goal, a process further retarded by the grid-settlement morphology and the reliance on the distant downtown for employment and most shopping.

Within the city, too, the streetcar sparked a spatial transformation. The ready availability and low fare of the electric trolley now provided every resident with access to the intracity circulatory system, thereby introducing truly "mass" transit to urban America in the final years of the nineteenth century. For nonresidential activities this new ease of movement among the city's various parts quickly triggered the emergence of specialized land-use districts for commerce, manufacturing, and transportation, as well as the continued growth of the multipurpose central business district (CBD) that had formed after mid-century. But the greatest impact of the streetcar was on the central city's social geography, because it made possible the congregation of ethnic groups in their own neighborhoods. No longer were these moderate-income masses forced to reside in the heterogeneous jumble of row-houses and tenements that ringed the factories. The trolley brought them the opportunity to "live with their own kind," allowing the sorting of discrete groups into their own inner-city social territories within convenient and inexpensive traveling distance of the workplace.

By World War I, the electric trolleys had transformed the tracked city into a full-fledged metropolis whose streetcar suburbs, in the larger cases, spread out more than 20 miles from the metropolitan center. It was at this point in time that intrametropolitan transportation achieved its greatest level of efficiency—that the bustling industrial city really "worked." How much closer the American metropolis might have approached optimal workability for all its residents, however, will never be known because the next urban transport revolution was already beginning to assert itself through the increasingly popular automobile. Americans took to cars as wholeheartedly as anything in the nation's long cultural history. Although Lewis Mumford and other scholars vilified the car as the destroyer of the city, more balanced assessments of the role of the automobile recognize its overwhelming acceptance for what it was—the long-awaited attainment of private mass transportation that offered users the freedom to travel whenever and wherever they chose. As cars came to the metropolis in ever greater numbers throughout the interwar decades, their major influence was twofold: to accelerate the deconcentration of population through the development of interstices bypassed during the streetcar era, and to push the suburban frontier farther into the countryside, again producing a compact, regular-shaped urban entity.

While it certainly produced a dramatic impact on the urban fabric by the eve of World War II, the introduction of the automobile into the American metropolis during the 1920s and 1930s came at a leisurely pace. The earliest flurry of auto adoptions had been in rural areas, where farmers badly needed better access to local service centers. In the cities, cars were initially used for weekend out-

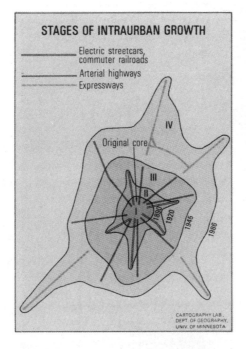

ings—hence the term *"Recreational Auto Era"*—and some of the earliest paved roadways were landscaped parkways along scenic water routes, such as New York's pioneering Bronx River Parkway and Chicago's Lake Shore Drive. But it was into the suburbs, where growth rates were

The ready availability and low fare of the electric trolley now provided every resident with access to the intracity circulatory system, thereby introducing truly "mass" transit to urban America.

now for the first time overtaking those of the central cities, that cars made a decisive penetration throughout the prosperous 1920s. In fact, the rapid expansion of automobile subur-

bia by 1930 so adversely affected the metropolitan public transportation system that, through significant diversions of streetcar and commuter-rail passengers, the large cities began to feel the negative effects of the car years before the auto's actual arrival in the urban center. By facilitating the opening of unbuilt areas lying between suburban rail axes, the automobile effectively lured residential developers away from densely populated traction-line corridors into the suddenly accessible interstices. Thus, the suburban homebuilding industry no longer found it necessary to subsidize privately-owned streetcar companies to provide low-fare access to trolley-line housing tracts. Without this financial underpinning, the modern urban transit crisis quickly began to surface.

The new recreational motorways also helped to intensify the decentralization of the population. Most were radial highways that penetrated deeply into the suburban ring and provided weekend motorists with easy access to this urban countryside. There they obviously were impressed by what they saw, and they soon responded in massive numbers to the sales pitches of suburban subdivision developers. The residential development of automobile suburbia followed a simple formula that was devised in the prewar years and greatly magnified in scale after 1945. The leading motivation was developer profit from the quick turnover of land, which was acquired in large parcels, subdivided, and auctioned off. Understandably, developers much preferred open areas at the metropolitan fringe, where large packages of cheap land could readily be assembled. Silently approving and underwriting this uncontrolled spread of residential suburbia were public policies at all levels of government: financing road construction, obligating lending institutions to invest in new homebuilding, insuring individual mortgages, and providing low-interest loans to FHA and VA clients.

Because automobility removed most of the pre-existing movement

(Boston Public Library)

Afternoon commuters converge at the tunnel leading out of central Boston, 1948.

constraints, suburban social geography now became dominated by locally homogeneous income-group clusters that isolated themselves from dissimilar neighbors. Gone was the highly localized stratification of streetcar suburbia. In its place arose a far more dispersed, increasingly fragmented residential mosaic to which builders were only too eager to cater, helping shape a kaleidoscopic settlement pattern by shrewdly constructing the most expensive houses that could be sold in each locality. The continued partitioning of suburban society was further legitimized by the widespread adoption of zoning (legalized in 1916), which gave municipalities control over lot and building standards that, in turn, assured dwelling prices that would only attract newcomers whose incomes at least equaled those of the existing local population. Among the middle class, particularly, these exclusionary economic practices were enthusiastically

Americans took to cars as wholeheartedly as anything in the nation's long cultural history.

supported, because such devices extended to them the ability of upper-income groups to maintain their social distance from people of lower socioeconomic status.

Nonresidential activities were also suburbanizing at an increasing rate during the Recreational Auto Era. Indeed, many large-scale manufacturers had decentralized during the streetcar era, choosing locations in suburban freight-rail corridors. These corridors rapidly spawned surrounding working-class towns that became important satellites of the central city in the emerging metropolitan constellation. During the interwar period, industrial employers accelerated their intraurban deconcentration, as more efficient horizontal fabrication methods replaced older techniques requiring multistoried plants-thereby generating greater space needs that were too expensive to satisfy in the high-density central city. Newly suburbaniz-

Central City-Focused Rail Transit

The widely dispersed distribution of people and activities in today's metropolis makes rail transit that focuses in the central business district (CBD) an obsolete solution to the urban transportation problem. To be successful, any rail line must link places where travel origins and destinations are highly clustered. Even more important is the need to connect places where people really want to go, which in the metropolitan America of the late twentieth century means suburban shopping centers, freeway-oriented office complexes, and the airport. Yet a brief look at the rail systems that have been built in the last 20 years shows that transit planners cannot—or will not—recognize those travel demands, and insist on designing CBD-oriented systems as if we all still lived in the 1920s.

One of the newest urban transit systems is Metrorail in Miami and surrounding Dade County, Florida. It has been a resounding failure since its opening in 1984. The northern leg of this line connects downtown Miami to a number of low- and moderate-income black and Hispanic neighborhoods, yet it carries only about the same number of passengers that used to ride on parallel bus lines. The reason is that the high-skill, service economy of Miami's CBD is about as mismatched as it could possibly be to the modest employment skills and training levels possessed by residents of that Metrorail corridor. To the south, the prospects seemed far brighter because of the possibility of connecting the system to Coral Gables and Dadeland, two leading suburban activity centers. However, both central Coral Gables and the nearby International Airport complex were bypassed in favor of a cheaply available, abandoned railroad corridor alongside U.S. 1. Station locations were poorly planned, particularly at the University of Miami and at Dadeland—where terminal location necessitates a dangerous walk across a six-lane highway from the region's largest shopping mall. Not surprisingly, ridership levels have been shockingly below projections, averaging only about 21,000 trips per day in early 1986. While Dade County's worried officials will soon be called upon to decide the future of the system, the federal government is using the Miami experience as an excuse to withdraw from financially supporting all construction of new urban heavy-rail systems. Unfortunately, we will not be able to discover if a well-planned, high-speed rail system that is congruent with the travel demands of today's polycentric metropolis is capable of solving traffic congestion problems. Hopefully, transportation policy-makers across the nation will heed the lessons of Miami's textbook example of how not to plan a hub-and-spoke public transportation network in an urban era dominated by the multicentered city.

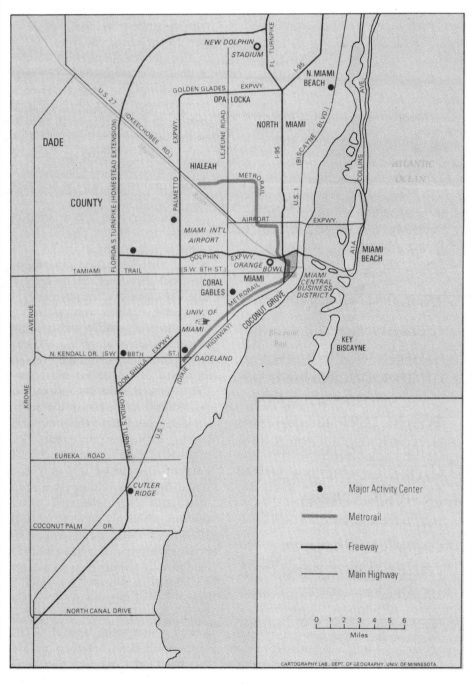

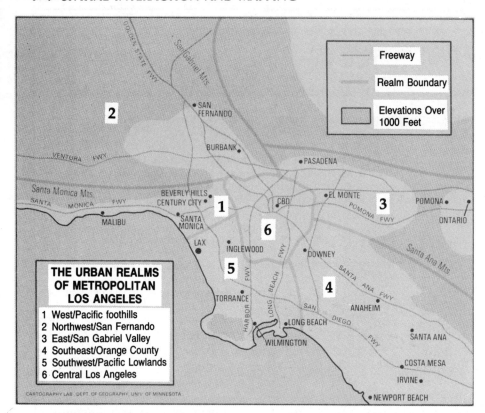

ing manufacturers, however, continued their affiliation with intercity freight-rail corridors, because motor trucks were not yet able to operate with their present-day efficiencies and because the highway network of the outer ring remained inadequate until the 1950s.

The other major nonresidential activity of interwar suburbia was retailing. Clusters of automobile-oriented stores had first appeared in the urban fringes before World War I. By the early 1920s the roadside commercial strip had become a common sight in many southern California suburbs. Retail activities were also featured in dozens of planned automobile suburbs that sprang up after World War I—most notably in Kansas City's Country Club District, where the nation's first complete shopping center was opened in 1922. But these diversified retail centers spread slowly before the suburban highway improvements of the 1950s.

Unlike the two preceding eras, the postwar Freeway Era was not sparked by a revolution in urban transportation. Rather, it represented the coming of age of the now pervasive automobile culture, which coincided with the emergence of the U.S. from 15 years of economic depression and war. Suddenly the automobile was no longer a luxury or a recreational di-

Retail activities were featured in dozens of planned automobile suburbs that sprang up after World War I—most notably in Kansas City's Country Club District, where the nation's first complete shopping center was opened in 1922.

version: overnight it had become a necessity for commuting, shopping, and socializing, essential to the successful realization of personal opportunities for a rapidly expanding majority of the metropolitan population. People snapped up cars as fast as the reviving peacetime automobile industry could roll them off the assembly lines, and a prodigious highway-building effort was launched, spearheaded by high-speed, limited-access expressways. Given impetus by the 1956 Interstate Highway Act, these new freeways would soon reshape every corner of urban America, as the more distant suburbs they engendered represented nothing less than the turning inside-out of the historic metropolitan city.

The snowballing effect of these changes is expressed geographically in the sprawling metropolis of the postwar era. Most striking is the enormous band of growth that was added between 1945 and the 1980s, with freeway sectors pushing the metropolitan frontier deeply into the urban-rural fringe. By the late 1960s, the maturing expressway system began to underwrite a new suburban co-equality with the central city, because it was eliminating the metropolitanwide centrality advantage of the CBD. Now any location on the freeway network could easily be reached by motor vehicle, and intraurban accessibility had become a ubiquitous spatial good. Ironically, large cities had encouraged the construction of radial expressways in the 1950s and 1960s because they appeared to enable the downtown to remain accessible to the swiftly dispersing suburban population. However, as one economic activity after another discovered its new locational flexibility within the freeway metropolis, nonresidential deconcentration sharply accelerated in the 1970s and 1980s. Moreover, as expressways expanded the radius of commuting to encompass the entire dispersed metropolis, residential location constraints relaxed as well. No longer were most urbanites required to live within a short distance of their job: the workplace had now become a locus of opportunity offering access to the best possible residence that an individual could afford anywhere in the urbanized area. Thus, the overall pattern of locally uniform, income-based clusters that had emerged in prewar automobile suburbia was greatly magnified in the Freeway Era, and such new social variables as age and lifestyle produced an ever more balkanized population mosaic.

The revolutionary changes in movement and accessibility introduced during the four decades of the Freeway Era have resulted in nothing less than the complete geographic restructuring of the metropolis. The single-center urban structure of the past has been transformed into a polycentric metropolitan form in which several outlying activity concentrations rival the CBD. These new "suburban downtowns," consisting of vast orchestrations of retailing, office-based business, and light industry, have become common features near the highway interchanges that now encircle every large central city. As these emerging metropolitan-level cores achieve economic and geographic parity with each other, as well as with the CBD of the nearby central city, they provide the totality of urban goods and services to their surrounding populations. Thus each metropolitan sector becomes a self-sufficient functional entity, or *realm*. The application of this model to the Los Angeles region reveals six broad realms. Competition among several new suburban downtowns for dominance in the five outer realms is still occurring. In wealthy Orange County, for example, this rivalry is especially fierce, but Costa Mesa's burgeoning South Coast Metro is winning out as of early 1986.

The new freeways would soon reshape every corner of urban America, as the more distant suburbs they engendered represented nothing less than the turning inside-out of the historic metropolitan city.

The legacy of more than two centuries of intraurban transportation innovations, and the development patterns they helped stamp on the landscape of metropolitan America, is suburbanization—the growth of the edges of the urbanized area at a rate faster than in the already-developed interior. Since the geographic extent of the built-up urban areas has, throughout history, exhibited a remarkably constant radius of about 45 minutes of travel from the center, each breakthrough in higher-speed transport technology extended that radius into a new outer zone of suburban residential opportunity. In the nineteenth century, commuter railroads, horse-drawn trolleys, and electric streetcars each created their own suburbs—and thereby also created the large industrial city, which could not have been formed without incorporating these new suburbs into the pre-existing compact urban center. But the suburbs that materialized in the early twentieth century began to assert their independence from the central cities, which were ever more perceived as undesirable. As the automobile greatly reinforced the dispersal trend of the metropolitan population, the distinction between central city and suburban ring grew as well. And as freeways eventually eliminated the friction effects of intrametropolitan distance for most urban functions, nonresidential activities deconcentrated to such an extent that by 1980 the emerging outer suburban city had become co-equal with the central city that spawned it.

As the transition to an information-dominated, postindustrial economy is completed, today's intraurban movement problems may be mitigated by the increasing substitution of communication for the physical movement of people. Thus, the city of the future is likely to be the "wired metropolis." Such a development would portend further deconcentration because activity centers would potentially be able to locate at any site offering access to global computer and satellite networks.

Further Reading

Jackson, Kenneth T. 1985. *Crabgrass Frontier: The Suburbanization of the United States.* New York: Oxford University Press.

Muller, Peter O. 1981. *Contemporary Suburban America.* Englewood Cliffs, N.J.: Prentice-Hall.

Schaeffer, K. H. and Sclar, Elliot. 1975. *Access for All: Transportation and Urban Growth.* Baltimore: Penguin Books.

BRIDGE to the PAST

The now-sunken land bridge that brought giant animals and people from Asia to the Americas was not what anyone had imagined.

by Scott Elias

Did the first people to arrive in the New World have to get their feet wet, or could they have crossed on dry land? Did they come by boat or on foot?

Such are the questions facing researchers studying the history of the peopling of the New World. Why? Because we know from ancient campsites that American ancestors lived in Siberia. And we know from similar campsites that such people made it to Alaska about 12,000 years ago.

What we don't quite understand is how they made the trip. If they'd come only a few thousand years earlier, answering how they got there would be easier, for the two land masses have not always been sliced apart by the frigid Bering and Chukchi seas as they are today. They were once connected by a swath of land called the Bering Land Bridge. Early immigrants could have walked across.

The trouble was that we thought the land bridge sank beneath the waves 14,500 years ago, 2,500 years *before* immigrants turned up on America's shores. This discrepancy poses some serious dilemmas. Did people really arrive 14,500 years ago? If so, why is there such a big gap in the archaeological record before their tools and fire hearths show up? The answer to these questions lay buried 180 feet beneath the sea.

My colleagues and I found a way to bring the land bridge to the surface—well, at least bits and pieces of it. Working at the University of Colorado's Institute of Arctic and Alpine Research, we looked at sediment cores yanked out of the Bering and Chukchi seabeds by drills connected to ships at the surface. From the cores, we pulled fossil plants, pollen and insects that have enabled us to reconstruct the lost land mass in unprecedented detail.

And the previous dates for the land bridge's drowning, we found, were much too old. Our new dates square with archeological evidence for a 12,000-year-old journey. We also know a lot more about the land that supported this trip. Rather than a broad expanse of grassland that was home to large herds of mammoths, arctic bison, horses and caribou, it now seems the land bridge was rather inhospitable to large grazing animals, so the animals and their predators did not linger there for long.

Imagine that you could start bailing water out of the world's oceans and lower them 400 feet. As the seas retreated across continental shelves worldwide, great swaths of land would appear. Essentially this is what happened 18,000 years ago. At that time, the vast continental ice sheets of the Ice Age held nearly five percent of the planet's water in their icy grasp, enough water to produce this huge dip in sea level—and expose this terrain.

Coastlines were radically altered by this dip, often shifted tens or even hundreds of miles seaward of their present positions. You could have walked from England to France, from the Asian mainland to Indonesia, from Australia to Papua New Guinea—and from eastern Siberia, from the Old World to the New.

One popular misconception about the land bridge is that it was a narrow isthmus between the two continents. In fact, the land bridge encompassed a huge region covering more than 579,000 square miles (about twice the size of Texas). In the midst of the last glaciation, 18,000 years ago, the distance between its northern and southern coasts was 1,125 miles, roughly the distance from Buffalo to Miami. The land bridge formed the center of the huge unglaciated region called Beringia. This included all of eastern Siberia, Alaska and the Yukon Territory, plus the intervening continental shelf regions, which are quite shallow. Beringia was an extremely unusual place: Nearly all the rest of the world's arctic regions were buried by miles of ice during most of the last 2 million years, but Beringia formed an ice-free refuge for arctic plants and animals.

What was this land like? For the past 20 years, researchers thought they had a pretty good idea. The vegetation along Beringia's frozen northern coast was similar, they thought, to that found in modern polar deserts of the high arctic. Imagine a land as dry as the Mojave Desert but as cold as Greenland. There was precious little vegetation growing on this barren landscape; few if any animals could have survived there. But in the south, things seemed different—and warmer.

Abundant remains of large grazing mammals such as woolly mammoths, arctic bison and horses found in Siberia, Alaska and the Yukon have led some scientists to propose that much of central and southern Beringia was a highly productive dry grassland, or steppe. Unlike the water-logged permafrost we see in arctic Alaska today, the landscape scientists envision must have been relatively dry, with deeply thawed soils in summer. These soils are thought to have nurtured a rich vegetation of grasses, herbs and sagebrush, mixed with tundra plants to form a unique vegetation type: steppe-tundra. Herds of grazing mammals would have cropped the grasses, stimulating more grass growth. The ecosystem has been compared to an arctic Serengeti Plain, where an abundant, diverse menagerie of grazing mammals and their predators roamed freely.

More support for the steppe view came from some studies of fossil pollen. Because these microscopic grains are nearly indestructible and are widely dispersed across the land, they provide an excellent record of changes in vegetation over thousands of years. Some areas that were near the center of the land

Imagine a land as dry as the Mojave Desert but as cold as Greenland.

bridge are still above sea level, such as St. Lawrence and St. Paul islands in the Bering Sea. In these areas, scientists found high percentages of sagebrush pollen, one of the dominant plants in the steppe vegetation of central Asia. In one such study, Paul Colinvaux, now at the Smithsonian Institution's Tropical Research Center in Panama, examined the pollen in Late Pleistocene lake sediments and found that cold, dry, steppe-like environments persisted there until 11,000 years ago.

So much for the types of pollen found in the upland regions of Beringia. What about the quantity? Pollen researchers place a great deal of importance on the amount of pollen that falls onto a landscape, as a measure of plant productivity. These researchers pointed out that the small quantities of pollen in typical samples from Beringia suggest sparse vegetation that would not have supported large numbers of grazers. We wanted to learn what the pollen studies, carried out in what were once highlands, could tell us [about] the ecology of the land that lies below the sea. No one had ever studied sediments from the vast lowlands of the land bridge itself.

So that's what we did. To conduct our "fieldwork" on the Bering Land Bridge, we traveled to Menlo Park, California, and the U.S. Geological Survey offices there. In the 1970s and 1980s, survey scientists Hans Nel-

EARTH: Rick Johnson

On the road.
Between Asia and America lie vast frigid seas (shaded areas top and middle) where once there was land. During the Pleistocene, which ended 10,000 years ago, Earth was in the grip of ice sheets holding so much water that sea level dropped. This exposed the Bering Land Bridge (bottom), the route early humans used to reach America from Asia at least 12,000 years ago.

son and Larry Phillips had collected cores from research ships cruising the shallow waters of the Bering and Chukchi seas. While their work had focused on the more remote geologic past of the seabed, the cores remained in storage awaiting studies of the uppermost layers. We chose 20 cores from the central and northern sectors of the land bridge. These cores contained organic-rich sediments laid down during the last 30,000 years of the last ice age.

The cores we studied all had layers containing peat. These organic sediments formed from decaying plant matter in wet depressions and were ideal for trapping the pollen, plant fragments and insects representative of the surrounding land. Susan Short, a pollen expert at the institute, found that all the samples had large amounts of shrub birch and sedge pollen, with lesser amounts of grass pollen and sphagnum moss spores. Shrub birch is a small plant growing no more than a few inches high—not the kind of plant typical of steppe grasslands. Short, in fact, found very little of the sagebrush pollen associated with steppes. Instead, the evidence pointed to shrub tundra with some grasses and sedges growing in moderately moist conditions.

Perhaps the closest modern analog to this type of environment is the North Slope of Alaska. Extending from the northern foothills of the Brooks Range to the coastal plains bordering the Arctic Ocean, this region is characterized by clumps of diminutive willows and birches growing with sedges in the peculiar mound-shaped form called tussocks. These mounds, perhaps two to three feet tall and four to five feet across, make the tundra nearly impossible to cross on foot. Imagine trying to hike over a landscape covered with giant, soggy pillows. Tussock tundra is dominated by plants that are at best poor food for grazers. Some of these species contain toxic compounds. Such vegetation cannot support large herds of grazing mammals for any length of time.

A similar story emerged from plant macro fossils—plant remains visible to the naked eye. Paleobotanist Hilary Birks, at the Botanical Institute of the University of Bergen, Norway, found that they too indicated moderately wet shrub tundra. The plant fragments also added details to our picture of the local habitats where the peats formed: Shallow ponds and lakes were choked with aquatic

A Mammoth (and Bison) Dispute

Land bridge dweller? "Blue Babe," a 36,000-year-old arctic bison found frozen near Fairbanks, Alaska, was so well-preserved some of its meat was still red.

Sea bed cores filled with peat and plant remnants aren't the only remains that testify to the nature of the Bering Land Bridge. There are also remains of animal life. And although the peat points to a largely inhospitable tundra, those who work with fossils of arctic animals say the bones tell a somewhat different story.

Paleontologist Dale Guthrie, of the Institute of Arctic Biology at the University of Alaska, in Fairbanks, studies fossil remains of mammoths, bison and other animals that once roamed the land bridge and the surrounding area. These animals have shown up in the fossil record of hundreds of well-documented sites ranging from England to western Canada. And they've always been associated with a sparse, arid grassland—an ecosystem similar to steppes today in central Asia. These animals were grazers, says Guthrie, and they needed a lot of grasses and sedges to eat; a barren tundra wouldn't do.

And so he doubts that a tundra such as Scott Elias describes (see main story) really did cover the entire land bridge. Indeed, he points out that the mammoth teeth have been dredged from the strait. And anywhere mammoth remains are found, a steppe environment is indicated. "The kind of vegetation that was there was all edible stuff, whereas modern tundra is not. It's virtually all toxic," Guthrie says.

Still, Guthrie does think that the lowest-lying portions of the land bridge could have been tundralike, particularly after about 14,000 years ago. "That is when the wetlands began to form all across the north." By this time, he points out, the Ice Age was drawing to a close, the planet was warming and sea level was rising. Water was reaching inland from the north and the south. The encroaching seas undoubtedly had a strong influence on the weather of the land that once joined the continents, bringing more moisture to the region. "If you look at the modern distribution of tundra," Guthrie notes, "it is a maritime phenomenon."

Perhaps, he suggests, this could explain another puzzle of the land bridge: What kept some animals from making the journey across? "It seems as though there was some sort of barrier out there to species that never made it," Guthrie says. Apparently, arctic bison, woolly mammoths, horses and some other mammals traversed the land bridge easily. Yet on the Siberian side, woolly rhinos have been found near the Bering Strait, but they did not cross. And on the Alaskan side, a number of species including the North American camel, the symbos (a type of musk-ox) and a long-legged ass made it to the Seward Peninsula but failed to traverse to Asia.

"We know there was no real physical barrier," Guthrie says. "It's quite possible that it was just the vegetation that created a barrier. You can make a strong case for that being a rather unique part of the whole steppe."

—Robert Anderson

plants and, above the water table, wet vegetation was dominated by sphagnum and other mosses.

As in the pollen data, no steppe or grassland plant macrofossils were found. My study of fossil insects produced more than two dozen fossil beetle species, whose hard exoskeletons are easily preserved in the peat. All of these turned out to be species found in moderately wet shrub tundra habitats today. None of the characteristic steppe-tundra beetles were found in my samples.

All this evidence from the seabed cores suggests that the land bridge was not a great place for large grazing animals such as mammoths or bison. It probably produced only enough grass to keep migrating herds alive on their way to grassy upland plains. So how did migrating herds of grazers and the predators and human hunting bands following them make it across the land bridge?

We'll never know for certain, but a few scenarios come to mind. The grazers may have had annual migrations from the steppes of Si-

How did migrating herds and hunters make it across the land bridge?

beria to those in Alaska and the Yukon. During their migration, the herds may have kept on the move as they crossed the lowland tundra of the land bridge.

At its narrowest point, the land bridge was only about 70 miles across from east to west. This distance could have been crossed in just a few days by large grazing herds of bison, woolly mammoth and horses. Alternatively, the land bridge may have been crossed in winter, when the whole region was probably snow covered. Perhaps the herds sustained themselves on their trek across the bridge by stopping off at patches of suitable grassy habitats, such as the highland region that is now St. Lawrence Island.

Our fossil plants and insects also gave us the answer to that other Beringian mystery: the date these migrations stopped. The land passage would come to an end when the Ice Age ended, and water from the melting continental ice sheets was released into the world's oceans, raising the sea over the last low-lying areas separating the continents. Prior to our core studies, the land bridge was thought to have been flooded as early as 14,500 years ago. But when we radiocarbon-dated plant and insect remains from the youngest layers of peat from the land bridge, we found that they were only 11,000 years old. The uppermost layers of peat that formed on the land bridge were probably swept away by the oncoming flood of seawater as the land bridge was inundated, so the final deluge must have happened even more recently.

We trust the new dates more than the previous 14,500-year dates because we now use different materials and markedly improved dating methods. The early dates were produced by conventional radiocarbon dating. Radiocarbon is formed in the upper atmosphere and then falls to earth in rain and snow, where it gets taken up by plants and animals. It is unstable, and half of the radiocarbon atoms decay to a stable form within 5,500 years. When a plant or animal dies, it stops taking in radiocarbon, so we can measure how long it has been since an organism died by measuring the amount of radiocarbon left in its remains.

The trouble is that the conventional radiocarbon-dating method requires relatively large samples of 10 to 15 grams, so researchers had to extract this amount of organic material from cores from the seabed. These sediments, however, contained minute particles of coal that were millions of years old, and that's enough ancient material to skew the results. The scientists attempted to compensate for this contamination, but they also warned that their dates should be treated with caution. Nevertheless, these dates were the only ones available for the last two decades, and were quoted and re-quoted in the scientific literature until they took on a mantle of reliability that the original authors never intended.

We benefited from the newer method of radiocarbon dating using an accelerator mass spectrometer. Its main advantage is that we can use small samples, less than a tenth of a gram, because the accelerator method counts the radioactive and stable carbon atoms individually. The seeds and insect parts that we picked out of our samples were sufficient, so we didn't have to use coal-contaminated sediments. Our new dates fit the generally accepted reconstruction of global sea level rise at the end of the last ice age.

The new dates also fit a lot better with the archeological evidence for the first humans in North America. The oldest human occupation on eastern Asia's Kamchatka Peninsula has been dated at 14,300 years ago, and another site near the Siberian arctic coast was occupied by people as early as 13,400 years ago. On the other side of the present Bering Strait, in Alaska, archeologists have found a few places where these migrants stopped to camp for a few days or weeks. They built stone fire circles and spent some time chipping stones to refurbish their tool kits.

One such "Paleoindian" camp that archaeologists Michael Kuntz and Richard Reanier recently found on the Alaskan North Slope is called the Mesa Site. Spear heads and darts were scattered about the camp along with bits of charcoal which yielded radiocarbon dates between 9,700 and 11,700 years ago. Other Paleoindian sites in Alaska were no more than 12,000 years old.

The evidence at the Mesa Site was a problem when scientists thought the land bridge had flooded more than 14,500 years ago. Archeologists were faced with the problem of finding older camps in North America or explaining—somehow—how people made the trek across in the absence of the land bridge. Now that we know the land bridge flooded soon after 11,000 years ago, everything fits; the first Americans moved across the land bridge during the warm period shortly before it was inundated.

The projectile points found at the Mesa Site are particularly important because they closely resemble some Paleoindian points found in the lower 48 states. These distinctive, fluted points, first discovered near Clovis, New Mexico, are generally believed to be the oldest stone tools from sites south of receding ice sheets; they date from 11,200 to 10,900 years ago. Other fluted projectile points, known as Folsom and Goshen points, date to 11,000 years ago.

These Mesa Site points, therefore, are a crucial link, tying together the history of Paleoindian migrations across the land bridge and south into unglaciated North America. By about 12,000 years ago, the vast Laurentide Ice Sheet had retreated eastward from the eastern slope of the Canadian Rockies. At the same time, the Cordilleran ice had withdrawn westward from the Rockies. This opened up an ice-free corridor connecting Alaska and the Yukon with the unglaciated south.

It wasn't the Promised Land, but to the northern hunters it must have looked pretty good. After all, the southern lands had trees. Perhaps hundreds of generations of people from Siberia and Alaska had never seen a tree, much less used tree limbs and trunks for fuel. They had managed to keep from freezing to death during long, arctic winters in a landscape where the best fuel for fires had been the pitiful little twigs of dwarf birch, willow, and arctic sage.

Article 26

County Buying Power, 1987-97

by Brad Edmonson/Map by Josh Galper

America's biggest cities lost some of their buying power in the last decade, according to Market Statistics estimates of consumer income. Suburbs and small Sunbelt cities picked up the gains.

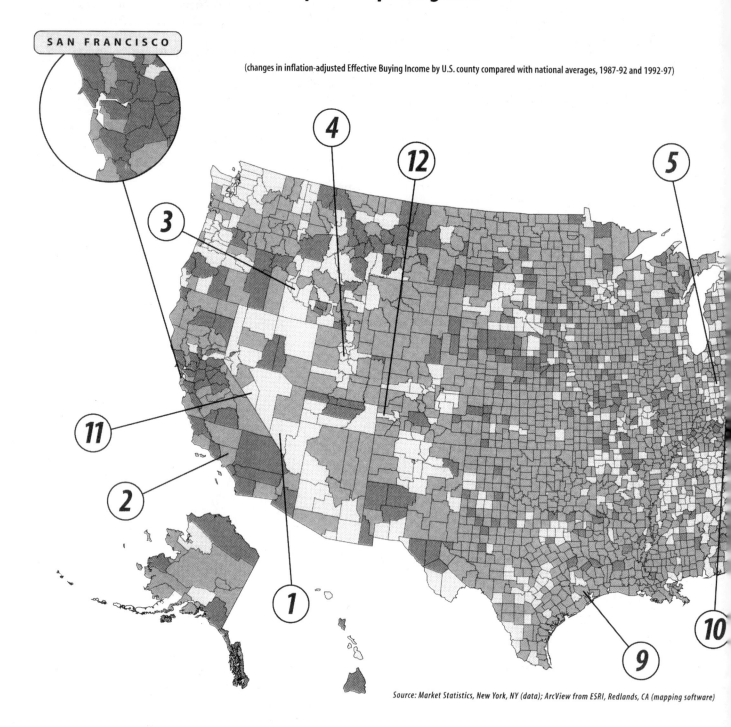

(changes in inflation-adjusted Effective Buying Income by U.S. county compared with national averages, 1987-92 and 1992-97)

Source: Market Statistics, New York, NY (data); ArcView from ESRI, Redlands, CA (mapping software)

Reprinted with permission from *American Demographics*, August 1998, pp. 28-29. © 1998 by American Demographics, Inc., Ithaca, NY. For subscription information, please call (800) 828-1133.

26. County Buying Power, 1987–1997

Think of the U.S. consumer market as one big economic pie, with America's 3,135 counties as the pieces. The entire pie grew 40 percent in size between 1987 and 1997, according to inflation-adjusted estimates of after-tax income. But the slice given to **(1) Las Vegas** (Clark County, NV) grew 172 percent, while the slice for **(2) Los Angeles County**, CA increased only 17 percent. Las Vegas's share of the national market grew between 1987 and 1992, then grew again between 1992 and 1997. Los Angeles is a two-time loser.

This map is based on Market Statistics' county-by-county estimates of "Effective Buying Income (EBI)," an exclusive measure of money income (wages and salaries; self-employment income; interest; dividends; rent; royalties; retirement payments; and welfare checks) minus personal tax payments (federal, state, and local income taxes; Social Security and other federal retirement payroll deductions; and residential property taxes). In the nation's hot spots, EBI has grown faster than the national average in both 1987–92 (26 percent) and 1992–97 (12 percent). In "turnaround" counties, EBI growth was slower than the national average in 1987–92, but faster than average in 1992–97. In "slowdown" counties, EBI growth was faster than average in 1987–92, but slower in 1992–97. And in "two-time losers," growth in EBI has been slower than the national average in both periods. Although the map shows every county in the U.S., those mentioned in this article are restricted to the 654 where aggregate EBI exceeded $1 million in 1997.

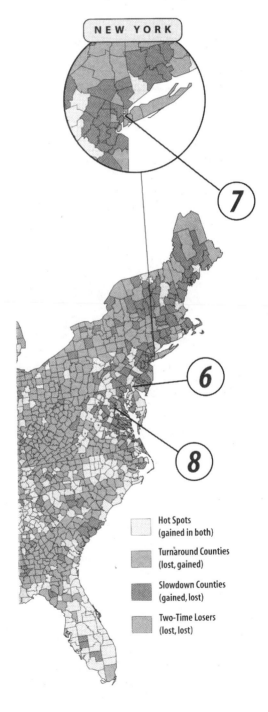

Hot Spots (gained in both)
Turnaround Counties (lost, gained)
Slowdown Counties (gained, lost)
Two-Time Losers (lost, lost)

The hottest counties in the U.S. these days are Rocky Mountain boom towns like **(3) Boise** (Ada County, ID) and **(4) Salt Lake**, UT. But the hot list is also dotted with outer-edge suburban counties like **(5) Hamilton**, IN, north of Indianapolis, and smaller eastern metros like **(6) Wilmington**, DE (New Castle County). It also includes the biggest downtown of them all, **(7) Manhattan** (New York County, NY). The other four boroughs of New York City are two-time losers.

> SOUTHERN CALIFORNIA SUFFERED A HUGE LOSS OF BUYING INCOME IN THE 1990s.

While Los Angeles lost share in both periods, buying income in other southern California counties grew faster than the U.S. in 1987–92. But the entire state hit the skids in 1992–97. Only two California counties gained share in both periods, and both were along the Interstate 80 corridor that leads to the nation's center of growth: Nevada. Slowdown counties also include many of the most affluent suburbs. That's why so many upscale retailers in places like **(8) Fairfax**, VA have been struggling.

The sharpest turnaround in the country happened in central **(9) Houston** (Harris County, TX), which lagged slightly behind national growth in 1987–92 (23 percent) but shot ahead in 1992–97 (18 percent). Most of the nation's turnaround counties are in the Ohio Valley and Great Plains, which were the first to recover from the recession of the early 1990s. In **(10) Louisville** (Jefferson County, KY), growth in EBI was 20 percent in 1987–92. Although it slowed to 14 percent in 1992–97, that was still faster than the nation as a whole.

Effective Buying Income is a shorthand way of measuring a market's potential, but the numbers may be misleading unless they are viewed in context. For example, EBI more than tripled in **(11) Esmeralda County**, NV over the last decade, as jobs were created in aerospace and mining. It also tripled in **(12) San Miguel County**, CO, as the rich and famous flocked to the ski slopes of Telluride. But Esmeralda has a total population of about 1,500 people, and San Miguel has just over 5,000 permanent residents. These days, the biggest boom towns are in the boondocks. In the real markets, slow growth is the rule.

Article 27

Puerto Rico, U.S.A.

by Brad Edmondson/Map by Josh Galper

After a century under U.S. rule, Puerto Rico still retains the culture of a West Indian island nation. But it has become a $12.2 billion market for U.S. goods with a strong middle class, and it may become the 51st state.

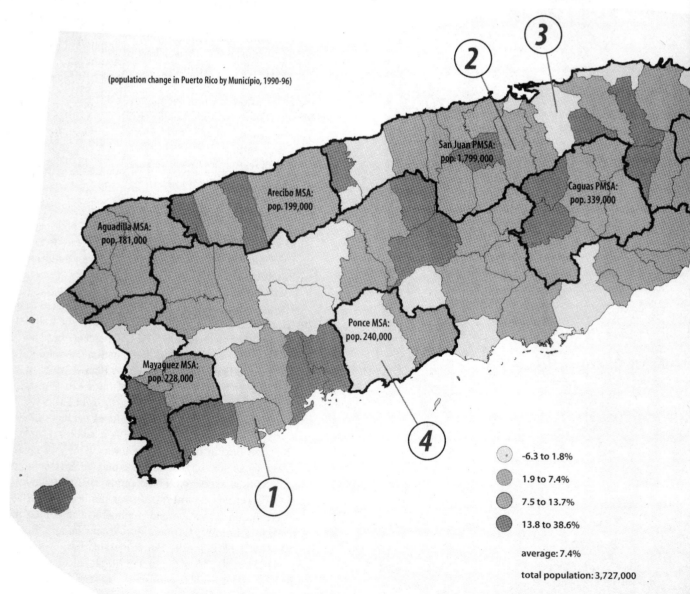

27. Puerto Rico, U.S.A.

This month, a celebration will be held on the beach in **(1) Guanica, Puerto Rico** (1996 population: 21,000) to commemorate the landing of American troops there on July 25, 1898. Spain ceded the island to the U.S. a century ago, but Puerto Rico is still caught between two worlds. It is a U.S. Commonwealth, but it has the culture of a West Indian island. It is a place where wealthy American tourists encounter vendors and beggars in a tropical port of call, and everyone in the scene is a U.S. citizen. Puerto Rico is also a notable market. U.S. businesses shipped $12.2 billion in goods to the island in 1996, and Puerto Ricans are the nation's second-largest Hispanic subgroup.

This map shows the change in population between 1990 and 1996 for Puerto Rico's 76 municipios, the equivalent of U.S. counties. It also shows the boundaries and 1996 population totals for the island's six metropolitan areas, which are home to 80 percent of its residents. Puerto Rico is densely populated: it is smaller in area than Connecticut (3,500 square miles, compared with 4,900), but it has more people (3,753,000, compared with 3,274,000). If Puerto Rico were to become the 51st state—a question that could soon be decided by island residents—it would be the 26th largest in population. Puerto Rico already uses the same currency as U.S. states do, follows many of the same laws, and has many of the same sources of marketing information.

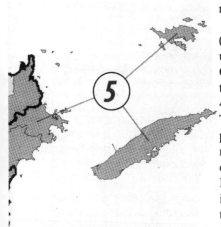

Along Highway 2 in the municipio of **(2) Bayamon** (pop. 232,000), are many of the same names that pop up on any big-city suburban strip: Taco Bell, Pizza Hut, Chase Manhattan, KFC, and Citicorp. In the tourist district of **(3) Old San Juan** (San Juan municipio, pop. 434,000), the names are Coach, Ralph Lauren, and Tommy Hilfinger. Puerto Rico was once known as "the poorhouse of the Caribbean," and it is still poor by mainland U.S. standards: in the 1990 census, 59 percent of residents had incomes below the poverty level. But Puerto Rico now has a healthy middle class, thanks in large part to a law that attracts U.S. businesses to the island by exempting them from federal taxes.

> **PUERTO RICANS ARE AMERICA'S SECOND-LARGEST HISPANIC GROUP.**

Middle-class homes and cars line suburban-style streets in the Bucana barrio of **(4) Ponce** (pop. 190,000), where tourist dollars flow freely at a Hilton casino. Another reason for the rapid rise in living standards is a rapid rise in education. The share of island residents aged 20 to 24 who are enrolled in college has risen from 18 percent in 1965 to nearly 60 percent today. The University of Puerto Rico's Ponce campus, founded in 1962, grants degrees in 23 fields. Still, higher wages in the states draw Puerto Ricans away. In 1991, median family income was about $10,000 on the island and $20,700 for Puerto Ricans on the mainland.

The mainland's influence on the island is easy to see in **(5) Ceiba** (pop. 17,700), where about 3,200 sailors are stationed at the Roosevelt Roads Naval Reservation. It is also easy to see on cable television in San Juan, which carries four stations from New York City. The eastern United States is to Puerto Rico what the southwestern United States is to Mexico: more than half (52 percent) of the 2,730,000 Puerto Ricans on the mainland live in New York or New Jersey, including about 500,000 in the Bronx. Another 9 percent live in Florida, 6 percent each in Pennsylvania and Massachusetts, and 5 percent each in Connecticut, Illinois, and California.

Some Puerto Ricans want to erase their remaining differences with the U.S. by becoming the 51st state. Earlier this year, the House of Representatives approved a bill that would allow residents to decide the island's status in a binding vote. The bill has not passed the Senate, however, and it may succumb to the same ambivalence many Puerto Ricans feel toward the mainland. "Puerto Rico treasures and defends its own identity," said Miguel Hernandez-Agosto, a senior statesman among Puerto Rico politicians, in *The New York Times*. "But it also treasures and defends its U.S. citizenship."

Article 28

Do We Still Need Skyscrapers?

**The Industrial Revolution made skyscrapers possible.
The Digital Revolution makes them (almost) obsolete**

by William J. Mitchell

Our distant forebears could create remarkably tall structures by exploiting the compressive strength of stone and brick, but the masonry piles they constructed in this way contained little usable interior space. At 146 meters (480 feet), the Great Pyramid of Cheops is a vivid expression of the ruler's power but inside it is mostly solid rock; the net-to-gross floor area is terrible. On a square base of 230 meters, it encloses the King's Chamber, which is just five meters across. The 52-meter spiraling brick minaret of the Great Mosque of Samarra does not have any interior at all. And the 107-meter stone spires of Chartres Cathedral, though structurally sophisticated, enclose nothing but narrow shafts of empty space and cramped access stairs.

The Industrial Revolution eventually provided ways to open up the interiors of tall towers and put large numbers of people inside. Nineteenth-century architects found that they could achieve greatly improved ratios of open floor area to solid construction by using steel and reinforced concrete framing and thin curtain walls. They could employ mechanical elevators to provide rapid vertical circulation. And they could integrate increasingly sophisticated mechanical systems to heat, ventilate and cool growing amounts of interior space. In the 1870s and 1880s visionary New York and Chicago architects and engineers brought these elements together to produce the modern skyscraper. Among the earliest full-fledged examples were the Equitable Building (1868–70), the Western Union Building (1872–75) and the Tribune Building (1873–75) in New York City, and Burnham & Root's great Montauk Building (1882) in Chicago.

These newfangled architectural contraptions found a ready market because they satisfied industrial capitalism's growing need to bring armies of office workers together at locations where they could conveniently interact with one another gain access to files and other work materials, and be supervised by their bosses. Furthermore, tall buildings fitted perfectly into the emerging pattern of the commuter city, with its high-density central business district, ring of low-density bedroom suburbs and radial transportation systems for the daily return journey. This centralization drove up property values in the urban core and created a strong economic motivation to jam as much floor area as possible onto every available lot. So as the 20th century unfolded, and cities such as New York and Chicago grew, downtown skylines sprouted higher while the suburbs spread wider.

But there were natural limits to this upward extension of skyscrapers, just as there are constraints on the sizes of living organisms. Floor and wind loads, people, water and supplies must ultimately be transferred to the ground, so the higher you go, the more of the floor area must be occupied by structural supports, elevators and service ducts.

Great Pyramid of Cheops
Built circa 2600 B.C.
Height 146 meters
Egypt

Minaret of Samarra
Built 9th century
Height 52 meters
Iraq

Chartres Cathedral
Built 13th century
Height 107 meters
France

Equitable Building
Built 1870
Height 43 meters
New York

Western Union Building
Built 1875
Height 70 meters
New York

Tribune Building
Built 1875
Height 79 meters
New York

Chrysler Building
Built 1930
Height 319 meters
New York

28. Do We Still Need Skyscrapers?

At some point, it becomes uneconomical to add additional floors; the diminishing increment of usable floor area does not justify the increasing increment of cost.

Urban planning and design considerations constrain height as well. Tall buildings have some unwelcome effects at ground level; they cast long shadows, blot out the sky and sometimes create dangerous and unpleasant blasts of wind. And they generate pedestrian and automobile traffic that strains the capacity of surrounding streets. To control these effects, planning authorities typically impose limits on height and on the ratio of floor area to ground area. More subtly, they may apply formulas relating allowable height and bulk to street dimensions—frequently yielding the stepped-back and tapering forms that so strongly characterize the Manhattan skyline.

The consequence of these various limits is that exceptionally tall buildings—those that really push the envelope—have always been expensive, rare and conspicuous. So organizations can effectively draw attention to themselves and express their power and prestige by finding ways to construct the loftiest skyscrapers in town, in the nation or maybe even in the world. They frequently find this worthwhile, even when it does not make much immediate practical sense.

There has, then, been an ongoing, century-long race for height. The Chrysler Building (319 meters) and the Empire State Building (381 meters) battled it out in New York in the late 1920s, adding radio antennas and even a dirigible mooring mast to gain the last few meters.

The contest heated up again in the 1960s and 1970s, with Lower Manhattan's World Trade Center twin towers (417 meters), Chicago's John Hancock tower (344 meters) and finally Chicago's gigantic Sears Tower (443 meters). More recently, Cesar Pelli's skybridge-linked Petronas Twin Towers (452 meters) in Kuala Lumpur have—for a while at least—taken the title of world's tallest building.

Along the way, there were some spectacular fantasy entrants as well. In 1900 Desiré Despradelle of the Massachusetts Institute of Technology proposed a 457-meter "Beacon of Progress" for the site of the Chicago World's Fair; like Malaysia's Petronas Towers of almost a century later, it was freighted with symbolism of a proud young nation's aspirations. Despradelle's enormous watercolor rendering hung for years in the M.I.T. design studio to inspire the students. Then, in 1956, Frank Lloyd Wright (not much more than five feet in his shoes and cape) topped it with a truly megalomaniac proposal for a 528-story, mile-high tower for the Chicago waterfront.

While this race has been running, though, the burgeoning Digital Revolution has been reducing the need to bring office workers together face-to-face, in expensive downtown locations. Efficient telecommunications have diminished the importance of centrality and correspondingly increased the attractiveness of less expensive suburban sites that are more convenient to the labor force. Digital storage and computer networks have increasingly supported decentralized remote access to data bases rather than reliance on cen-

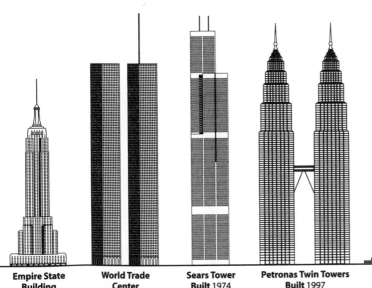

Empire State Building Built 1931 Height 381 meters New York

World Trade Center Built 1972 Height 417 meters New York

Sears Tower Built 1974 Height 443 meters Chicago

Petronas Twin Towers Built 1997 Height 452 meters Kuala Lumpur, Malaysia

Microsoft HQ Started in 1986 Height 20 meters Redmond, Wash.

Beacon of Progress Proposed 1900 Never built Height 457 meters Planned for Chicago

Mile High Tower Proposed 1956 Never built Height 1,609 meters Planned for Chicago

tralized paper files. And businesses are discovering that their marketing and public-relations purposes may now be better served by slick World Wide Web pages on the Internet and Superbowl advertising spots than by investments in monumental architecture on expensive urban sites.

We now find, more and more, that powerful corporations occupy relatively unobtrusive, low- or medium-rise suburban office campuses rather than flashy downtown towers. In Detroit, Ford and Chrysler spread themselves amid the greenery in this way—though General Motors has bucked the trend by moving into the lakeside Renaissance Center. Nike's campus in Beaverton, Ore., is pretty hard to find, but www.nike.com is not. Microsoft and Netscape battle it out from Redmond, Wash., and Mountain View, Calif., respectively, and—though their logos, the look and feel of their interfaces, and their Web pages are familiar worldwide—few of their millions of customers know or care what the headquarters buildings look like. And—a particularly telling straw in the wind—Sears has moved its Chicago workforce from the great Loop tower that bears its name to a campus in far-suburban Hoffman Estates.

Does this mean that skyscrapers are now dinosaurs? Have they finally had their day? Not quite, as a visit to the fancy bar high atop Hong Kong's prestigious Peninsula Hotel will confirm. Here the washroom urinals are set against the clear plate-glass windows so that powerful men can gaze down on the city while they relieve themselves. Obviously this gesture would not have such satisfying effect on the ground floor. In the 21st century, as in the time of Cheops, there will undoubtedly be taller and taller buildings, built at great effort and often without real economic justification, because the rich and powerful will still sometimes find satisfaction in traditional ways of demonstrating that they're on top of the heap.

WILLIAM J. MITCHELL is dean of the School of Architecture and Planning at the Massachusetts Institute of Technology.

INDIAN GAMING IN THE U.S.

DISTRIBUTION, SIGNIFICANCE AND TRENDS

Dick G. Winchell, John F. Lounsbury and Lawrence M. Sommers

Introduction: The widespread distribution of Indian gaming

Legalized gaming is the most rapidly growing industry in the United States in recent years. Close to $600 billion was wagered in 1996, an increase of over 54 percent from 1993. Although some forms of gaming showed a decrease—charitable bingo, greyhound racing and jai alai—there was a significant increase in wagering in casinos, lotteries and on Indian reservations. Indian gaming increased over 125 percent in this three year period compared to a wagering increase in casinos of 48 percent. Although recent entrants to the gaming industry, Indian bingo and more recently tribal casinos were developed rapidly during the 1980s and '90s. In 1993, Indian gaming accounted for $28.96 billion or 7.4 percent of the total money wagered, and grew to $65.18 billion or 11.1 percent by 1996 as shown in Table 1, "Money Wagered on Various Types of Legal Gambling." This growth has been controversial in a number of states because it has been approved only for Indian lands without the potential for development off-reservation, causing complaints from many non-Indian public and private interests. The spread of Indian gaming has led to consideration of legalization of gambling activities for casinos in many states throughout the nation, and has often brought states into conflict with the tribes.

In 1997, 142 Indian tribes or communities operated 281 casinos and bingo halls in 24 states. The widespread distribution of Indian gaming includes major tribal casinos in the east as well as those in the west, where there are more reservations. There are many more tribes preparing to enter into negotiations with the states or actually constructing facilities for casinos based on the success of current tribal casinos. The rapid expansion of Indian gaming dates back to 1988 when the Indian Gaming Regulatory Act (IGRA) was passed by Congress. This article examines the distribution, significance, historical evolution and major issues in recent Indian gaming growth in the U.S. Tribal gaming in the states of Washington, Arizona, Michigan and Connecticut will be used to illustrate the characteristics and selected issues resulting from this trend toward increased Indian gaming.

Background: Expansion of gaming in the United States

The increase nationally in gaming activities is seen in the gross revenues in 1996—what's left over after payouts and prizes—which was $41.8 billion, an increase of 20 percent in 3 years. In 1996, gross gaming revenues were $47.6 billion, about the same as the total amount of money spent on all the following major forms of entertainment combined—recorded music ($12.5 billion), video games ($7.1 billion), theme parks ($7.2 billion), movies ($5.9 billion), spectator sports ($6.3 billion), and non-sport live entertainment ($9.2 billion). Once con-

Jamestown Clallam tribe totem Sequim, Wash.
Photo: D. Winchell

Table 1

Money Wagered on Various Types of Legal Gambling

Activity	1996 (in billions)	Percent of Total
Casinos	$438.68	74.8
Indian Reservations	$65.18	11.1
Lotteries	$42.93	7.3
Horse Racing	$14.10	2.4
Card Rooms	$9.86	1.7
Charitable Games	$5.68	1.0
Charitable Bingo	$4.04	0.7
Bookmaking	$2.61	0.5
Greyhound Racing	$2.31	0.4
Jai Alai	$0.24	0.1
Totals	**$586.52**	**100.0**

Source: International Gaming and Wagering Business Magazine, "1996 Gross Annual Wager," August, 1997, p. 8.

sidered an immoral and reckless activity, gaming is now increasingly seen as an acceptable form of recreation, and a potential source of tax revenue for state governments. Further, as the type of hotels and theme parks built recently in Las Vegas indicate, attempts are being made to provide entertainment for the whole family.

At this time, casino gaming far surpasses other forms of legalized gaming, accounting for about three-fourths of the total dollars wagered. Gaming in 1996 employed nearly 350,000 workers in the ten largest gaming states, including an estimated 42,500 employed in Indian gaming. Lotteries now operate in 37 states and the District of Columbia. Collectively, they generate millions of dollars in revenue which is generally earmarked for education in state budgets. Many states hungry for revenue and afraid of higher taxes favor lotteries as an easier alternative. In some states, such as Michigan, competition by casinos has decreased the lottery intake.

The amount of money wagered varies greatly from one state to another as shown in Table 2 "Gaming by State—1996." This variation is the result of differences in gaming regulations, the nature of gaming facilities and the population of the state and the region. In 1996, there were 31 states in which gross wagering exceeded $1 billion. In order of amounts wagered, the top five were Nevada, New Jersey, Mississippi, Louisiana and Illinois. Nevada has had well-established full-scale casinos since 1933, and New Jersey since 1978. Las Vegas, Reno, Lake Tahoe and Laughlin in Nevada, and Atlantic City in New Jersey have elaborate casino/resort hotels that attract people from all parts of the country and abroad, while other "boom towns" have grown rapidly in Nevada as discussed in a 1991 *Focus* article by two of the present authors. The Mississippi Gaming Control Act was enacted in 1990, and nine counties have subsequently approved dockside casino gaming. Thirty casinos operated in 1997 with a yearly total gross revenue of nearly $2 billion. These casinos are drawing people from all over the state and adjacent states as well. Illinois and California, with popular race tracks and lotteries in existence for many years, have large populations from which to obtain customers.

The states in which gross wagering is less than $100 million are Hawaii, Tennessee and Utah, which had no legal gaming in 1996, and North Carolina, Vermont and Wyoming. The latter states have small populations or do not have elaborate gaming complexes at this time.

Indian gaming historical background

There are over 500 American Indian Tribes which are officially recognized by the United States Government, with extremes in size from those with no tribal land base to the Navajo Nation with approximately 25,000 square miles; and extremes in population from tribes with less than ten tribal members to over 150,000 Navajo tribal members. Generally, Indian reservations were established in rural areas, and in recent decades the expansion of urban areas, especially in the west, has resulted in many reservations now being surrounded by or adjoining cities including San Diego and Palm Springs, California; Phoenix and Tucson, Arizona; and Seattle, Washington. For the most part, reservation economies suffer from persistent poverty, and lacking independent tax bases, tribal governments have remained dependent upon federal resources.

To alleviate this poverty, to try to build the local tribal economies and tribal governments, and to respond to the sovereignty of tribes to control their own communities, several tribes sought to introduce gambling as tribal enterprises in the 1960s and 1970s. By the 1980s several Supreme Court rulings provided tribes with the rights to establish bingo operations, where legal for other organizations by state law. Efforts by tribes to expand these operations resulted in the passage of the Indian Gaming Regulatory Act (IGRA) in 1988, which permits tribes to operate a wide range of gaming operations including casinos with some restrictions.

Many American Indian tribes had gambling activities as integral functions of community interaction prior to European contact. Stick games, foot races, and ball games are just a few of the traditional Indian gambling activities which often provided social opportunities for interaction and exchange. During the period of contact with Europeans, many traditional tribal members were quick to adopt European forms of gambling, especially dice games. Europeans also introduced the non-Indian moral attitude that gambling is evil, especially when done by Indians. Laws were passed to control gambling, and this argument is still used to try to restrict or limit Indian gaming "for their own good."

Not only gambling but even those traditional ceremonials which involved giveaways, such as the potlatches of the American Northwest tribes, were made illegal in the reservation period of the late 1800s up to the 1960s. During the 1960s and 1970s some tribes began to openly carry out traditional gaming activities, as well as bingo operations for profit. By the late 1970s a number of tribal governments operated bingo facilities in states where such activities were legal for other organizations, and some developments, such as the Seminole Bingo Hall in Florida and the Cabazon Band facility near San Diego, established high stakes bingo for non-Indian markets. When the states tried to stop these two operations, the courts ruled in favor of the tribes.

The Supreme Court left standing a lower court in the Seminole case which established that, according to the *Congressional Quarterly Almanac*, "Florida ...and by implication other states... could not regulate bingo on Indian reservations if the game was legal elsewhere in the state." This was reinforced by a decision on the Cabazon Band when the legality of tribal high stakes

Table 2
Gaming by State—1996

Gross wagering in the USA (amount players spent on gaming), and gross revenues (the amounts wagered minus the winnings returned to the players) in 1996[1] (in millions).

State	Wagering	Revenue
Ala.	264.0	63.7
Alaska	264.9	58.1
Ariz.	760.4	219.6
Ark.	267.9	55.8
Calif.	14,416.2	2,393.2
Colo.	8,003.1	654.1
Conn.	1,134.7	406.6
Del.	2,169.3	284.2
D.C.	234.0	117.9
Fla.	3,986.9	1,510.7
Ga.	1,713.8	860.5
Idaho	145.9	49.1
Ill.	26,099.2	2,312.3
Ind.	8,050.6	788.8
Iowa	11,835.3	773.8
Kan.	367.6	128.0
Ky.	1,689.3	501.3
La.	28,667.5	2,070.9
Maine	272.4	94.6
Md.	2,318.6	739.4
Mass.	3,783.5	1,121.6
Mich.	2,254.9	865.8
Minn.	1,807.8	417.4
Miss.	35,075.1	1,894.4
Mo.	11,748.8	831.3
Mont.	2,328.8	249.3
Neb.	493.3	137.2
Nev.	222,253.9	7,452.8
N.H.	477.7	135.9
N.J.	89,412.0	4,860.8
N.M.	220.1	64.4
N.Y.	6,535.2	2,431.3
N.C.	34.7	8.4
N.D.	551.3	67.2
Ohio	3,677.4	1,347.3
Okla.	278.1	61.2
Ore.	3,871.8	540.1
Pa.	2,959.9	1,141.1
R.I.	1,221.7	193.6
S.C.	1,446.5	419.3
S.D.	2,778.4	241.0
Texas	4,739.4	1,821.5
Vt.	90.0	34.5
Va.	1,151.3	489.0
Wash.	1,554.0	480.7
W.Va.	1,131.0	185.4
Wis.	664.2	247.1
Wyo.	29.2	7.6
Total	**514,234.5**	**41,829.9**

[1]Does not include gaming on oceangoing cruise ships or on Indian reservations. Note: Hawaii, Tennessee and Utah have no legal gambling.

Source: International Gaming & Wagering Business. "1996 Gross Annual Wager." August, 1997, p. 20, 39.

bingo was established in *California v. Cabazon Band of Mission Indians*. As summarized in the *Congressional Quarterly Almanac*:

> Thus, after the 1987 Cabazon ruling, "Games on the reservation still had to abide by federal law and by state criminal laws. If a particular form of gambling, such as *roulette*, was prohibited altogether under state law it was also illegal on Indian reservations. But 45 states permitted bingo, while regulating where the games could be played and what prizes could be offered. On a reservation, those restrictions did not apply. The state could ban bingo outright, but then other groups that used the games for fund raising would be shut out as well.

After the Florida decision in 1982, many tribes started to develop bingo operations, and by 1986 over 100 tribes operated bingo facilities.

IGRA: Indian Gaming Regulatory Act

The Indian Gaming Regulatory Act of 1988 (PL. 1000–497) was an attempt to reach a compromise between state efforts to restrict tribal gambling, particularly Class III gaming, and tribal efforts to seek gaming operations as a means of badly needed economic development. The IGRA first established definitions for different types of gambling, and, provided for Class I gaming, "traditional ceremonial gaming or social games for prizes of limited value" which come under the sole control of the tribes; Class II gaming including bingo, lotto, and certain card games (but not blackjack, chemin de fer or baccarat), which are subject to oversight by a five-member National Indian Gaming Commission (NIGC); and Class III gaming which, again quoting the *CQA*, includes "casino gambling, slot machines, horse and dog racing, and jai alai and which are prohibited unless they are legal in the state and the state and tribe enter into a compact for their operation."

Problems with the Act have centered upon the slow time frame for establishing the NIGC, for its delays in the adoption of regulations for tribal gaming, and for the complex requirement for compacts between tribes and states to authorize tribal gaming. Although the Act provides for tribes to establish their own gaming authority if states do not negotiate in good faith, states argue they cannot be forced to sign an agreement with the tribes. This provision of the Act both reduces tribal sovereign powers and has led states to delay signing compacts or to require extensive control over tribal gaming. U.S. Senator Pete Domenici stated in Congressional hearings that:

> ...we didn't have a commission (National Indian Gaming Commission) in

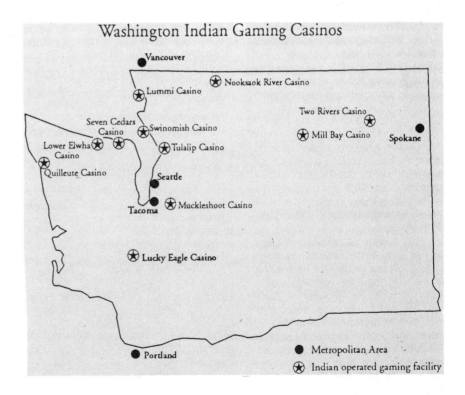

Map by: Michigan State Cartography Lab

a timely manner and we do not even have regulations and regulatory distinctions between Class II and Class III. So to the extent that some of us are frustrated about the statute (IGRA) we drew ... (we are still in the state of) determining what is criminal and what isn't (in Class III gaming under IGRA).

Some of the problems may be resolved with the empowerment of the NIGC, which has only been fully operationalized for a few years. In the NIGC structure, the state's role might again be diminished, particularly if the NIGC's role includes enforcement capabilities. Despite not having the Commission or its regulations in place, many tribes have pursued compacts for Class III gaming, opened additional Class II gaming facilities, and have sought to create their own regulations which will assure the success of their programs. Particularly for those tribes which entered the Class III gaming early, their programs have been extremely successful.

The 1988 IGRA, as it now reads, is itself receiving criticism from various sources. Some feel that it is too vague and, as a consequence, 49 state governors have petitioned Congress to clarify the act seeking tighter regulations and clearer negotiating guidelines. The tribes, on the other hand, want less intrusion. It is a political battle pitting sovereign state rights against sovereign tribal rights. It also is an economic battle over entertainment dollars—Indian-operated casinos versus state lotteries, race tracks, privately owned casinos, charitable games and bingo. Billions of dollars are at stake.

Another criticism is that the Bureau of Indian Affairs does not have the expertise and labor to provide the necessary oversight, and that as a consequence, Indian-operated casinos could become havens for money laundering and organized crime. Although states have often fallen back on the argument that organized crime and unrestricted gaming by Indians is a dangerous problem, both the FBI and Justice Department say there is no evidence of substantial organized crime in Indian gaming at this time.

The Indian Affairs Committee of Congress has worked to establish a minimum set of standards for regulating Indian gaming, and the National Indian Gaming Commission (NIGC) has been formed to serve as the regulatory agency over tribal gaming, working in partnership with other regulatory bodies, mostly tribal gaming commissions as noted in Indian Country Today. Numerous bills to change the IGRA are introduced each year, but so far none have resolved the complex issues which generally pit state interests against tribal interests. The states won a partial victory in a 1996 Supreme Court Decision, Seminole Tribe of Florida v. Florida, which upheld a lower court ruling that the tribe could not sue the state to force them to negotiate to permit specific gaming activities on their reservation, as noted in Facts on File.

Photo by: D. Winchell

Totem poles symbolize traditional designs of the Jamestown Clallam Tribe, and welcome visitors at the entrance of the tribe's Seven Cedars Casino near Sequim, WA.

The significance of Class III Indian gaming developments

The operation of Class III Gaming Facilities has provided a whole new arena for tribal economic development and self-sufficiency. Gambling operations have brought employment opportunities to tribal members, revenues to tribal gov-

Table 3
Washington State Indian Gaming Operations

Tribe or Indian Community	Casino	Location
Chehalis Reservation	Lucky Eagle Casino	Olympia Area
Colville Confederated Tribes	Mill Bay Casino (Completed 1995)	Lake Chelan
Jamestown Clallam Res.	Seven Cedars Casino	Sequim
Lower Elwha Tribe	Bingo/Casino	Port Angeles Area
Lummi Tribal Council	Lummi Casino	Bellingham Area
Muckleshoot Reservation	Muckleshoot Casino	Greater Tacoma Area
Nooksack Indian Tribe	Nooksack River Casino	Greater Seattle Area
Quileute Reservation	Quileute Casino	La Push
Spokane Tribe	Two Rivers Casino	Greater Spokane Area
Swinomish Tribe	Bingo, Casino	LaConner Area
Tulalip Tribe	Casino	Everette Area

Table 4
Arizona Indian Gaming Operations

Tribe or Indian Community	Casino	Location
Ak Chin	Harrah's Ak Chin Casino	Greater Phoenix Area
Cocopah	Cocopah Bingo/Casino	Greater Yuma Area
Mohave-Apache	Fort McDowell Gaming Center	Greater Phoenix Area
Gila River	Gila River Casino	Greater Phoenix Area
Pascus Yaqui	Casino of The Sun	Greater Tucson Area
Quechan	Quechan Bingo	Greater Yuma Area
San Carlos Apache	Apache Gold Casino	Globe-San Carlos
Tohono O'odham	Desert Diamond Casino	Greater Tucson Area
Tonto-Apache	Mazatzal Casino	Payson
White Mountain Apache	Hon Dah Casino	Pinetop
Yavapai-Prescott	Yavapai Gaming Center	Prescott

ernments through taxes or as profits to the tribes, and have produced capital for investment and development not just for casinos but for other community programs, activities and facilities.

Tribal casino development has often taken advantage of tribal locations near urban centers, with the largest tribal casino at Ledyard, Connecticut, a short distance from New York City, as shown in Figure 3, "Foxwoods Casino/Hotel, Connecticut." Other early tribal casinos include Lummi and Tulalip in the state of Washington, which draw from the Seattle and Vancouver, British Columbia urban areas; and tribal facilities near Minneapolis, San Diego, Palm Springs, Phoenix and Upper Peninsula, Michigan.

A special issue of *Indian Country Today* reported the degree of success by tribes in Class III gaming. The Cabazon Band of California grossed $50 million in 1992, with tax revenues to the tribe of over $530,000. The article stated that "Gaming revenues have enabled the Cabazons to become self-sufficient, with a true and functioning tribal government and court system."

The Oneida Tribal gaming operation in Wisconsin has had a $650 million annual impact on the surrounding area, generating over $43 million in revenues and employing over 1,100 people, approximately 75 percent of whom are Native Americans. The Mashantucket Pequot Tribe in Connecticut provided over 10,000 jobs through gaming, while the Mille Lacs Band of the Chippewa in Minnesota eliminated unemployment.

The distribution of Class III gaming facilities correlates closely with the location of reservations themselves, but the largest and most successful have been linked to urban populations or to recreation/resort amenities in rural areas. Some facilities, however, have been able to create their own destination attraction based on the existence of a casino alone. The following case studies in Connecticut, Washington, Arizona and Michigan illustrate the significance of Indian reservation gaming developments and the associated issues, changes, and trends in these four representative states.

Mashantucket Pequots: the landmark Connecticut case

Perhaps no other Indian-operated gaming facility has received as much publicity and notoriety in recent years as the Foxwoods Casino on the Mashantucket Pequot reservation near Ledyard, Connecticut. It wasn't until 1983 that the small tribe (179 enrolled members) obtained federal recognition. In 1989, the spotlight focused on the tribe as it entered into a bitter two-year battle with the state of Connecticut to open a full-scale casino based on the claim that charitable "Las Vegas" nights were legal in the state. After months of highly publicized confrontations, the U.S. Supreme Court refused to hear the state's challenge to the tribal plans. This case served as a legal landmark, and opened the door to casino developments on Indian lands nationwide.

In February, 1992, the Foxwoods Casino opened and expanded rapidly. It now is the largest casino in the western hemisphere, larger than the MGM Grand in Las Vegas. Located about half way between Boston and New York, the casino had a total handle of nearly $8 billion and a gross revenue of nearly $600 million in slot machine income alone. The facilities include two hotels, a virtual reality 1,500-seat theater, and a sporting events complex. The tribe is expanding its existing 1,794 acre development to over 9,000 acres with another 1,100 room hotel, two golf courses, and a theme park. They are investing in a variety of other local businesses such as shipbuilding, spas and gravel quarries. As is often the case in areas undergoing rapid land use changes, neighboring communities oppose the expansion, fearing the loss of the region's rural character. No doubt the conflict will be resolved over a period of time.

The tribe, now numbering over 300 members, has become a major political

Table 5
Michigan Gaming Operations

Tribe or Indian Community	Casino	Location
Lower Peninsula		
Isabella Indian Reservation	Soaring Eagle Casino	Near Mt. Pleasant
Grand Traverse Band of Ottawa & Chippewa Indians	Leelanau Sands Casino	Suttons Bay, North of Traverse City
Upper Peninsula		
Sault Ste. Marie Tribe of Chippewa Indians	Kewadin Shores Casino	St. Ignace
Sault Ste. Marie Tribe of Chippewa Indians	Vegas Kewadin Plus Casino	Sault Ste. Marie, Christmas, Hessel & Manistique
Bay Mills Indian Community	King's Club Casino	Near Brimley
Potawatomi Tribe	Chip-In Casino	Near Harris
Keeweenaw Bay Indian Community	Ojibwa Casino	Near Baraga
Lac Vieux Desert Tribe	Lac Vieux Desert	Watersmeet

and economic force in the state and region. The tribe employs more than 10,000 workers and pumps more than $100 million a year from slot machine revenue into state coffers. It has pledged $10 million to the Smithsonian Institution for a Museum of the American Indian—the largest single donation ever received by the Smithsonian; and has poured millions into lobbying and campaign contributions.

A second Indian casino, the Mohegan Sun operated by the Mohegan Tribe, opened in 1996 just 10 miles from the Foxwoods. This facility is much smaller than the Foxwoods, but nonetheless is a growing competitor.

The State of Washington and the Lummi Indian Nation Casino

Legalized gaming by American Indian tribes within the state of Washington was initiated even before the adoption of the IGRA in 1988. By summer, 1994, six tribes operated casinos as shown in Table 3, "Washington State Indian Gaming Operations" and Figure 4, "Washington Indian Gaming Casinos," while three additional tribes had casinos under development, two of which have been completed and are now open. Fifteen tribes had bingo operations. Total employment in 1996 from tribal gaming has been significant for Washington tribes, with over 6,000 jobs created on reservations since 1990 according to studies by author Winchell. Gaming employment in 1994 represented approximately 10 percent of total reservation employment in the state of Washington, and completion of new casinos increased this employment rate threefold by 1996. In addition, tribal governments have received revenues from gaming which has allowed expansion of tribal services and programs. Many tribes have developed special programs

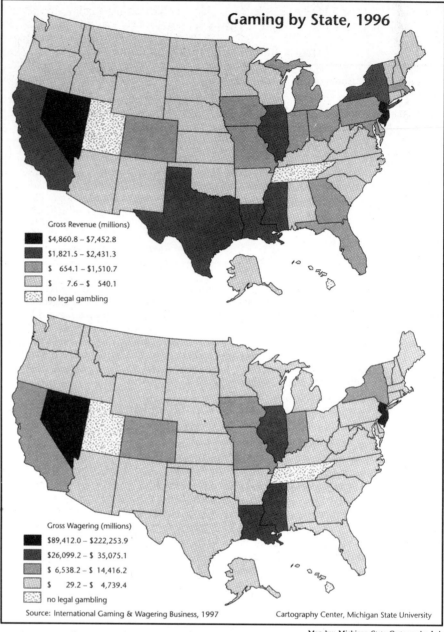

> Gaming is good for economic development only if you can import the gamblers.

for gaming-related job training, but have also increased support for general education and established programs to directly address potential negative problems such as gaming addiction and substance abuse for tribal members. Tribal leaders report that negative impacts are not as great as anticipated from casino gaming, and most tribes have sought non-Indian clients as their target. To better illustrate the impacts of gaming in Washington, the development of the Lummi Casino is described below.

The Lummi tribe in northwest Washington between Seattle, Washington and Vancouver, British Columbia, took advantage of its proximity to two major urban centers and the growing autonomy of tribes to initiate bingo and other gaming activities in the early 1980s. The Lummi tribe began gaming activities in 1983, when it converted a community center into a blackjack gaming room, and used a bank loan to pave the road onto the reservation so patrons could get to the facility. A federal judge closed the tribe's "casino" within two months for violation of state laws, but when the IGRA was passed in 1988, it allowed previously operating facilities to be "grandfathered into operation" without being required to complete a compact with the state.

The Lummi tribe reopened its casino in 1992 with a facility which held 42 blackjack tables, and which employed 413 people, of whom about 70 percent were tribal members. The Lummi Casino grossed over $3 million in revenues in its first year of operation, and the success of this casino was the stimulus for other tribes within the state to pursue Class III gaming.

29. Indian Gaming in the U.S.

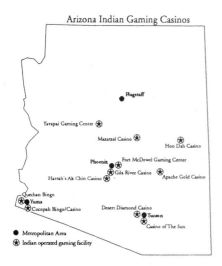

Map by: Michigan State Cartography Lab

The Lummi Casino operated under a separate management board which the tribe established, and the tribe took an active role to create positive employment opportunities for tribal members, and established a dealer training program and a gaming/resort degree program on the reservation in conjunction with the Northwest Indian Community College. It established rules that prevented tribal elected leaders and casino employees from gambling at the facility, and used some of its profits to establish its own security force. The casino also provided funds for two tribal policemen and a police car to offset the impact of increased traffic through the reservation.

The tribal government, in addition to creating jobs and training programs for tribal members, received special program support and tribal revenues of approximately $1 million a year. This dramatically contributed to the effectiveness of its programs and services.

According to interviews with the casino manager and tribal officials, the social impacts were more positive than negative. There seems to have actually been a decrease in social problems as a result of more people being employed. There was a strong recognition of potential problems, however, and the tribe established a special counseling and Gambler's Anonymous program run by American Indian counselors. The casino is clearly oriented to tourists, and has been especially successful in capturing a strong market from Canadians in the Vancouver area, less than an hour's drive away. This orientation toward non-Indians has helped lessen the negative social impact.

The Lummi Casino is also an example of the volatility in the Indian gaming industry. A new casino/hotel operated by the Nooksack River Tribe in conjunction with Harrah's only 30 miles away had some impact on the Lummi Casino, but more significantly Canada liberalized its gaming laws, and Canadians no longer need to travel across the border for gambling. By mid-year 1997, plans were being carried out to close the Lummi Casino. The Lummi Casino did produce short-term employment and revenues, and many tribal employees have been able to find employment in other nearby casinos. The tribe was able to pay off its capital expenses over the short but successful operation of the casino, and make considerable gains for the tribe.

The Arizona case

Tourism has been a major part of Arizona's economy for several decades. The climate, scenery, resort hotels, golf courses and national-level sporting events attract millions of visitors each year. In addition, legalized gaming has been in existence in the state for many years in the form of horse and greyhound racing, lotteries, and bingo halls. Indian reservations make up 28 percent of the state's land area, and some are located in close proximity to the populous urban areas of Phoenix, Tucson, Yuma and Flagstaff. Considering this mix of circumstances, it is not surprising that Indian-operated gaming would develop shortly after the passage of the 1988 Indian Gaming Regulatory Act.

As of January 1995, sixteen of Arizona's twenty-one tribes had signed gaming compacts with the state. Thirteen of these tribes or communities operated casinos in 1997, which generate tribal revenues of millions of dollars. The spatial distribution of the thirteen shows a distinct relationship to the metropolitan areas of Phoenix, Tucson, and Yuma or to major tourist areas as shown in Table 4, "Arizona Indian Gaming Operations" and Figure 5, "Arizona Indian Gaming Casinos."

The establishment and rise of Indian gaming has not developed without opposition. The horse and greyhound racing industries have not welcomed the competition, and understandably so. Inasmuch as these industries have made large contributions to most legislators' campaign funds over many years, their influence is felt in the state legislature. Also, other groups have voiced concerns regarding the impact of increased traffic congestion and crime on neighboring communities. After months of confrontations that included lawsuits, use of federal mediators, the temporary banning of charitable "casino" or "Las Vegas" nights, standoffs with federal agents and the granting of special tax benefits to the racetracks, many issues were resolved to some degree. However, other areas of dispute between the Indian tribes and the state remain. Foremost was the manner of determining which types of gaming are legal. Also, there was confusion over the jurisdiction regarding legal matters related to casinos. Whether state or tribal courts have the ultimate

Photo by: R. D. Winchell

The Seven Cedars Casino includes a gaming area, restaurant, a native arts gallery, arts and crafts sales and a large showroom for dining, dancing and performances.

167

authority was not clearly defined. In November, 1994, the tribes that operate casinos established the Arizona Indian Gaming Association to provide a united front to challenge the state gaming agency concerning these disputes and other issues that might arise.

It is likely that one of the most recent developments may have the most far-reaching effect on the Phoenix metropolitan area. The Salt River Pima-Maricopa Indians recently voted overwhelmingly in favor of going into the casino business. The reservation abuts Scottsdale, Mesa and Tempe, and within a five-mile radius there are several of the area's largest resort hotels. The establishment of a large high quality casino on Indian lands will have a major impact on Scottsdale and Tempe, for better or worse. A collaborative effort between the communities is essential to maximize the economic benefits and to head off whatever adverse impacts that will arise.

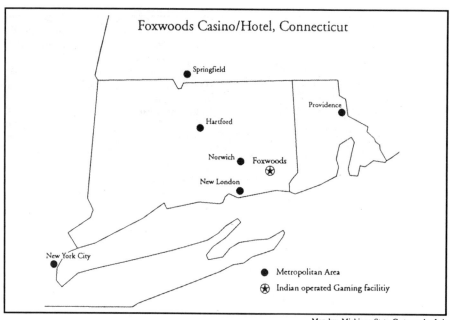

Map by: Michigan State Cartography Lab

The Michigan case

Indian commercial gaming in Michigan, as in most other states, started with bingo and moved gradually into casino type operations. The initial casino efforts were small and created much controversy over their legality. The first Indian-owned casino was a minuscule operation in a remodeled garage near Baraga in the Upper Peninsula in 1984. Indian casino gaming mushroomed in number and size with the signing of a compact between the tribes and the state in August of 1993. By 1995, six Indian operated casino complexes were in operation in the Upper Peninsula as shown in Table 5 and Figure 6, "Michigan Indian Gaming Operations." Growth has taken place at a rapid rate and is characterized by major casino floor space expansion, the building and enlargement of adjoining motels, hotels, and restaurants and the possibility of marinas and golf courses.

The nature of gaming activities has evolved rapidly from bingo halls to casinos and from blackjack and slot machines to all table and video games. In the compact with the state, the tribes agreed to give the state eight percent and the local government two percent of the video poker and slot machine profits. This was done in order to be able to keep those highly profitable machines in their casinos.

The locations of Indian gaming in Michigan present a marketing problem, in that only the Soaring Eagle Casino near Mt. Pleasant, operated by the Saginaw Chippewa Indian Tribe, is within easy daily driving distance of the major population concentrations of the southern one-third of Michigan, especially the Grand Rapids, Saginaw Bay, Flint, Lansing and Detroit urban areas. The other seven casinos are in thinly populated but highly popular summer and winter tourist areas of Michigan. The Leelanau Sands Casino, on Suttons Bay, about 20 miles north of the important Traverse City tourist area, is a good example of a location possessing excellent summer water-related activities for tourists as well as the winter attractions of ski areas, such as the nearby Sugarloaf Ski Resort. This casino is operated by the Grand Traverse Band of the Ottawa and Chippewa Indians and includes over 20,000 square feet of gambling space in two buildings, including all types of casino gaming, two restaurants and a 51-unit motel. Profits from this casino gaming complex have been used for improving Indian housing, building a health center and a tribal center, as well as plowing funds into casino expansion and reducing debt obligations. Some Michigan tribes are also distributing some gaming profits to official tribal members on a quarterly basis.

Much controversy has existed over whether casino gaming should be developed in the Detroit area. State referendums on casinos in Detroit were defeated until the Casino Windsor opened in 1994, just across the Detroit River in Windsor, Ontario, Canada. This facility draws thousands of dollars and thousands of Michiganians daily, mostly from Detroit. It is estimated that 80 percent of the Casino Windsor customers are from Detroit. As a result, a referendum on casino gambling passed easily in 1996.

The 1996 referendum permits three casinos to be developed in Detroit. Much discussion took place in the Michigan legislature on how to adequately control the developments and the amount of required financial returns to the state. Agreements were reached in July, 1997. The mayor of Detroit screened applications from major gaming enterprises, casino locations are being decided,

Photo by: R. D, Winchell

Seven Cedars Casino electronic sign advertising "The Platters," Sequim, WA.

29. Indian Gaming in the U.S.

Photo by: L. Sommers

Soaring Eagle Casino and Hotel near Mount Pleasant, MI. This huge establishment, operated by the Isabella Indian Reservation, is one of the largest between the U.S. east coast and Las Vegas, NV.

including consideration of the mayor's recommendation of a 100-acre site in northwest Detroit, and construction will begin in 1998. Casino gambling alone will not solve Detroit's economic problems, but it will provide considerable employment and reduce the flow of gaming money to Ontario.

Gaming on Indian reservation lands is having a major economic and social impact on Michigan Indians and the areas in which Indian operated casinos exist. About half of the workers in the casinos are Indians. Before casino gaming, more than half of the Michigan Indians were unemployed or on welfare. This figure has been drastically reduced and some Indians and Indian tribes are now much more economically self-sufficient. The economic impact on the areas where casinos have been built is considerable in terms of employment and the increase in income of nearby businesses supplying goods and services to the casinos and their patrons.

Income from Indian gaming has been used to enlarge casinos and associated buildings. The Roaring Eagle Casino near Mt. Pleasant is a good example. A major casino, one of the largest between the east coast and Las Vegas, has been built, and a hotel with over 500 beds was completed in 1997. Considerable funds also have been funneled to improve community, economic, and social conditions. Existing housing for Indians was renovated and new homes are being built. Other funds are being used to build health facilities, social service and substance abuse centers, day-care centers, elderly housing, and to improve and fund education.

Some Indian strategies to broaden the long-term impacts of gaming are demonstrated by investments in industry and activities other than gaming that will benefit Indians as well as others. The overall result is that there has been marked improvement in the economic and social well-being of Michigan Indians in areas where casino gaming is located. Indians are also playing a more important role in the policies and decisions of local and state government. Problems such as drug and alcohol abuse exist, as in the rest of society, but the tribes are assuming more responsibility and making inputs which augur well for their communities as well as for their contributions to the social and economic health of Michigan.

Political, legal, economic and social questions

The advent of gaming on Indian lands has resulted in political and legal issues—federal, state and local—that will take years to resolve. At the federal level, several members of Congress wish to change the 1988 IGRA, although at this time there is no clear consensus as to what these changes should be. Other members of Congress would repeal the Act, while still others are satisfied with it as it now stands. The tribes involved in gaming obviously resist more federal or state intrusion into their sovereignty. In any case, any bill introduced in Congress regarding changes or amendments to the 1988 Act is bound to be argued for a long period of time.

At the state level, it becomes a battle of state sovereign rights versus tribal sovereign rights. States resist Indian gaming because it diverts gamblers from state-sponsored games and other forms of recreation, and adds to the state's cost of roads, police and community services. Perhaps the major issue is who has the ultimate say as to which games can be offered and the magnitude of those operations. There is a great diversity among the states regarding this issue and the manner of negotiations with tribal units. Unfortunately, it is likely that many of these issues will be taken to the courts for decision.

At the local level, counties, townships and municipalities near Indian gaming facilities are concerned with rural-to-urban changes in lifestyle, and confusion over jurisdiction regarding enforcement of crimes associated with the increase of people and traffic such as speeding, drunken driving, fraud and car burglary. Again, many of these issues across the country will likely end up in long drawn-out court cases.

In addition to political issues, economic and social impacts of Indian gaming are of concern in many areas. Economically, other forms of recreation, particularly charitable bingo and card games, horse and greyhound racing, nightclubs, and bars do not welcome the competition. Whether or not a given Indian gaming operation is a boon to the

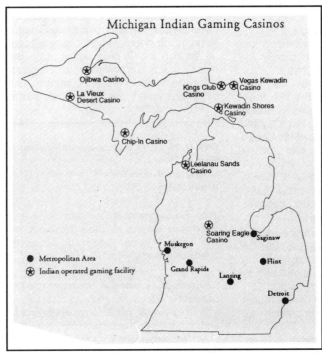

Map by: Michigan State Cartography Lab

4 ❖ SPATIAL INTERACTION AND MAPPING

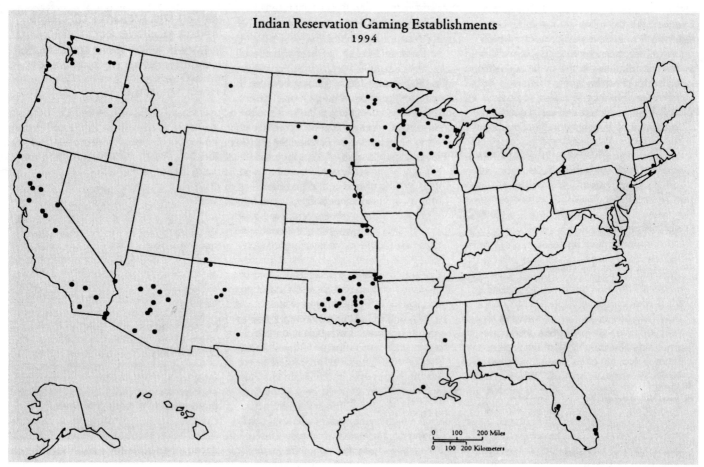

Map by: Michigan State Cartography Lab

local area depends on whether it is a *basic* or *non-basic* economic activity. Basic economic activities generate local or regional growth by exporting goods or services beyond the borders of the locality or area. Conversely, nonbasic economic activities serve the local area markets only: they bring in or import money from the outside. In this context, gaming is good for economic development only if you can import the gamblers. There are gaming operations that are non-basic in nature which have not greatly benefited the local region, such as in Colorado, but many have proven to be basic to their region and have produced tremendous economic growth. The Foxwoods in Connecticut is a prime example.

From the social standpoint, the economic gains have drastically helped the living standards of many tribes. Increased income, employment, new schools, water and sewer facilities, roads, social centers and the like have exceeded the wildest dreams of many tribes in various parts of the country. Conversely, there are cases of increased crime, unwise spending and investments, and breakdowns of tribal cultures. Overall, in the majority of cases, the positive economic and social gains have greatly outweighed the negative factors, at least in the short run.

Another major issue is whether Indian gaming developments will be able to compete in the gaming industry over the long run. As Indian reservation casinos increase in number, size and complexity of gaming, and the number of non-Indian casinos increase, the competition for customers will also intensify. It remains to be seen whether the nationwide increase in casino gaming can continue to be successful in its various forms and levels. It is likely that some developments will fail as has been the case in Las Vegas. As that city has evolved to megacasinos and theme parks, some of the smaller and less appealing casinos have failed to remain competitive. The Indian gaming industry undoubtedly will also go through this kind of evolution. In the meantime it has provided the single-greatest boon to reservation economies and tribal government resources since Indian

> In the majority of cases, the positive economic and social gains have greatly outweighed the negative factors.

Dick G. Winchell *is Professor of Planning at Eastern Washington University. His interests include rural development and ethnicity in America. Email: DWINCHELL@ewu.edu*

John F. Lounsbury *is Professor Emeritus of Geography at Arizona State University. He and Larry Sommers wrote "Border Boom Towns of Nevada," FOCUS 41(4), in 1991.*

Lawrence M. Sommers *is Professor in the Department of Geography, Michigan State University. His interests include applied geography and economic development. He and John Lounsbury wrote "Border Boom Towns of Nevada," FOCUS 41(4), in 1991.*

reservations were first established in the United States.

References and Further Readings

Devereaux, Francis and Asmara Habte. 1997. The Planning Impacts of Tribal ming in Washington. The Western Planner.

Eadington, W. (ed.) 1990. Indian Gaming and the Law. Reno: Institute for the Study of Gambling and Commercial Gaming, University of Nevada, Reno.

Facts on File. Supreme Court: Indian Gaming Act Overturned. April 4, 1996. v. 56, n 2882, p. 220.

Indian Country Today. 1992. Gaming: Winner's Circle Special Edition. November 5, 1992. 38 pp.

1996. Bloom Is Off American Indian Gaming Rose. 1994, 1997. July 14–21, p. A5.

International Gaming and Wagering Business. 1994, 1997. New York.

Meyer-Arendt, K. and R. Hartmann (eds.). 1998. Casino Gambling in America: Origins, Trends and Impacts. Cognizant C.C.: New York.

Minnesota Indian Gaming Association. 1992. Economic Benefits of Tribal Gaming in Minnesota. Minneapolis, Minnesota: Minnesota Indian Gaming Association Report, March, 1992 (also reprinted in Senate Select Committee, 1992, part 2, pp. 273–298).

The New York Times Magazine. 1994. Gambling Nation. The New York Times Magazine. July 17, 1994, pp. 34–61.

Senate Select Committee on Indian Affairs. 1992. Implementation of Indian Gaming Regulatory Act. Hearing before the Select Committee on Indian Affairs United States Senate. March 18, 1992. Parts 1, 2, 3 and 4. Washington, D.C. U.S. Government Printing Office.

Sommers, Lawrence M. and John F. Lounsbury. 1991. Border Boom Towns of Nevada. FOCUS 41 (4), Winter, 1991, pp. 12–18.

White, R. 1990. The Rebirth of Native America. New York: Henry Holt and Co.

Winchell, Dick G. and Francis Devereaux. 1993. The Impact of Gaming on the Colville Confederated Tribes. Nespelem, WA.: The Planning Department, Colville Confederated Tribes.

Legal Cases

California v. Cabazon Band of Mission Indians,] 480 U.S. 208 (1987).

Confederated Tribes v. State of Washington, 938 F. 2nd 146(9th Cir. 1990).

Seminole Tribe of Florida v. Florida. March 27, 1996.

Legislation

Public Law 100–497 Indian Gaming Regulatory Act (IGRA), 25 U.S.C. sec. 2701, et. seq. and 18 U.S.C. sec. 1166–1168, enacted October, 1988.

For Poorest Indians, Casinos Aren't Enough

By PETER T. KILBORN

PINE RIDGE, S.D.—The image of the Indian gambling casino as money machine evaporates among the destitute Oglala Sioux of the Pine Ridge Reservation. The casino here consists of three double-wide trailers stuck together on cinder blocks in a rutted lot with two rusty oil drums outside for trash.

The Oglalas earn $1 million a year from gambling, about what the Pequot tribe in Connecticut earns in half a day at Foxwoods, the nation's richest Indian casino. The Oglalas tried to build a bigger, permanent casino, but construction stopped in December with just the foundation laid because the contractor wanted more money. The dispute could take years to resolve, and fissures have formed in the foundation's concrete walls.

This two-million-acre reservation is a pit of fissured dreams. It is the site of the Wounded Knee massacre, where the Army killed some 200 braves, women and children in 1890, and was the home of the Oglala band of Red Cloud and Crazy Horse.

It is as poor as America gets.

"It's very bleak," said John Yellow Bird Steele, 51, president of the Oglala Sioux Tribal Council.

In recent years, with casinos going up on one reservation after another, many Americans have come to believe that Indian reservations are boom towns. Indeed, with Indian casinos bringing in about $6 billion last year, compared with $100 million in 1988, a few tribes are doing well.

A visit to Pine Ridge is a striking reminder that most reservations remain places of bone-crushing poverty. And things are likely to get worse as the Government cuts some of the welfare payments that are crucial to their economies.

At least half of the 1.3 million American Indians who live on reservations are poor, the Bureau of Indian Affairs says, and 49 percent of the reservations' labor force is unemployed—10 times the national unemployment rate. The sickest economy is in Pine Ridge, the second-largest reservation, after the Navajo reservation in the Southwest.

Here, just 1 in 4 adults has a job. People subsist almost entirely on Federal, state and tribal government checks: paychecks and welfare checks, retirement and disability checks. Now welfare and housing aid is being cut.

The casino, for all its travails, sets the standard for success. Alone among the businesses sponsored by the tribe, it makes a little money.

Nature has wrought a topographical masterpiece here, with expanses of butte-broken prairie and the Badlands' sinister white spires and menacing cliffs. But this immense terrain yields no oil or gas, coal or gold, crops or lumber.

Its one readily exploitable natural resource is grass. The Bureau of Indian Affairs says 20,205 cattle, 753 bison and about 5,000 horses are grazing reservation lands. But that business, too, has stalled. Cattle prices have dropped to 60 cents a pound, compared with 90 cents four years ago.

About 25,000 people—no one has a reliable count—live on the reservation in shacks, trailers and small, shoebox-like ranch houses subsidized by the Government, the lots of many cluttered with the hulks of dead cars. No car here usually means no job because of the vast distances between home and work.

Five years ago, Mr. Steele aspired to build three casinos with adjoining motels and restaurants. But that idea has faltered, as have all other missions to rescue the economy—a ranch, a moccasin factory and a beef-packing plant. The prospects for yet another venture, a herd of choice cows to produce beef for Pequot casino restaurants, dimmed in May when the council suspended from his post as treasurer the man who had conceived it.

The Census Bureau found eight years ago that Shannon County, S.D., which includes most of the reservation, was the poorest of the nation's 3,143 counties. Using different criteria in its latest survey, done four years ago, the bureau said Shannon County was the fourth-poorest.

"It's like living at the bottom of a well," said Milo Yellow Hair, vice president of the tribe. "The great white father looks down and says, 'Here's a few dollars.'"

Now those dollars are dwindling. One new blow is the demise of the old welfare system. Wilbur Campbell, manager of the local office of the South Dakota Department of Social Services, said that through April, 128 of the 896 reservation mothers on welfare had lost a portion of their cash assistance because they had not gone to work. For a mother with one child, he said, that meant the check was cut to $78 a month, from $250.

Another group of welfare recipients could begin losing out soon. The Bureau of Indian Affairs allots $38 a month in general assistance to the poor who do not qualify for family as-

30. For Poorest Indians, Casinos Aren't Enough

sistance. Robert D. Ecoffey, superintendent of the bureau's Pine Ridge agency, said that in the last couple of years, Congress had chopped his general-assistance budget to $714,000, from $1.7 million. He has not cut anyone off yet, he said, because he has been scrounging up the money from other accounts.

Housing, primitive as it is, is also being squeezed. Paul Iron Cloud, director of the tribe's housing authority, said Federal money for home renovation had been cut by 20 percent last year. The wait for Government-subsidized homes, about the only kind built here, stretches for four to five years.

Families double and triple up; Mr. Steele has 4 adults and 18 children in his house. Now, Mr. Iron Cloud said, changes in Federal policy could freeze out the poorest families, who pay no rent and get free utilities.

Poverty is taking an ever-bigger toll on health. Pine Ridge's infant mortality rate is almost three times the national average. Alcoholism, the scourge of indigent Indians, is as entrenched as ever. Doctors say that for the first time, they are spotting adult-onset diabetes, which usually strikes people past 40, in much younger patients.

"I just saw a 12-year-old who has it," said Dr. David Mulder, the reservation hospital's chief of staff.

"That's really frightening—they're already insulin-resistant," he said.

Mr. Yellow Hair explained: "A six-pack of soda costs less than a gallon of milk. So mothers get the soda and put it in their babies' bottles and shake it up to get the fizz out. It starts there."

Kibbe Conti, the hospital's nutritionist, said that for these Northern Plains Indians, who have a genetic predisposition for diabetes, "pop is poison."

The blame for these conditions goes all around. The roots of the troubles lie in political pandering among tribal leaders, who must run for election every two years, and in corruption, the 19th-century treaties that shunted the Oglalas off richer lands, the cuts in state and Federal support, and discrimination.

In early May, for example, the Justice Department penalized the nearby First National Bank of Gordon, Neb., for charging the Pine Ridge Sioux interest rates that were five percentage points higher than the rate paid by white customers off the reservation.

But the Indians also blame themselves for their problems.

Baptiste Poirier, 45, is an anomaly here. A tribe member and, like his wife, Pat, 39, a recovering alcoholic, he owns Big Bat's, a bustling convenience store, restaurant and Texaco gasoline station at the four-way stop in the center of the village of Pine Ridge.

Mr. Poirier and his family also own a propane distribution business, a small, 150-head herd of cattle and, over the Nebraska line in Chadron, another convenience store. The Poiriers and their two sons, both students at an Indian college, live in a 16-by-55-foot trailer with a single concession to conspicuous consumption, a hot tub.

"It's so hard to do business here," he said. "For every one Indian trying to get ahead, there's 20 trying to pull him down."

In the nine years since he opened Big Bat's, Mr. Poirier said, last year was his worst. The store lost money. "You know what it was over?" he asked. "Employee theft. Merchandise and cash." Nationally, stores lose about 3 percent to employee theft, he said. "For me, it was well over 10 percent," he said.

Scarce as jobs are here, Mr. Poirier said his employee turnover rate was 50 percent a year; a typical worker stays only six months. "Most of it is alcohol-related," he said. "There's so much hopelessness. They get drunk. They don't show up for work. I've got to let them go."

"We don't have a business class, a middle class," Mr. Poirier said. "We have a political class. It's a welfare government. When you've got political people in charge, they're going to take care of the voters."

The tribe is the biggest employer, through services and business ventures. Among the ventures that have not panned out is the tribe's own ranching operation, Farm and Ranch Enterprise, which once received a Federal development grant of $5.5 million. "That money was lost," Mr. Yellow Hair said.

The ranch, with a herd of only 340 cattle, has little water, so the tribe installed a windmill-powered water pump. "But they put the windmill in a gulley, a place where there's no wind," said Melvin Lee Houston, an Oglala official.

The farm survives on a $100,000 annual tribal subsidy.

The three-year-old casino is off to a happier start. Its $1 million in profits, although only $38 per capita and far from enough to offset the decline in the Government's aid, is distributed across the reservation, mostly to the elderly.

It has also created jobs, including 129 for residents of the reservation.

Thirty-four-year-old Amy Rodriquez, a pit boss who oversees the casino's four blackjack tables, earns $10.92 an hour, a very high wage around here. "I'm doing a lot better," she said. "I have a steady job. I have a car now."

But the tribe's great expectations for casino gambling, like those of many tribes, are withering.

Here, one obstacle is the construction dispute. A bigger obstacle is the reservation's isolation. The nearest sizeable source of customers is Rapid City, S.D., 80 miles to the north. "The reality is," Mr. Yellow Hair said, "there is no population."

For a short time recently, the reservation's hopes shifted to a newer enterprise.

The tribe's treasurer, Chuck Jacobs, 41, proposed buying a herd of cattle that were not only free of growth hormones and other additives but also had an impressive genetic record of yielding choice beef. He would lease the bulls to reservation ranchers and, bypassing most middlemen, sell the beef to the Pequots and other wealthy tribes for their restaurants.

A former all-Indian rodeo champion, Mr. Jacobs has a degree in regional planning from the University of Massachusetts and was awarded a MacArthur Foundation fellowship 10 years ago for analyses of reservation economies.

Mr. Jacobs prepared an inch-thick plan including reams of correspondence showing the support of potential customers. In a close vote last December, the tribal council authorized Mr. Jacobs to proceed. He bought a foundation herd of 70 cows and 17 bulls for $187,000.

Today, the cattle are grazing at the Farm and Ranch Enterprise site, but what will become of them is unclear. Some tribal leaders, alarmed at the cost and the realization that the cattle would not be yielding much beef for some years, now deride the purchase as the "golden cows."

The tribal council voted 8 to 8 earlier this month to suspend Mr. Jacobs. Mr. Steele, who defeated Mr. Jacobs among other candidates in the race for the presidency last year, broke the tie by voting against him.

"People have a lot of questions about the viability of the whole project," Mr. Steele said. Mr. Ecoffey, of the Bureau of Indian Affairs, said, "He was the kind of guy we need to be leaning on."

Unit 5

Unit Selections

31. **Before the Next Doubling,** Jennifer D. Mitchell
32. **The End of Cheap Oil,** Colin J. Campbell and Jean H. Laherrère
33. **Reseeding the Green Revolution,** Charles Mann
34. **How Much Food Will We Need in the 21st Century?** William H. Bender
35. **The Changing Geography of U.S. Hispanics, 1850–1990,** Terrence Haverluk
36. **'Hispanics' Don't Exist,** Linda Robinson
37. **Russia's Population Sink,** Toni Nelson
38. **Vanishing Languages,** David Crystal
39. **Risky Business: Who Will Pay for the Growing Costs of Global Change?** Options

Key Points to Consider

❖ How are you personally affected by the population explosion?

❖ Give examples of how economic development adversely affects the environment. How can such adverse effects be prevented?

❖ How do you feel about the occurrence of starvation in developing world regions?

❖ What might it be like to be a refugee?

❖ In what forms is colonialism present today?

❖ How is Earth a system?

❖ For how long are world systems sustainable?

❖ What is your scenario of the world in the year 2010?

 **Links**   www.dushkin.com/online/

27. **African Studies WWW (U.Penn)**
    http://www.sas.upenn.edu/African_Studies/AS.html
28. **Human Rights and Humanitarian Assistance**
    http://info.pitt.edu/~ian/resource/human.htm
29. **Hypertext and Ethnography**
    http://www.umanitoba.ca/faculties/arts/anthropology/tutor/ aaa_presentation.new.html
30. **Research and Reference (Library of Congress)**
    http://lcweb.loc.gov/rr/
31. **Space Research Institute**
    http://arc.iki.rssi.ru/Welcome.html

These sites are annotated on pages 6 and 7.

Population, Resources, and Socioeconomic Development

The final unit of this anthology includes discussions of several important problems facing humankind. Geographers are keenly aware of regional and global difficulties. It is hoped that their work with researchers from other academic disciplines and representatives of business and government will help bring about solutions to these serious problems.

Probably no single phenomenon has received as much attention in recent years as the so-called population explosion. World population continues to increase at unacceptably high rates. The problem is most severe in the less developed countries, where in some cases, populations are doubling in less than 20 years.

The human population of the world will pass the 6 billion mark in 1999. It is anticipated that population increase will continue well into the twenty-first century, despite a slowing in the rate of population growth globally since the 1960s. The first three articles in this section deal with related issues of population growth, declining petroleum reserves, and the need for a new Green Revolution. The next article asks what world food needs will be in the next century. The next two articles deal with aspects of the Hispanic population. "Russia's Population Sink" reports on the dilemma of death rates exceeding birth rates in this postcommunist country. "Vanishing Languages" details the loss of many languages due to disease and economic relocations. The last article in this unit cites greatly increased cost levels associated with natural and human-induced disasters, an increasing problem as we near the twenty-first century.

Before the Next Doubling

Nearly 6 billion people now inhabit the Earth—almost twice as many as in 1960. At some point over the course of the next century, the world's population could double again. But we don't have anything like a century to prevent that next doubling; we probably have less than a decade.

by Jennifer D. Mitchell

In 1971, when Bangladesh won independence from Pakistan, the two countries embarked on a kind of unintentional demographic experiment. The separation had produced two very similar populations: both contained some 66 million people and both were growing at about 3 percent a year. Both were overwhelmingly poor, rural, and Muslim. Both populations had similar views on the "ideal" family size (around four children); in both cases, that ideal was roughly two children smaller than the actual average family. And in keeping with the Islamic tendency to encourage large families, both generally disapproved of family planning.

But there was one critical difference. The Pakistani government, distracted by leadership crises and committed to conventional ideals of economic growth, wavered over the importance of family planning. The Bangladeshi government did not: as early as 1976, population growth had been declared the country's number one problem, and a national network was established to educate people about family planning and supply them with contraceptives. As a result, the proportion of couples using contraceptives rose from around 6 percent in 1976 to about 50 percent today, and fertility rates have dropped from well over six children per woman to just over three. Today, some 120 million people people live in Bangladesh, while 140 million live in Pakistan—a difference of 20 million.

Bangladesh still faces enormous population pressures—by 2050, its population will probably have increased by nearly 100 million. But even so, that 20 million person "savings" is a colossal achievement, especially given local conditions. Bangladeshi officials had no hope of producing the classic "demographic transition," in which improvements in education, health care, and general living standards tend to push down the birth rate. Bangladesh was—and is—one of the poorest and most densely populated countries on earth. About the size of England and Wales, Bangladesh has twice as many people. Its per capita GDP is barely over $200. It has one doctor for every 12,500 people and nearly three-quarters of its adult population are illiterate. The national diet would be considered inadequate in any industrial country, and even at current levels of population growth, Bangladesh may be forced to rely increasingly on food imports.

All of these burdens would be substantially heavier than they already are, had it not been for the family planning program. To appreciate the Bangladeshi achievement, it's only necessary to look at Pakistan: those "additional" 20 million Pakistanis require at least 2.5 million more houses, about 4 million more tons of grain each year, millions more jobs, and significantly greater investments in health care—or a significantly greater burden of disease. Of the two nations, Pakistan has the more robust economy—its per capita GDP is twice that of Bangladesh. But the Pakistani economy is still primarily agricultural, and the size of the average farm is shrinking, in part because of the expanding population. Already, one fourth of the country's farms are under 1 hectare, the standard minimum size for economic viability, and Pakistan is looking increasingly towards the international grain markets to feed its people. In 1997, despite its third consecutive year of near-record harvests, Pakistan attempted to double its wheat imports but was not able to do so because it had exhausted its line of credit.

And Pakistan's extra burden will be compounded in the next generation. Pakistani women still bear an average of well over five children, so at the current birth rate, the 10 million or so extra couples would produce at least 50 million children. And these in turn could bear nearly 125 million children of their own. At its current fertility rate, Pakistan's population will double in just 24 years—that's more than twice as fast as Bangladesh's population is growing. H. E. Syeda Abida Hussain, Pakistan's Minister of Population Welfare, explains the problem bluntly: "If we achieve success in lowing our population growth substantially, Pakistan has a future. But if, God forbid, we should not—no future."

The Three Dimensions of the Population Explosion

Some version of Mrs. Abida's statement might apply to the world as a whole. About 5.9 billion people currently inhabit the Earth. By the middle of the next century, according to U.N. projections, the population will probably reach 9.4 billion—and all of the net increase is likely to occur in the developing world. (The total population of the industrial countries is expected to decline slightly over the next 50 years.) Nearly 60 percent of the increase will occur in Asia, which will grow from 3.4 billion people in 1995 to more than 5.4 billion in 2050. China's population will swell from 1.2 billion to 1.5 billion, while India's is projected to soar from 930 million to 1.53 billion. In the Middle East and North Africa, the population will probably more than double, and in sub-Saharan Africa, it will triple. By 2050, Nigeria alone is expected to have 339 million people—more than the entire continent of Africa had 35 years ago.

Despite the different demographic projections, no country will be immune to the effects of population growth. Of course, the countries with the highest growth rates are likely to feel the greatest immediate burdens—on their educational and public health systems, for instance, and on their forests, soils, and water as the struggle to grow more food intensifies. Already some 100 countries must rely on grain imports to some degree, and 1.3 billion of the world's people are living on the equivalent of $1 a day or less.

But the effects will ripple out from these "front-line" countries to encompass the world as a whole. Take the water predicament in the Middle East as an example. According to Tony Allan, a water expert at the University of London, the Middle East "ran out of water" in 1972, when its population stood at 122 million. At that point, Allan argues, the region had begun to draw more water out of its aquifers and rivers than the rains were replenishing. Yet today, the region's population is twice what it was in 1972 and still growing. To some degree, water management now determines political destiny. In Egypt, for example, President Hosni Mubarak has announced a $2 billion diversion project designed to pump water from the Nile River into an area that is now desert. The project—Mubarak calls it a "necessity imposed by population"—is designed to resettle some 3 million people outside the Nile flood plain, which is home to more than 90 percent of the country's population.

Elsewhere in the region, water demands are exacerbating international tensions; Jordan, Israel, and Syria, for instance, engage in uneasy competition for the waters of the Jordan River basin. Jordan's King Hussein once said that water was the only issue that could lead him to declare war on Israel. Of course, the United States and the western European countries are deeply involved in the region's antagonisms and have invested heavily in its fragile states. The western nations have no realistic hope of escaping involvement in future conflicts.

Yet the future need not be so grim. The experiences of countries like Bangladesh suggest that it is possible to build population policies that are a match for the threat. The first step is to understand the causes of population growth. John Bongaarts, vice president of the Population Council, a non-profit research group in New York City, has identified three basic factors. (See figure on the next page.)

Unmet demand for family planning. In the developing world, at least 120 million married women—and a large but undefined number of unmarried women—want more control over their pregnancies, but cannot get family planning services. This unmet demand will cause about one-third of the projected population growth in developing countries over the next 50 years, or an increase of about 1.2 billion people.

Desire for the large families. Another 20 percent of the projected growth over the next 50 years, or an increase of about 660 million people, will be caused by couples who may have access to family planning services, but who choose to have more than two children. (Roughly two children per family is the "replacement rate," at which a population could be expected to stabilize over the long term.)

Population momentum. By far the largest component of population growth is the least commonly understood. Nearly one-half of the increase projected for the next 50 years will occur simply because the next reproductive generation—the group of people currently entering puberty or younger—is so much larger than the current reproductive generation. Over the next 25 years, some 3 billion people—a number equal to the entire world population in 1960—will enter their reproductive years, but only about 1.8 billion will leave that phase of life. Assuming that the couples in this reproductive bulge begin to have children at a fairly early age, which is the global norm, the global population would still expand by 1.7 billion, even if all of those couples had only two children—the longterm replacement rate.

Meeting the Demand

Over the past three decades, the global percentage of couples using some form of family planning has increased dramatically—from less than 10 to more than 50 percent. But due to the growing population, the absolute number of women not using family planning is greater today than it was 30 years ago. Many of these women fall into that first category above—they want the services but for one reason or another, they cannot get them.

Sometimes the obstacle is a matter of policy: many governments ban or restrict valuable methods of contraception. In Japan, for instance, regulations discourage the use of birth control pills in favor of condoms, as a public health measure against sexually transmitted diseases. A study conducted in 1989 found that some 60 countries required a husband's permission before a woman can be sterilized; several required a husband's consent for all forms of birth control.

Elsewhere, the problems may be more logistical than legal. Many developing countries lack clinics and pharmacies in rural areas. In some rural areas of sub-Saharan Africa, it takes an average of two hours to reach the nearest contraceptive provider. And often contraceptives are too expensive for most people. Sometimes the products or services are of such poor quality that they are not simply ineffective, but dangerous. A woman who has been injured by a badly made or poorly inserted IUD may well be put off by contraception entirely.

In many countries, the best methods are simply unavailable. Sterilization is often the only available nontraditional option, or the only one that has gained wide acceptance. Globally, the procedure accounts for about 40 percent of contraceptive use and in some countries the fraction is much higher: in the Dominican Republic and India, for example, it stands at 69 percent. But women don't generally resort to sterilization until well into their childbearing years, and in some countries, the procedure isn't permitted until a woman reaches a certain age or bears a certain number of children. Sterilization is therefore no substitute for effective temporary methods like condoms, the pill, or IUDs.

There are often obstacles in the home as well. Women may be prevented from seeking family planning services by disapproving husbands or in-laws. In Pakistan, for example, 43 percent of husbands object to family planning. Frequently,

Population of Developing Countries, 1950–95, with Projected Growth to 2050

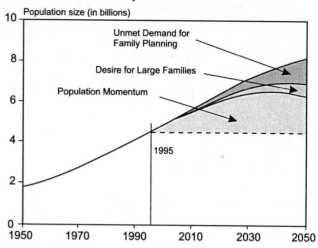

Source: U.N., *World Population Prospects: the 1996 Revision* (New York: forthcoming); and John Bongaarts, "Population Policy Options in the Developing World," *Science*, 11 February 1994.

such objections reflect a general social disapproval inculcated by religious or other deeply-rooted cultural values. And in many places, there is a crippling burden of ignorance: women simply may not know what family planning services are available or how to obtain them.

Yet there are many proven opportunities for progress, even in conditions that would appear to offer little room for it. In Bangladesh, for instance, contraception was never explicitly illegal, but many households follow the Muslim custom of *purdah*, which largely secludes women in their communities.

Since it's very difficult for such women to get to family planning clinics, the government brought family planning to them: some 30,000 female field workers go door-to-door to explain contraceptive methods and distribute supplies. Several other countries have adopted Bangladesh's

approach. Ghana, for instance, has a similar system, in which field workers fan out from community centers. And even Pakistan now deploys 12,000 village-based workers, in an attempt to reform its family planning program, which still reaches only a quarter of the population.

Reducing the price of contraceptives can also trigger a substantial increase in use. In poor countries, contraceptives can be an extremely price-sensitive commodity even when they are very cheap. Bangladesh found this out the hard way in 1990, when officials increased contraceptive prices an average of 60 percent. (Under the increases, for example, the cheapest condoms cost about 1.25 U.S. cents per dozen). Despite regular annual sales increases up to that point, the market slumped immediately: in 1991, condom sales fell by 29 percent and sales of the pill by 12 percent. The next year, prices were rolled back; sales rebounded and have grown steadily since then.

Additional research and development can help broaden the range of contraceptive options. Not all methods work for all couples, and the lack of a suitable method may block a substantial amount of demand. Some women, for instance, have side effects to the pill; others may not be able to use IUDs because of reproductive tract infections. The wider the range of available methods, the better the chance that a couple will use one of them.

Planning the Small Family

Simply providing family planning services to people who already want them won't be enough to arrest the population juggernaut. In many countries, large families are still the ideal. In Senegal, Cameroon, and Niger, for example, the average woman still wants six or seven children. A few countries have tried to legislate such desires away. In India, for example, the Ministry of Health and Family Welfare is interested in promoting a policy that would bar people who have more than two children from political careers, or deny them promotion if they work within the civil service bureaucracy. And China's well-known policy allows only one child per family.

But coercion is not only morally questionable—it's likely to be ineffective because of the backlash it invites. A better starting point for policy would be to try to understand why couples want large families in the first place. In many developing countries, having lots of children still seems perfectly rational: children are a source of security in old age and may be a vital part of the family economy. Even when they're very young, children's labor can make them an asset rather than a drain on family income. And in countries with high child mortality rates, many births may be viewed as necessary to compensate for the possible deaths (of course, the cumulative statistical effect of such a reaction is to *over*-compensate).

Religious or other cultural values may contribute to the big family ideal. In Pakistan, for instance, where 97 percent of the population is Muslim, a recent survey of married women found that almost 60 percent of them believed that the number of children they have is "up to God." Preference for sons is another widespread factor in the big family psychology: many large families have come about from a perceived need to bear at least one son. In India, for instance, many Hindus believe that they need a son to perform their last rites, or their souls will not be released from the cycle of births and rebirths. Lack of a son can mean abandonment in this life too. Many husbands desert wives who do not bear sons. Or if a husband dies, a son is often the key to a woman's security: 60 percent of Indian women over 60 are widows, and widows tend to rely on their sons for support. In some castes, a widow has no other option since social mores forbid her from returning to her birth village or joining a daughter's family. Understandably, the fear of abandonment prompts many Indian women to continue having children until they have a son. It is estimated that if son preference were eliminated in India, the fertility rate would decline by 8 percent from its current level of 3.5 children per woman.

Yet even deeply rooted beliefs are subject to reinterpretation. In Iran, another Muslim society, fertility rates have dropped from seven children per family to just over four in less than three decades. The trend is due in some measure to a change of heart among the government's religious authorities, who had become increasingly concerned about the likely effects of a population that was growing at more than 3 percent per year. In 1994, at the International Conference on Population and Development (ICPD) held in Cairo, the Iranian delegation released a "National Report on Population" which argued that according to the "quotations from prophet Mohammad . . . and verses of [the] holy Quran, what is standing at the top priority for the Muslims' community is the social welfare of Muslims." Family planning, therefore, "not only is not prohibited but is emphasized by religion."

Promotional campaigns can also change people's assumptions and behavior, if the campaigns fit into the local social context. Perhaps the most successful effort of this kind is in Thailand, where Mechai Viravidaiya, the founder of the Thai Population and Community Development Association, started a program that uses witty songs, demonstrations, and ads to encourage the use of contraceptives. The program has helped foster widespread awareness of family planning throughout Thai society. Teachers use population-related examples in their math classes; cab drivers even pass out condoms. Such efforts have paid off: in less than three decades, contraceptive use among married couples has risen from 8 to 75 percent and population growth has slowed from over 3 percent to about 1 percent—the same rate as in the United States.

Better media coverage may be another option. In Bangladesh, a recent study found that while local journalists recognize the importance of family planning, they do not understand population issues well enough to cover them effectively and objectively. The study, a collaboration between the University Research Corporation of Bangladesh and Johns Hopkins University in the United States, recommended five ways to improve coverage: develop easy-to-use information for journalists (press releases, wall charts, research summaries), offer training and workshops, present awards for population journalism, create a forum for communication between journalists and family planning professionals, and establish a population resource center or data bank.

Often, however, the demand for large families is so tightly linked to social conditions that the conditions themselves must be viewed as part of the problem. Of course, those conditions vary greatly from one society to the next, but there are some common points of leverage:

Reducing child mortality helps give parents more confidence in the future of the children they already have. Among the most effective ways of reducing mortality are child immunization programs, and the promotion of "birth spacing"—lengthening the time between births. (Children born less than a year and a half apart are twice as likely to die as those born two or more years apart.)

Improving the economic situation of women provides them with alternatives to child-bearing. In some countries, officials could reconsider policies or customs that limit women's job opportunities or other economic rights, such as the right to inherit property. Encouraging "micro-leaders" such as Bangladesh's Grameen Bank can also be an effective tactic. In Bangladesh, the Bank has made loans to well over a million villagers—mostly impoverished women—to help them start or expand small businesses.

Improving education tends to delay the average age of marriage and to further the two goals just mentioned. Compulsory school attendance for children undercuts the economic incentive for larger families by reducing the opportunities for child labor. And in just about every society, higher levels of education correlate strongly with small families.

Momentum: The Biggest Threat of All

The most important factor in population growth is the hardest to counter—and to understand. Population momentum can be easy to overlook because it isn't directly captured by the statistics that attract the most attention. The global growth rate, after all, is dropping: in the mid-1960s, it amounted to about a 2.2 percent annual increase; today the figure is 1.4 percent. The fertility rate is dropping too: in 1950, women bore an average of five children each; now they bear roughly three. But despite these continued declines, the absolute number of births won't taper off any time soon. According to U.S. Census Bureau estimates, some 130 million births will still occur annually for the next 25 years, because of the sheer number of women coming into their child-bearing years.

The effects of momentum can be seen readily in a country like Bangladesh, where more than 42 percent of the population is under 15 years old—a typical proportion for many poor countries. Some 82 percent of the population growth projected for Bangladesh over the next half century will be caused by momentum. In other words, even if from now on, every Bangladeshi couple were to have only two children, the country's population would still grow by 80 million by 2050 simply because the next reproductive generation is so enormous.

The key to reducing momentum is to delay as many births as possible. To understand why delay works, its helpful to think of momentum as a kind of human accounting problem in which a large number of births in the near term won't be balanced by a corresponding number of deaths over the same period of time. One side of the population ledger will contain those 130 million annual births (not all of which are due to momentum, of course), while the other side will contain only about 50 million annual deaths. So to put the matter in a morbid light, the longer a substantial number of those births can be delayed, the longer the death side of the balance sheet will be when the births eventually occur. In developing countries, according to the Population Council's Bongaarts, an average 2.5-year delay in the age when a woman bears her first child would reduce population growth by over 10 percent.

One way to delay childbearing is to postpone the age of marriage. In Bangladesh, for instance, the median age of first marriage among women rose from 14.4 in 1951 to 18 in 1989, and the age at first birth followed suit. Simply raising the legal age of marriage may be a useful tactic in countries that permit marriage among the very young. Educational improvements, as already mentioned, tend to do the same thing. A survey of 23 developing countries found that the median age of marriage for women with secondary education exceeded that of women with no formal education by four years.

Another fundamental strategy for encouraging later childbirth is to help women break out of the "sterilization syndrome" by providing and promoting high-quality, temporary contraceptives. Sterilization might appear to be the ideal form of contraception because it's permanent. But precisely because it is permanent, women considering sterilization tend to have their children early, and then resort to it. A family planning program that relies heavily on sterilization may therefore be working at cross purposes with itself: when offered as a primary form of contraception, sterilization tends to promote early childbirth.

What Happened to the Cairo Pledges?

At the 1994 Cairo Conference, some 180 nations agreed on a 20-year reproductive health package to slow population growth. The agreement called for a progressive rise in annual funding over the life of the package; according to U.N. estimates, the annual price tag would come to about $17 billion by 2000 and $21.7 billion by 2015. Developing countries agreed to pay for two thirds of the program, while the developed countries were to pay for the rest. On a global scale, the package was fairly modest: the annual funding amounts to less than two weeks' worth of global military expenditures.

Today, developing country spending is largely on track with the Cairo agreement, but the developed countries are not keeping their part of the bargain. According to a recent study by the U.N. Population Fund (UNFPA), all forms of developed country assistance (direct foreign aid, loans from multilateral agencies, foundation grants, and so on) amounted to only $2 billion in 1995. That was a 24 percent increase over the previous year, but preliminary estimates indicate that support declined some 18 percent in 1996 and last year's funding levels were probably even lower than that.

The United States, the largest international donor to population programs, is not only failing to meet its Cairo commitments, but is toying with a policy that would undermine international family planning efforts as a whole. Many members of the U.S. Congress are seeking reimposition of the "Mexico City Policy" first enunciated by President Ronald Reagan at the 1984 U.N. population conference in Mexico City, and repealed by the Clinton administration in 1993. Essentially, a resurrected Mexico City Policy would extend the current U.S. ban on funding abortion services to a ban on funding any organization that:

- funds abortions directly, or
- has a partnership arrangement with an organization that funds abortions, or
- provides legal services that may facilitate abortions, or
- engages in any advocacy for the provision of abortions, or
- participates in any policy discussions about abortion, either in a domestic or international forum.

The ban would be triggered even if the relevant activities were paid for entirely with non-U.S. funds. Because of its draconian limits even on speech, the policy has been dubbed the "Global Gag Rule" by its critics, who fear that it could stifle, not just abortion services, but many family planning operations involved only incidentally with abortion. Although Mexico City proponents have not managed to enlist enough support to reinstate the policy, they have succeeded in reducing U.S. family planning aid from $547 million in 1995 to $385 million in 1997. They have also imposed an unprecedented set of restrictions that meter out the money at the rate of 8 percent of the annual budget per month—a tactic that Washington Post reporter Judy Mann calls "administrative strangulation."

If the current underfunding of the Cairo program persists, according to the UNFPA study, 96 million fewer couples will use modern contraceptives in 2000 than if commitments had been met. One-third to one-half of these couples will resort to less effective traditional birth control methods; the rest will not use any contraceptives at all. The result will be an additional 122 million unintended pregnancies. Over half of those pregnancies will end in births, and about 40 percent will end in abortions. (The funding shortfall is expected to produce 16 million more abortions in 2000 alone.) The unwanted pregnancies will kill about 65,000 women by 2000, and injure another 844,000.

Population funding is always vulnerable to the illusion that the falling growth rate means the problem is going away. Worldwide, the annual population increase had dropped from a high of 87 million in 1988 to 80 million today. But dismissing the problem with that statistic is like comforting someone stuck on a railway crossing with the news that an oncoming train has slowed from 87 to 80 kilometers an hour, while its weight has increased. It will now take 12.5 years instead of 11.5 years to add the next billion people to the world. But that billion will surely arrive—and so will at least one more billion. Will still more billions follow? That, in large measure, depends on what policymakers do now. Funding alone will not ensure that population stabilizes, but lack of funding will ensure that it does not.

The Next Doubling

In the wake of the Cairo conference, most population programs are broadening their focus to include improvements in education, women's health, and women's social status among their many goals. These goals are worthy in their own right and they will ultimately be necessary for bringing population under control. But global population growth has gathered so much momentum that it could simply overwhelm a development agenda. Many countries now have little choice but to tackle their population problem in as direct a fashion as possible—even if that means temporarily ignoring other social problems. Population growth is now a global social emergency. Even as officials in both developed and developing countries open up their program agendas, it is critical that they not neglect their single most effective tool for dealing with that emergency: direct expenditures on family planning.

The funding that is likely to be the most useful will be constant, rather than sporadic. A fluctuating level of commitment, like sporadic condom use, can

end up missing its objective entirely. And wherever it's feasible, funding should be designed to develop self-sufficiency—as, for instance, with UNFPA's $1 million grant to Cuba, to build a factory for making birth control pills. The factory, which has the capacity to turn out 500 million tablets annually, might eventually even provide the country with a new export product. Self-sufficiency is likely to grow increasingly important as the fertility rate continues to decline. As Tom Merrick, senior population advisor at the World Bank explains, "while the need for contraceptives will not go away when the total fertility rate reaches two—the donors will."

Even in narrow, conventional economic terms, family planning offers one of the best development investments available. A study in Bangladesh showed that for each birth prevented, the government spends $62 and saves $615 on social services expenditures—nearly a tenfold return. The study estimated that the Bangladesh program prevents 890,000 births a year, for a net annual savings of $547 million. And that figure does not include savings resulting from lessened pressure on natural resources.

Over the past 40 years, the world's population has doubled. At some point in the latter half of the next century, today's population of 5.9 billion could double again. But because of the size of the next reproductive generation, we probably have only a relatively few years to stop that next doubling. To prevent all of the damage—ecological, economic, and social—that the next doubling is likely to cause, we must begin planning the global family with the same kind of urgency that we bring to matters of trade, say, or military security. Whether we realize it or not, our attempts to stabilize population—or our failure to act—will likely have consequences that far outweigh the implications of the military or commercial crisis of the moment. Slowing population growth is one of the greatest gifts we can offer future generations.

Jennifer D. Mitchell is a staff researcher at the Worldwatch Institute.

The End of Cheap Oil

Global production of conventional oil will begin to decline sooner than most people think, probably within 10 years

by Colin J. Campbell and Jean H. Laherrère

In 1973 and 1979 a pair of sudden price increases rudely awakened the industrial world to its dependence on cheap crude oil. Prices first tripled in response to an Arab embargo and then nearly doubled again when Iran dethroned its Shah, sending the major economies sputtering into recession. Many analysts warned that these crises proved that the world would soon run out of oil. Yet they were wrong.

Their dire predictions were emotional and political reactions; even at the time, oil experts knew that they had no scientific basis. Just a few years earlier oil explorers had discovered enormous new oil provinces on the north slope of Alaska and below the North Sea off the coast of Europe. By 1973 the world had consumed, according to many experts' best estimates, only about one eighth of its endowment of readily accessible crude oil (so-called conventional oil). The five Middle Eastern members of the Organization of Petroleum Exporting Countries (OPEC) were able to hike prices not because oil was growing scarce but because they had managed to corner 36 percent of the market. Later, when demand sagged, and the flow of fresh Alaskan and North Sea oil weakened OPEC's economic stranglehold, prices collapsed.

The next oil crunch will not be so temporary. Our analysis of the discovery and production of oil fields around the world suggests that within the next decade, the supply of conventional oil will be unable to keep up with demand. This conclusion contradicts the picture one gets from oil industry reports, which boasted of 1,020 billion barrels of oil (Gbo) in "proved" reserves at the start of 1998. Dividing that figure by the current production rate of about 23.6 Gbo a year might suggest that crude oil could remain plentiful and cheap for 43 more years—probably longer, because official charts show reserves growing.

Unfortunately, this appraisal makes three critical errors. First, it relies on distorted estimates of reserves. A second mistake is to pretend that production will remain constant. Third and most important, conventional wisdom erroneously assumes that the last bucket of oil can be pumped from the ground just as quickly as the barrels of oil gushing from wells today. In fact, the rate at which any well—or any country—can produce oil always rises to a maximum and then, when about half the oil is gone, begins falling gradually back to zero.

From an economic perspective, when the world runs completely out of oil is thus not directly relevant: what matters is when production begins to taper off. Beyond that point, prices will rise unless demand declines commensurately. Using several different techniques to estimate the current reserves of conventional oil and the amount still left to be discovered, we conclude that the decline will begin before 2010.

Digging for the True Numbers

We have spent most of our careers exploring for oil, studying reserve figures and estimating the amount of oil left to discover, first while employed at major oil companies and later as independent consultants. Over the years, we have come to appreciate that the relevant statistics are far more complicated than they first appear.

Consider, for example, three vital numbers needed to project future oil production. The first is the tally of how much oil has been extracted to date, a figure known as cumulative production. The second is an estimate of reserves, the amount that companies can pump out of known oil fields before having to abandon them. Finally, one must have an educated guess at the quantity of conventional oil that remains to be discovered and exploited. Together they add up to ultimate recovery, the total number of barrels that will have been extracted when production ceases many decades from now.

The obvious way to gather these numbers is to look them up in any of several publications. That approach works well enough for cumulative production statistics because companies meter the oil as it flows from their wells. The record of production is not perfect (for example, the two billion barrels of Kuwaiti oil wastefully burned by Iraq in 1991 is usually not included in official statistics), but errors are relatively easy to spot and rectify. Most experts agree that the industry had removed just over 800 Gbo from the earth at the end of 1997.

Getting good estimates of reserves is much harder, however. Almost all the publicly available statistics are taken from surveys conducted by the *Oil and Gas Journal* and *World Oil*. Each year these two trade journals query oil firms and governments around the world. They then publish whatever production and reserve numbers they receive but are not able to verify them.

The results, which are often accepted uncritically, contain systematic errors. For one, many of the reported figures are unrealistic. Estimating reserves is an inexact science to begin with, so petroleum engineers assign a probability to their assessments. For example, if, as geologists estimate, there is a 90 percent chance that the Oseberg field in Norway contains 700 million barrels of recoverable oil but only a 10 percent chance that it will yield 2,500 million more barrels, then the lower figure should be cited as the so-called P90 estimate (P90 for "probability 90 percent") and the higher as the P10 reserves.

In practice, companies and countries are often deliberately vague about the likelihood of the reserves they report, preferring instead to publicize whichever figure, within a P10 to P90 range, best suits them. Exaggerated estimates can, for instance, raise the price of an oil company's stock.

The members of OPEC have faced an even greater temptation to inflate their reports because the higher their reserves, the more oil they are allowed to export. National companies, which have exclusive oil rights in the main OPEC countries, need not (and do not) release detailed statistics on each field that could be used to verify the country's total reserves. There is thus good reason to suspect that when, during the late

5 ❖ POPULATION, RESOURCES, AND SOCIOECONOMIC DEVELOPMENT

COURTESY OF THE SOCIETY OF EXPLORATION GEOPHYSICISTS

FLOW OF OIL starts to fall from any large region when about half the crude is gone. Adding the output of fields of various sizes and ages (*bottom curves at right*) usually yields a bell-shaped production curve for the region as a whole. M. King Hubbert (*left*), a geologist with Shell Oil, exploited this fact in 1956 to predict correctly that oil from the lower 48 American states would peak around 1969.

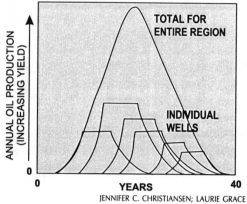

JENNIFER C. CHRISTIANSEN; LAURIE GRACE

1980s, six of the 11 OPEC nations increased their reserve figures by colossal amounts, ranging from 42 to 197 percent, they did so only to boost their export quotas.

Previous OPEC estimates, inherited from private companies before governments took them over, had probably been conservative, P90 numbers. So some upward revision was warranted. But no major new discoveries or technological breakthroughs justified the addition of a staggering 287 Gbo. That increase is more than all the oil ever discovered in the U.S.—plus 40 percent. Non-OPEC countries, of course, are not above fudging their numbers either: 59 nations stated in 1997 that their reserves were unchanged from 1996. Because reserves naturally drop as old fields are drained and jump when new fields are discovered, perfectly stable numbers year after year are implausible.

Unproved Reserves

Another source of systematic error in the commonly accepted statistics is that the definition of reserves varies widely from region to region. In the U.S., the Securities and Exchange Commission allows companies to call reserves "proved" only if the oil lies near a producing well and there is "reasonable certainty" that it can be recovered profitably at current oil prices, using existing technology. So a proved reserve estimate in the U.S. is roughly equal to a P90 estimate.

Regulators in most other countries do not enforce particular oil-reserve definitions. For many years, the former Soviet countries have routinely released wildly optimistic figures—essentially P10 reserves. Yet analysts have often misinterpreted these as estimates of "proved" reserves. *World Oil* reckoned reserves in the former Soviet Union amounted to 190 Gbo in 1996, whereas the *Oil and Gas Journal* put the number at 57 Gbo. This large discrepancy shows just how elastic these numbers can be.

Using only P90 estimates is not the answer, because adding what is 90 percent likely for each field, as is done in the U.S., does not in fact yield what is 90 percent likely for a country or the entire planet. On the contrary, summing many P90 reserve estimates always understates the amount of proved oil in a region. The only correct way to total up reserve numbers is to add the mean, or average, estimates of oil in each field. In practice, the median estimate, often called "proved and probable," or P50 reserves, is more widely used and is good enough. The P50 value is the number of barrels of oil that are as likely as not to come out of a well during its lifetime, assuming prices remain within a limited range. Errors in P50 estimates tend to cancel one another out.

We were able to work around many of the problems plaguing estimates of conventional reserves by using a large body of statistics maintained by Petroconsultants in Geneva. This information, assembled over 40 years from myriad sources, covers some 18,000 oil fields worldwide. It, too, contains some dubious reports, but we did our best to correct these sporadic errors.

According to our calculations, the world had at the end of 1996 approximately 850 Gbo of conventional oil in P50 reserves—substantially less than the 1,019 Gbo reported in the *Oil and Gas Journal* and the 1,160 Gbo estimated by *World Oil*. The difference is actually greater than it appears because our value represents the amount most likely to come out of known oil fields, whereas the larger number is supposedly a cautious estimate of proved reserves.

For the purposes of calculating when oil production will crest, even more critical than the size of the world's reserves is the size of ultimate recovery—all the cheap oil there is to be had. In order to estimate that, we need to know whether, and how fast, reserves are moving up or down. It is here that the official statistics become dangerously misleading.

Diminishing Returns

According to most accounts, world oil reserves have marched steadily upward over the past 20 years. Extending that apparent trend into the future, one could easily conclude, as the U.S. Energy Information Administration has, that oil production will continue to rise unhindered for decades to come, increasing almost two thirds by 2020.

Such growth is an illusion. About 80 percent of the oil produced today flows

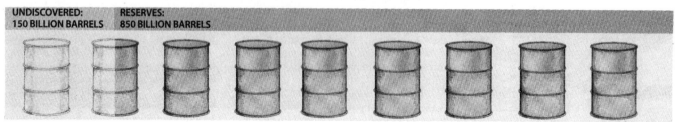

EARTH'S CONVENTIONAL CRUDE OIL is almost half gone. Reserves (defined here as the amount as likely as not to come out of known fields) and future discoveries together will provide little more than what has already been burned.

UNDISCOVERED: 150 BILLION BARRELS

RESERVES: 850 BILLION BARRELS

32. End of Cheap Oil

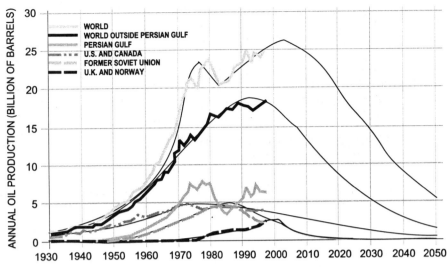

GLOBAL PRODUCTION OF OIL, both conventional and unconventional, recovered after falling in 1973 and 1979. But a more permanent decline is less than 10 years away, according to the authors' model, based in part on multiple Hubbert curves (*thin lines*). U.S. and Canadian oil topped out in 1972; production in the former Soviet Union has fallen 45 percent since 1987. A crest in the oil produced outside the Persian Gulf region now appears imminent.

LAURIE GRACE; SOURCE: JEAN H. LAHERRÈRE

from fields that were found before 1973, and the great majority of them are declining. In the 1990s oil companies have discovered an average of seven Gbo a year; last year they drained more than three times as much. Yet official figures indicated that proved reserves did not fall by 16 Gbo, as one would expect—rather they expanded by 11 Gbo. One reason is that several dozen governments opted not to report declines in their reserves, perhaps to enhance their political cachet and their ability to obtain loans. A more important cause of the expansion lies in revisions: oil companies replaced earlier estimates of the reserves left in many fields with higher numbers. For most purposes, such amendments are harmless, but they seriously distort forecasts extrapolated from published reports.

To judge accurately how much oil explorers will uncover in the future, one has to backdate every revision to the year in which the field was first discovered—not to the year in which a company or country corrected an earlier estimate. Doing so reveals that global discovery peaked in the early 1960s and has been falling steadily ever since. By extending the trend to zero, we can make a good guess at how much oil the industry will ultimately find.

We have used other methods to estimate the ultimate recovery of conventional oil for each country [*see box*], EARTH's CONVENTIONAL CRUDE OIL, and we calculate that the oil industry will be able to recover only about another 1,000 billion barrels of conventional oil. This number, though great, is little more than the 800 billion barrels that have already been extracted.

It is important to realize that spending more money on oil exploration will not change this situation. After the price of crude hit all-time highs in the early 1980s, explorers developed new technology for finding and recovering oil, and they scoured the world for new fields. They found few: the discovery rate continued its decline uninterrupted. There is only so much crude oil in the world, and the industry has found about 90 percent of it.

Predicting the Inevitable

Predicting when oil production will stop rising is relatively straightforward once one has a good estimate of how much oil there is left to produce. We simply apply a refinement of a technique first published in 1956 by M. King Hubbert. Hubbert observed that in any large region, unrestrained extraction of a finite resource rises along a bell-shaped curve that peaks when about half the resource is gone. To demonstrate his theory, Hubbert fitted a bell curve to production statistics and projected that crude oil production in the lower 48 U.S. states would rise for 13 more years, then crest in 1969, give or take a year. He was right: production peaked in 1970 and has continued to follow Hubbert curves with only minor deviations. The flow of oil from several other regions, such as the former Soviet Union and the collection of all oil producers outside the Middle East, also follows Hubbert curves quite faithfully.

The global picture is more complicated, because the Middle East members of OPEC deliberately reined back their oil exports in the 1970s, while other nations continued producing at full capacity. Our analysis reveals that a number of the largest producers, including Norway and the U.K., will reach their peaks around the turn of the millennium unless they sharply curtail production. By 2002 or so the world will rely on Middle East nations, particularly five near the Persian Gulf (Iran, Iraq, Kuwait, Saudi Arabia and the United Arab Emirates), to fill in the gap between dwindling supply and growing demand. But once approximately 900 Gbo have been consumed, production must soon begin to fall. Barring a global recession, it seems most likely that world production of conventional oil will peak during the first decade of the 21st century.

Perhaps surprisingly, that prediction does not shift much even if our estimates are a few hundred billion barrels high or low. Craig Bond Hatfield of the University of Toledo, for example, has conducted his own analysis based on a 1991 estimate by the U.S. Geological Survey of 1,550 Gbo remaining—55 percent higher than our figure. Yet he similarly concludes that the world will hit maximum oil production within the next 15 years. John D. Edwards of the University of Colorado publish-

PRODUCED: 800 BILLION BARRELS

How Much Oil Is Left to Find?

We combined several techniques to conclude that about 1,000 billion barrels of conventional oil remain to be produced. First, we extrapolated published production figures for older oil fields that have begun to decline. The Thistle field off the coast of Britain, for example, will yield about 420 million barrels (*a*). Second, we plotted the amount of oil discovered so far in some regions against the cumulative number of exploratory wells drilled there. Because larger fields tend to be found first—they are simply too large to miss—the curve rises rapidly and then flattens, eventually

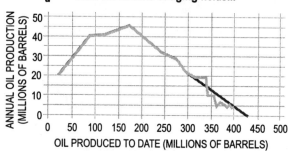

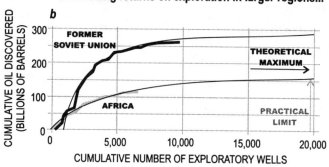

ed last August one of the most optimistic recent estimates of oil remaining: 2,036 Gbo. (Edwards concedes that the industry has only a 5 percent chance of attaining that very high goal.) Even so, his calculations suggest that conventional oil will top out in 2020.

Smoothing the Peak

Factors other than major economic changes could speed or delay the point at which oil production begins to decline. Three in particular have often led economists and academic geologists to dismiss concerns about future oil production with naive optimism.

First, some argue, huge deposits of oil may lie undetected in far-off corners of the globe. In fact, that is very unlikely. Exploration has pushed the frontiers back so far that only extremely deep water and polar regions remain to be fully tested, and even their prospects are now reasonably well understood. Theoretical advances in geochemistry and geophysics have made it possible to map productive and prospective fields with impressive accuracy. As a result, large tracts can be condemned as barren. Much of the deepwater realm, for example, has been shown to be absolutely nonprospective for geologic reasons.

What about the much touted Caspian Sea deposits? Our models project that oil production from that region will grow until around 2010. We agree with analysts at the USGS World Oil Assessment program and elsewhere who rank the total resources there as roughly equivalent to those of the North Sea—that is, perhaps 50 Gbo but certainly not several hundreds of billions as sometimes reported in the media.

A second common rejoinder is that new technologies have stead-ily increased the fraction of oil that can be recovered from fields in a basin—the so-called recovery factor. In the 1960s oil companies assumed as a rule of thumb that only 30 percent of the oil in a field was typically recoverable; now they bank on an average of 40 or 50 percent. That progress will continue and will extend global reserves for many years to come, the argument runs.

Of course, advanced technologies will buy a bit more time before production starts to fall [see "Oil Production in the 21st Century," by Roger N. Anderson*]. But most of the apparent improvement in recovery factors is an artifact of reporting. As oil fields grow old, their owners often deploy newer technology to slow their decline. The falloff also allows engineers to gauge the size of the field more accurately and to correct previous underestimation—in particular P90 estimates that by definition were 90 percent likely to be exceeded.

Another reason not to pin too much hope on better recovery is that oil companies routinely count on technological progress when they compute their reserve estimates. In truth, advanced technologies can offer little help in draining the largest basins of oil, those onshore in the Middle East where the oil needs no assistance to gush from the ground.

Last, economists like to point out that the world contains enormous caches of unconventional oil that can substitute for crude oil as soon as the price rises high enough to make them profitable. There is no question that the resources are ample: the Orinoco oil belt in Venezuela has been assessed to contain a staggering 1.2 trillion barrels of the sludge known as heavy oil. Tar sands and shale deposits in

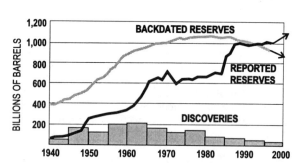

LAURIE GRACE; SOURCE: PETROCONSULTANTS, *OIL AND GAS JOURNAL* AND U.S. GEOLOGICAL SURVEY

GROWTH IN OIL RESERVES since 1980 is an illusion caused by belated corrections to oil-field estimates. Back-dating the revisions to the year in which the fields were discovered reveals that reserves have been falling because of a steady decline in newfound oil (*bottom bars*).

Canada and the former Soviet Union may contain the equivalent of more than 300 billion barrels of oil [see "Mining for Oil," by Richard L. George *]. Theoretically, these unconventional oil reserves could quench the world's thirst for liquid fuels as conventional oil passes its prime. But the industry will be hard-pressed for the time and money needed to ramp up production of unconventional oil quickly enough.

Such substitutes for crude oil might also exact a high environmental price. Tar sands

reaching a theoretical maximum: for Africa, 192 Gbo. But the time and cost of exploration impose a more practical limit of perhaps 165 Gbo (*b*). Third, we analyzed the distribution of oilfield sizes in the Gulf of Mexico and other provinces. Ranked according to size and then graphed on a logarithmic scale, the fields tend to fall along a parabola that grows predictably over time. (*c*). (Interestingly, galaxies, urban populations and other natural agglomerations also seem to fall along such parabolas.) Finally, we checked our estimates by matching our projections for oil production in large areas, such as the world outside the Persian Gulf region, to the rise and fall of oil discovery in those places decades earlier (*d*).

—*C.J.C. and J.H.L.*

...by extrapolating the size of new fields into the future...

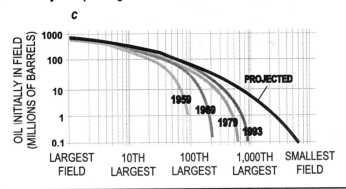

...and by matching production to earlier discovery trends.

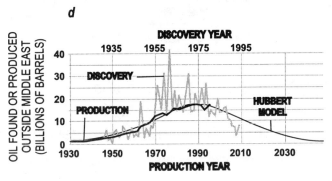

LAURIE GRACE; SOURCE: JEAN H. LAHERRÈRE

typically emerge from strip mines. Extracting oil from these sands and shales creates air pollution. The Orinoco sludge contains heavy metals and sulfur that must be removed. So governments may restrict these industries from growing as fast as they could. In view of these potential obstacles, our skeptical estimate is that only 700 Gbo will be produced from unconventional reserves over the next 60 years.

On the Down Side

Meanwhile global demand for oil is currently rising at more than 2 percent a year. Since 1985, energy use is up about 30 percent in Latin America, 40 percent in Africa and 50 percent in Asia. The Energy Information Administration forecasts that worldwide demand for oil will increase 60 percent (to about 40 Gbo a year) by 2020.

The switch from growth to decline in oil production will thus almost certainly create economic and political tension. Unless alternatives to crude oil quickly prove themselves, the market share of the OPEC states in the Middle East will rise rapidly. Within two years, these nations' share of the global oil business will pass 30 percent, nearing the level reached during the oil-price shocks of the 1970s. By 2010 their share will quite probably hit 50 percent.

The world could thus see radical increases in oil prices. That alone might be sufficient to curb demand, flattening production for perhaps 10 years. (Demand fell more than 10 percent after the 1979 shock and took 17 years to recover.) But by 2010 or so, many Middle Eastern nations will themselves be past the midpoint. World production will then have to fall.

With sufficient preparation, however, the transition to the post-oil economy need not be traumatic. If advanced methods of producing liquid fuels from natural gas can be made profitable and scaled up quickly, gas could become the next source of transportation fuel [see "Liquid Fuels from

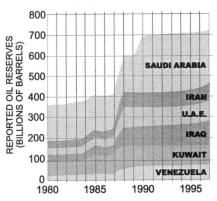

SUSPICIOUS JUMP in reserves reported by six OPEC members added 300 billion barrels of oil to official reserve tallies yet followed no major discovery of new fields.

Natural Gas," by Safaa A. Fouda *]. Safer nuclear power, cheaper renewable energy, and oil conservation programs could all help postpone the inevitable decline of conventional oil.

Countries should begin planning and investing now. In November a panel of energy experts appointed by President Bill Clinton strongly urged the administration to increase funding for energy research by $1 billion over the next five years. That is a small step in the right direction, one that must be followed by giant leaps from the private sector.

The world is not running out of oil—at least not yet. What our society does face, and soon, is the end of the abundant and cheap oil on which all industrial nations depend.

The Authors

COLIN J. CAMPBELL and JEAN H. LAHERRÈRE have each worked in the oil industry for more than 40 years. After completing his Ph.D. in geology at the University of Oxford, Campbell worked for Texaco as an exploration geologist and then at Amoco as chief geologist for Ecuador. His decade-long study of global oil-production trends has led to two books and numerous papers. Laherrère's early work on seismic refraction surveys contributed to the discovery of Africa's largest oil field. At Total, a French oil company, he supervised exploration techniques worldwide. Both Campbell and Laherrère are currently associated with Petroconsultants in Geneva.

Further Reading

UPDATED HUBBERT CURVES ANALYZE WORLD OIL SUPPLY. L. F. Ivanhoe in *World Oil*, Vol. 217, No. 11, pages 91–94; November 1996.

THE COMING OIL CRISIS. Colin J. Campbell. Multi-Science Publishing and Petroconsultants, Brentwood, England, 1997.

OIL BACK ON THE GLOBAL AGENDA. Craig Bond Hatfield in *Nature*, Vol. 387, page 121; May 8, 1997.

Editor's note: All of the articles mentioned in this article can be found in Scientific American, March 1998.

Reseeding the Green Revolution

High-yielding varieties of wheat, rice, and maize helped double world grain production. A repeat performance is now needed, and that will require a new commitment to agricultural research

Searing images of Ethiopian children with bloated bellies and flies clinging to their faces spurred the world in 1984 to combat that nation's devastating famine. To publicize their plight, which was exacerbated by the country's raging civil war, UNESCO filmed actresses Liv Ullman and the late Audrey Hepburn touring camps full of starving children. Pop icons Michael Jackson and Lionel Richie raised millions for the cause by gathering U.S. musicians to sing "We Are the World." In the United Kingdom, rock star Bob Geldof launched a similar campaign, called Band Aid. His arrival in Addis Ababa with a planeload of food sealed Ethiopia's reputation as the epitome of a country incapable of feeding itself.

Things have changed. The civil war ended in 1991, and since then Ethiopia has almost doubled its production of grain. Last year it exported about 200,000 tons of grain to neighboring Kenya, which was hard hit by drought. "Even people in Africa can't believe that Ethiopia is *exporting food*—do you have any idea what a change that is?" asks Nobel Peace Prize–winning plant geneticist Norman Borlaug of CIMMYT, the Mexican cereals-research center.

Borlaug was a captain of the Green Revolution—the potent combination of higher-yielding grain varieties, greatly increased use of chemical fertilizer, and the techniques to demonstrate their use to poor farmers—that enabled much of Asia and Latin America to achieve agricultural self-sufficiency in the 1960s and 1970s. While acknowledging that many Ethiopians still go hungry because of poverty and poor food distribution networks, he says the nation's turnaround shows that Green Revolution ideas can help Africa feed itself in the 1990s. "If African political leaders put agriculture high on their order of priorities, rather than military hardware, and if foreign assistance programs are reasonably well funded, we could see some really dramatic improvements in the next 5 to 6 years."

But Borlaug—like many agricultural researchers—is less optimistic about the prospects for rapidly boosting crop yields in the rest of the world. The Green Revolution has already transformed agriculture in most of Asia, Europe, and the Americas, with enormous impact on human well-being. Globally speaking, a child is less likely to be malnourished today than ever before. "The problem is that population growth hasn't stopped,

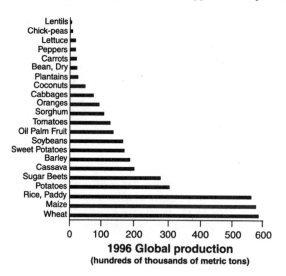

1996 Global production
(hundreds of thousands of metric tons)

SOURCE: FAO

so the Green Revolution has to happen all over again," says agricultural economist Lester R. Brown, president of Worldwatch Institute, an environmental advocacy group in Washington, D.C. "And that won't be easy."

By the year 2030, the United Nations predicts, today's world population of 5.9 billion will likely jump to 7.1 billion. The impact of 1.2 billion additional mouths will be compounded by affluence, which drives up consumption of meat, requiring high volumes of grain for animal feed. "So you're not just increasing the load on what land we have, you're multiplying it," says William C. Paddock, a retired Iowa State University plant pathologist who has written about potential food shortfalls since the 1960s.

The surge in demand will occur even as evidence suggests that the Green Revolution is petering out. In recent years grain yields have stopped rising as fast, and plant scientists agree that they are facing physical limits as they try to coax plants to produce ever more of their weight in grain. Supplies of fresh water are growing scarce. Soil quality is deteriorating. There is little unplanted arable land left to exploit. "We're running out of gas at the time we most need it," Paddock says.

Will humankind ultimately be able to feed itself? Interviews with plant breeders, crop physiologists, and botanical geneticists in Africa, Asia, Europe, and North America suggest that the answer is yes. Life can be sustained, and at a relatively healthy level. But, they caution, it can happen only if the world engages in a gigantic, multiyear, multibillion-dollar scientific effort—a kind of agricultural "person-on-the-Moon project," says Kenneth G. Cassman, a crop physiologist at Nebraska State University, Lincoln, who specializes in rice.

To feed the world, Cassman and other researchers say, scientists will have to bring modern agricultural methods to areas where they are not now used, as Borlaug and others

33. Reseeding the Green Revolution

are trying to do in Africa. At the same time, they will have to squeeze more productivity from every piece of arable land where the Green Revolution has already taken place by developing "high-precision farming" techniques that optimize every step from seed to harvest.

Unfortunately, there is little evidence that society—or science—will embrace this heroic task in time. "The stakes are huge," Cassman says, "but when you look at it in terms of the global agenda of science, most people aren't even aware of it. People are more concerned about the impact of an asteroid." Worldwide funding for agricultural science is flagging, and several major research institutions are laying off staff. As a result, Cassman says, "I think that science could feed the world, but I'm quite worried that it won't be allowed to."

Defusing a bomb

Ecologist Paul Ehrlich, author of *The Population Bomb*, forecast in 1969 that within a decade, Japan would starve and a horde of famished Chinese would invade Russia. "Most of the people who are going to die in the greatest cataclysm in the history of man have already been born," he warned.

Back then, according to the U.N. Food and Agricultural Organization (FAO), 56% of the human race lived in nations with average per capita food supplies of 2200 calories per day or less, a level barely enough to get by. As human numbers climbed relentlessly, population seemed destined to outstrip food production in the classic Malthusian scenario. Instead, grain harvests soared. By 1992–1994, FAO estimates, the percentage of the world's population living at or below 2200 calories/day had fallen to 10%. Ehrlich hadn't included the Green Revolution in his apocalyptic scenario.

It began in 1943, when Borlaug, funded by the Rockefeller Foundation and the Mexican Ministry of Agriculture, headed a program to breed high-yielding wheat varieties that resisted stern rust, a fungus that then plagued Latin American agriculture. The program set up two labs separated by 10 degrees of latitude and 2600 meters in altitude. By simultaneously testing wheat strains in both stations, the program developed high-yielding, rust-resistant hybrids that were insensitive to climatic variables such as temperature and day length.

The new varieties produced so much grain, in fact, that they "lodged"—that is, the plants became top-heavy and fell over, ruining the crop. So the researchers sought out wheat strains with shorter, stouter stalks, screening the entire U.S. Department of Agriculture (USDA) wheat collection before learning of some "dwarf" varieties in Japan. Except for its size, however, the Japanese wheat was unpromising. It produced unusable grain, was often sterile, and was so susceptible to disease that the first year's experimental crop was wholly lost to rust. After seven more years, Borlaug's team introduced the dwarfing genes without the undesirable characteristics. And there was an unexpected bonus: The dwarfing genes ultimately had synergistic effects on yield, increasing harvests to as much as 8 tons per hectare (t/ha)—a staggering increase from the previous average of 0.75 t/ha.

In 1960, the Rockefeller Foundation, the Ford Foundation, the U.S. Agency for International Development, and the Filipino government launched a similar campaign for rice. The International Rice Research Institute (IRRI), based in Los Baños, the Philippines, duplicated the Mexican success by breeding high-yielding, disease-resistant strains of rice and crossing them with dwarf varieties to prevent lodging. Meanwhile, Borlaug's program—renamed the International Center for Maize and Wheat Improvement, but known by its Spanish acronym, CIMMYT—adapted the new wheat strains for Pakistan and India. Largely because of the introduction of hybrid wheat and rice, says USDA, farmers around the world raised grain yields by an average of 2.1% a year between 1950 and 1990, almost tripling grain harvests during that period.

Recently, though, the rate of increase has slowed. Since the harvest of 1989–1990, world grain yields have risen by just 0.5% per year. As a result, cereal stockpiles plunged from 383 million tons in 1992 to an estimated 281 million tons in 1997 "well below... the minimum necessary to safeguard world food security," according to the May/June FAO "Food Outlook" report. And cereal stocks held by developing countries have declined for 3 years in a row.

To Brown, the declining stocks are "clear signs" that humankind is running into "a fundamental biological phenomenon—the S-shaped growth curve—where enormous increases in productivity hit a wall and level out." He believes the shortfalls are the harbinger of a widening, long-term, and nearly unavoidable "gap between the demand and supply of grain."

Others dismiss these gloomy predictions. According to Timothy Roche, grain chair at the USDA Production Estimates and Crop Assessment Division, the primary cause of the slippage in world productivity gains was the collapse of the Soviet Union, where grain harvests fell from 180 million tons in 1989–1990, the last year of Communism, to an estimated 116 million tons in 1996–1997. Far from signaling the approach of ecological limits, Roche says, the drop in the ex-Communist states "is 100% economics." Indeed, had Soviet yields remained at the levels attained in 1989—an average year until today, global grain yields would have risen at an annual clip of 1.6%, more than triple the apparent rate. And many agriculturists expect that as the economies stabilize in the former Soviet Union, yields will gradually bounce back.

Although 1.6% per year still represents a decline, economist Nikos Alexandratos, chief of FAO's global perspective studies in Rome, says it may not be a problem. "One does not need to continue the growth at the same rate we have in the past for the simple reason that today a much higher proportion of the world population is well fed," says Alexandratos. Because production in rich countries doesn't need to increase, he says, "the aggregate will not grow as fast [as it did] in the past. If you observe slower growth today than yesterday, it could be good"—an indication of success, not failure.

Spreading the revolution

To Borlaug, worrying about the Green Revolution slowing is less important than recognizing that it has never been fully applied to some of the world's poorest areas, especially Africa. To be sure, Africa is an especially hard case—much of its soil is eroded, nitrogen-deprived, and lacking in organic matter. With sufficient water, fertilization could overcome many of these deficiencies. But arid Africa does not have the water. "It's more difficult there in every way," says Nebraska's Cassman.

Still, Borlaug believes that Africa has been "unjustly neglected." Indeed, the International Crops Research Institute for the Semi-Arid Tropics (ICRISAT), in Hyderabad, India—the equivalent of IRRI and CIMMYT for African staples like sorghum and millet—was not founded until 1972, more than a decade after its fellows. And even when research was done, Borlaug says, the world made "no major effort to move the technology to farmers' fields."

Dismayed by the slow progress, the late Ryoichi Sasakawa, a Japanese industrialist-philanthropist, asked Borlaug in 1986 to come out of retirement and bring the Green Revolution to Africa. With Borlaug and former President Jimmy Carter on board, Sasakawa created Sasakawa Global 2000 (SG2000), hoping to do in Africa what the Rockefeller Foundation did in Latin America 40 years before. Today, Borlaug says, SG2000 has set up between 350,000 and 400,000 demonstration plots where Green Revolution approaches are compared to traditional, current practices. "What this has shown is that you can always at least double the yields—and frequently triple them—and in some cases quadruple them by the application of the best package of technology that you can put together," he says. "very simple steps can have a dramatic impact." Africans rarely use commercial fertilizer, for example, and the first high-yield sorghum was only brought to the continent in 1991.

A recent, dramatic success story has been Ethiopia, says Marco Quiñones, an SG2000 agronomist in Addis Ababa. Until 1991, the ruling military cabal favored heavy industry

Saving Sorghum by Boiling the Wicked Witchweed

Civil war, genocide, corruption, and political incompetence have conspired to keep entire regions of Africa on the brink of famine, earning it the sad reputation as the hungry continent. But even if these dreadful socioeconomic problems were alleviated, African agriculture would still suffer from a host of more traditional problems—insects, birds, and plant diseases. Indeed, one of the greatest sources of crop losses in Africa is not war or corruption, but three species of the parasitic plant *Striga hermonthica*.

Commonly known as witchweed, *Striga* feeds on the roots of cereals and legumes in much of Africa and South Asia. Estimates of crop losses caused by *Striga* range from 15% to 40% of Africa's total cereal harvest; many areas lose two-thirds or more of their crops every year. Gebisa Ejeta, an agronomist at Purdue University, says *Striga*—which attacks maize, sorghum, and millet, Africa's three most important cereals—has long been the "strongest biological constraint to crop production" in the continent.

All efforts to control what Ejeta calls "this scourge" failed until he and his Purdue colleague, the late Larry Butler, developed the first *Striga*-resistant sorghum. Introduced in 1995, the new varieties are now grown in such desperately poor places as Chad, Mali, Niger, Rwanda, and the Sudan. According to Marco Quiñones, an agronomist in Addis Ababa, Ethiopia, the Ejeta-Butler sorghum was so successful that it spread throughout Sudan despite a civil war. Farmers in neighboring Ethiopia wanted *Striga*-resistant sorghum badly enough to smuggle the seeds across the hostile border. "They are growing sorghum in areas that were abandoned to *Striga* for years," Quiñones says. "The change is enormous."

From a biological perspective, *Striga* is a fascinating problem. Once established, witchweed is almost impossible to eradicate—the United States has spent millions in an attempt to contain a single small outbreak in the Carolinas. Each plant produces 40,000 to 100,000 seeds, although production of half a million seeds has been observed. The seeds, smaller than grains of sand, lie dormant in the soil for as long as 20 years, germinating only when stimulated by a specific chemical exudate given off by the root of the host plant. After germination, a second host exudate triggers the development of a root-like organ called a haustorium, which the parasite uses to penetrate the host and siphon away nutrients. Dozens of plants can attack the same host, stunting or killing it. Although *Striga* eventually grows into an 80-cm plant with bright pink or red flowers, it wreaks most harm while still invisibly underground. "Before farmers know they have the parasite in their farms," Ejeta says, "the damage has been done."

S. hermonthica and *S. asiatica* parasitize cereals; *S. gesneroides* targets cowpeas and tobacco. But all three rapidly adapt to new hosts—one reason that *Striga* losses are growing. Barley and the Ethiopian grain teff, once perceived as immune, are now attacked. Pearl millet was introduced to eastern Sudan, where sorghum crops had been wiped out by *S. hermonthica*, the largest and most virulent species. Within a few years *Striga* was wiping out millet, too.

In the past, African farmers shifted their planting from one plot to another, rotating crops with long fallow periods between. "If a problem came up with *Striga*," Ejeta says, "they had the luxury of leaving the field alone for a few years." With populations rising, farmers now often stay put, cropping the same land—ideal conditions for *Striga*.

Having left his native Ethiopia in 1974 just before a military coup deposed Emperor Haile Selassie, Ejeta finished his doctorate in the United States and then went to the Sudan office of the International Crops Research Institute for the Semi-Arid Tropics (ICRISAT), an agricultural think tank headquartered in Hyderabad, India. At ICRISAT, he developed Hageen Dura-1, the first commercial sorghum hybrid in sub-Saharan Africa, and introduced it to Sudanese farmers. Although the new variety was drought-tolerant and yielded as much as 150% more than traditional varieties, it was plagued by witchweed. When Ejeta came to Purdue in 1984, he was determined to do something about this parasite.

Believing the interactions among parasite, host, and environment are too complex to control in field conditions, Ejeta decided to unravel the basic biology of *Striga*. He teamed up with Butler, a biochemist, hoping that understanding the specific biochemical signals between host and parasite would allow farmers to disrupt them. In 1992, Bupe Siame, a Zambian graduate student working with Ejeta and Butler, identified the exudate—sorgolactone—that activated germination of *Striga* seeds in sorghum, maize, and millet.

Dale Hess, another Purdue graduate student, meanwhile developed a simple agar assay that separated sorghum genotypes on the basis of their ability to germinate *Striga*. He spread *Striga* seeds evenly over a petri dish and placed a growing sorghum seed in the center. By measuring at intervals the distance of the furthest germinated *Striga* seed from the sorghum, the researchers determined the host's level of sorgolactone.

Using this technique, Hess and Ejeta discovered that one sorghum line, SRN-39, had a recessive gene that limited sorgolactone production. "The ability to produce this chemical compound was under simple Mendelian control," Ejeta says. "So we were able to extract the gene through conventional plant breeding and put it into eight varieties of cultivated sorghum." Field testing took place in Niger, Sudan, and Mali. (Field studies of *Striga* are usually performed in Africa, for fear of letting witchweed escape into new territory.) In the United States, Ejeta and Butler mapped the gene on the molecular linkage map for sorghum, the first step in cloning the *Striga*-resistance gene and transferring it to other crops.

By 1999, Ejeta believes, some 200,000 farms should be growing *Striga*-resistant sorghum in 12 African countries. "We were lucky," he says. "Usually varieties perform well in one environment but fail in others. Ours seem to be doing well throughout arid Africa."

The success was marred by Butler's unexpected death after surgery for prostate cancer last February. "To get these kind of breakthroughs—first in Sudan with the hybrid sorghum and then this one—I've been very fortunate to have it happen," Ejeta says. "My greatest sadness is that Larry isn't here to cherish it with me."

— C.M.

---

over farming. Partly because of urgings by Carter—who helped convince the prime minister to view SG2000 demonstration plots in 1994—the new government has emphasized agriculture, lending money for improved seed and fertilizer to the country's millions of small private farms.

As a result, Quiñones says, Ethiopian grain production went from less than 6 million tons in the 1994–1995 harvest season to an estimated 11.7 million tons in the 1996–1997 harvest season. "And that's with fertilizer levels typically less than half of what they are in the United States, ... with many farmers still not having access to improved varieties of maize and sorghum, and no improved varieties of teff [a traditional Ethiopian grain] yet available." Quiñones believes further increases will occur as farmers adopt new innovations such as parasite-resistant sorghum (see box above, *Saving Sorghum by Foiling the Wicked Witchweed*).

Researchers in Africa emphasize, however, that the continent's social and economic problems remain a major obstacle to development. SG2000 refuses to work in countries without stable governments, which

33. Reseeding the Green Revolution

bars it from many nations. It didn't begin operations in Mozambique, for instance, until the 1995–1996 season, a year after the end of a disastrous civil war. Based in the capital city of Maputo, the foundation selected 40 1-ha farms for demonstration plots, half near the border with Zimbabwe, half in the far North. It provided each farm with 100 kilograms of fertilizer and disease-resistant white maize seed. Harvests from the region near the Zimbabwe border usually averaged less than 1 t/ha, according to Wayne Haag, the SG2000 representative in Maputo. "With just this little bit of fertilizer and better seed," he says, "their yields were over 3 t/ha."

In the North, though, the results were less beneficial, but not because the improved seed and fertilizer failed to produce—the farms averaged 4.7 t/ha. The northern farmers were unable to take advantage of the surplus. The cash-strapped Mozambican government didn't fulfill its promise to buy excess production for about $120/t. Worse, the poor condition of local roads prevented farmers from transporting their produce. The northern area ended up awash in maize; with stockpiles rotting, the price fell to a ruinous $40/t. Meanwhile, a drought hit southern Mozambique, which paid $160/t to import maize from South Africa. "It made economic sense," Haag says. "If you figured in paying the transportation costs, which are fairly high due to the poor infrastructure, and the high interest rates, it probably would have cost nearly $200/t to take the grain produced in the North to the South."

In Haag's view, such woes demonstrate that the success of the African Green Revolution will depend on investment in infrastructure. But, he says, agencies like the International Monetary Fund are demanding that African governments "follow a very tight, austere public-financing policy, so there's no money." At the same time, foreign aid is being cut back, especially from U.S. sources. Says Haag: "Years ago, to start the Green Revolution there was a lot of external assistance. Now there's almost none and people stand back saying 'Africa is hopeless.' Well, Africa is not hopeless. You give the farmers here a chance and they respond magnificently."

Growing pains

If African nations make the necessary investments in agriculture—and Borlaug, for one, is confident that they eventually will—the region's poor soils and lack of water still make it unlikely that, even with the Green Revolution, Africa will ever produce enough surplus food to help meet the growing planetwide need. That task, says Takeshi Horie, an agronomist at Kyoto University, requires regions that have better soils, water, and climate. "To feed the world, we will have to take areas that have already increased yields greatly—Japan, California, Europe—and make them repeat it a second time," says Horie. "It will be a big job."

Just how big a job will depend in part on how fast demand for grains is likely to grow—a topic of considerable debate. In 1994, Lester Brown of Worldwatch caused an international uproar by predicting that China's growing appetite for grain would set off a worldwide economic convulsion. Brown argued that China is losing arable land and exhausting its water supply, while explosive economic growth and a shift in consumption from rice and wheat to meat will drive up its demand for grain. Brown forecast that China would have to import "massive quantities" of rice, wheat, and maize. This extraordinary demand would "trigger unprecedented rises in world food

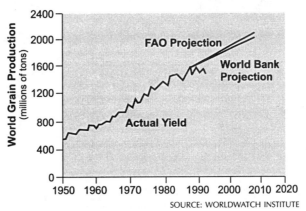

Yield of dreams? Worldwatch's Brown says World Bank and FAO predictions about future yields are overly optimistic.

prices," he said, tipping the world food balance "from surplus to scarcity" and leading to mass starvation.

Most agricultural economists agree that China's appetite for grain will surge, but they believe—surprisingly—that the world can increase production fast enough to satisfy it, and with relative ease. World Bank analysts Donald O. Mitchell and Merlinda D. Ingco, for example, predicted in a widely cited 1993 study that future yields would "continue along the path of past growth." And Alexandratos's 1996 model for FAO argues that eliminating malnutrition in the world's poorest nations by 2010 "would certainly not tax the capacity of the world."

Agricultural scientists, although tending to dismiss Brown's scenario as overly apocalyptic, are considerably less sanguine than the economists are. "The ones that predict higher yields forever, I keep asking them, 'How are we going to do this?' " says Thomas R. Sinclair, an environmental horticulturist at the University of Florida-Gainesville's Agricultural Research Center.

With his colleagues, Sinclair has been assessing the maximum potential yields of individual wheat and maize plants—the harvest index, as it is known. "To grow corn," he points out, "you have to have leaves, stalks, and roots, so there's got to be mass committed to what you don't harvest." The question is how small the nongrain proportion needs to be. "At the beginning of this century," he says, "many crops had harvest indexes on the order of 0.25 of their weight in grain, and now many crops are approaching 0.5." Sinclair says the index can't rise much higher. "Maybe you could go up to 0.6 or 0.65," Sinclair says, "but beyond that you can't have a viable plant."

Nor, he believes, can farmers keep dumping ever-greater quantities of fertilizer on their fields. Maximum yields at IRRI experimental stations declined from 10 t/ha in the tropical dry season to 6 t/ha as overuse of fertilizer reduced the level of easily decomposable organic compounds in the soil, in turn reducing its nitrogen-supplying capacity. With less ability to hold nitrogen, over-fertilized soils let it wash into rivers and groundwater, polluting them. Partly for this reason, fertilizer use in Europe—where run-off is a problem—declined from 169 kg/ha in 1988 to 116 kg/ha in 1993, the latest year for which FAO statistics are available.

These fundamental physical constraints mean that researchers can no longer easily apply the old Green Revolution paradigm—breed shorter plants with more grain per stalk, provide lots of fertilizer, and watch yields triple—to wheat, rice, and maize, the main cereal crops. "Producing higher yields will no longer be like unveiling a new model of a car," Nebraska's Cassman says. "We won't be pulling off the sheet and there it is, a twofold yield increase."

An example is the "new plant type" rice under development at IRRI. Cultivated rice grows as a clump of almost 30 stemlike "panicles" that bear the flowers and grain. But only half the panicles produce grain, so Gurdev S. Khush, the principal plant breeder at IRRI's base in the Philippines, and coworkers selected for those and thickened their stems. IRRI scientists hope that the new plant type, which should be in field tests within 3 years, will push the current 0.55 harvest index of rice to 0.6 or 0.65—a 10% to 20% increase, not the 200% to 300% increases of the past.

But even this modest rise may be unattainable. Cassman and his collaborators reviewed the literature on the new plant type at the end of 1993. "We did not find a strong scientific paper trail, based on published data, that would justify or support the supposition that there's an untapped 25% yield potential," he says. (For another approach to

191

5 ❖ POPULATION, RESOURCES, AND SOCIOECONOMIC DEVELOPMENT

Cashing in on Seed Banks' Novel Genes

In 1970, an epidemic of Southern corn leaf blight ravaged farms throughout North America, causing the biggest economic losses ever recorded for a single crop in a single year. Nothing seemed able to stop the fungus that caused the blight—until scientists discovered that a wild variety of maize was resistant to it. By crossing the wild and cultivated maizes, researchers created resistant varieties, saving thousands of farmers from ruin.

The blight spread so rapidly because 70% of the maize in the United States had the same genetic susceptibility to the disease. This stark evidence of the dangers of genetic uniformity led to an international effort to conserve crop diversity. Today, collections hold more than 6 million germ plasm samples, mostly seeds, covering some 100 crop species and their wild relatives. But as Cornell University plant breeders Steven Tanksley and Susan McCouch contend..., plant breeders have failed to exploit seed banks.

"The embarrassing, paradoxical fact is that we've made this major investment in biodiversity but—except for corn blight—hardly ever used it," says Tanksley. "Seed banks are supposed to be storehouses of important genetic traits, but breeders pay practically no attention to them." Ronald Phillips, chief scientist for the competitive grants program at the U.S. Department of Agriculture, says the article "provides a great service by alerting people to the value of hidden genes in seed banks and by pointing out the newer methods for their detection."

In part, breeders have ignored seed banks because conventional breeding involves eliminating all but the most desirable traits. As a result, breeders tend to regard seed banks as botanical *salons des refusés*: storehouses of rejected traits. But Tanksley's own work shows how useful these storehouses can be. Tanksley began hunting through seed banks for novel genes 6 years ago, eventually teaming up with his Cornell colleague McCouch and Jiming Li and Longping Yuan of the National Hybrid Rice Research and Development Center, in Hunan, China. This group now has bred rice that may yield 20% to 40% more than conventional high-yielding strains—all by capturing genes from uncultivated rice varieties that themselves show little obvious sign of being useful.

Using methods pioneered by Tanksley with tomatoes, the researchers crossed a weedy, unpromising wild Malaysian rice (*Oryza rufipogon*) with cultivated Asian rice (*Oryza sativa*), hoping that the wild species might have some unknown beneficial traits. Of the 300 test plants they bred by crossing these species, about 15% outyielded the cultivated strain, a few by as much as 50%. At Cornell, Tanksley and McCouch genetically mapped the high-yielders and found two wild genes that seemed to be responsible for the increased yield. Such a finding, McCouch says, "flies in the face of traditional breeding, where the best parents give the best children. Here, Steve [Tanksley] was taking parents with poor phenotypes and using them to improve yields in elite varieties."

Although Tanksley believes that the hybrid rice will be useful, he is most pleased by the larger implications of the new method. Maintaining but not using seed banks, he says, "was like having this huge bank account in Switzerland, but nobody had given us the password, so we couldn't tap into it. The genes that passed the muster of evolution for millions of years are sitting there, waiting to be used. And now maybe we can start using them."

—C.M.

---

increasing the rice-harvest index, see box above, *Cashing in on Seed Banks' Novel Genes.*)

Because increasing the harvest index will be difficult, progress will lie in combining a variety of approaches—breeding strains that better resist disease, tolerate acidic or metallic soils, or provide better nutrition. Borlaug is particularly excited about the aluminum-tolerant breeds of corn, soybeans, rice, wheat, and pasture grass now being tried on the highly leached Brazilian cerrado. Hopeful of finding useful new genes, U.S. cereal geneticists are proposing a federally funded project to spend more than $100 million on mapping the genomes of wheat, rice, and corn (*Science,* 27 June, p. 1960). But Cassman cautions that these "good and useful" efforts "will be expensive, compared to the past," and are unlikely "to shoot up yields overnight." Increasing crop yields, in his estimation, will be "incremental, tortuous, and slow."

Given that current varieties are approaching their biological limits, Russell Muchow of the Commonwealth Scientific and Industrial Research Organization, in Brisbane, Australia, believes that "the big opportunities lie not in raising maximum yields but in getting actual yields closer to the maximum." To find places where crop production falls below the maximum potential yield, Kyoto University's Horie factors in variables such as water availability, temperature, the length of day, and the harvest index. In the temperate California desert, Horie has calculated, the potential rice harvest is 19.3 t/ha. "The actual yield is only about 8 t/ha or 9 t/ha, so there is room for improvement by yield management," Horie says. Japan and Australia, he says, have similar possibilities. But in other areas, like China's Yunan province, Horie sees little opportunity. "The farmers get something like 13 to 15 tons per hectare, a very high yield. But there is no deficiency of nutrients or water, no insect damage, and it's carefully managed, so the yield is very close to potential."

Maximizing yields, researchers believe, ultimately requires an expensive global effort to wring the last bit of productivity from plant genomes and employ "high-precision farming" techniques to realize the gains in the field. Martin Kropff, a theoretical ecologist at Wageningen Agricultural University in the Netherlands, says that lengthening the grain-filling period between flowering and maturity of the crop is one key. Temperate environments have cooler nights, naturally providing a longer grain-filling period. "[That's] one reason why the U.S. Midwest has such high yields," says Kropff. Hotter places like Africa may be able to partly overcome their disadvantage if breeders create new varieties with longer grain-filling periods, but they will have to be precisely managed. "Having a longer grain-filling period will depend on supplying the nitrogen at exactly the right time and in exactly the right amount," he says.

R&D is starving

Most of the optimistic forecasts of farm production depend crucially on a single variable: investment in R&D. "Science is not a panacea," says Per Pinstrup-Andersen, director-general of the International Food Policy Research Institute (IFPRI). "It will take more than that. But without it, we won't make it."

Modelers bank on R&D not only providing future productivity rises, but on maintaining current agricultural conditions. Biological systems are constantly changing, as pests and diseases evolve, soil conditions change from irrigation and cropping practices, and people heat up the earth with carbon dioxide. "You have to run harder and harder just to stand still in agriculture," says IFPRI researcher Philip G. Pardey, "It's not only a matter of generating more input, it's a matter of running to keep what we have now."

The trend, however, is worrisome. Public agricultural research funding has been declining for years, according to a new analysis by Pardey, Julian M. Alston of the University of California at Davis, and Johannes Roseboom of the International Service for National Agricultural Research at The Hague. In 1985 dollars, the three researchers reported at a 10 August international gathering of agricultural economists, global research spending doubled between 1971 and 1991, from $7.3 billion to $15 billion. But the average annual rate of increase fell from 4.4% in 1971–1981 to 2.8% in 1981–1991. A continuation of this trend, warns economist Pierre Crosson, a senior fellow at Resources for the Future, a Washington, D.C.–based think tank, "would pose a major threat to the achievement of a successful supply response to [the] rising world demand for food."

Privately funded research will not come to the rescue. Although private money funded 53% of all agricultural research in 1993, the last year for which data are available, the IFPRI researchers calculate that just 12% of the money went to direct crop improvement. For self-pollinating crops like wheat and rice, Pardey explains, industry has trouble recouping its investment in new varieties, because farmers only purchase the seed once. (Industry is more interested in corn, which in the United States is mostly grown from sterile hybrid seed.) Economics thus drives private R&D to focus on drugs, pesticides, food processing, and mechanization.

The slowdown has especially hit international R&D. The principal vehicle for such research is the Consultative Group on International Agricultural Research (CGIAR), a group of 43 public- and private-sector donors that supports 17 research centers, including CIMMYT, IRRI, IFPRI, and ICRISAT. Although CGIAR funding accounts for just 2% of all agricultural R&D, its catalytic role is disproportionately important, especially in the Third World. The chief focus of many national research institutions in poor nations is developing local adaptations to CGIAR technologies. Today, CGIAR is a victim of donor fatigue. Since 1993, its budget has remained roughly constant, at about $315 million. But because that money is parceled out to an increasing number of institutions, budgets on a per-organization level have fallen. IRRI, for instance, lost almost a quarter of its $30 million budget in the last 2 years; it recently laid off 550 people, half its staff. "There is so little investment to use science to solve poor people's problems," Pinstrup-Andersen says.

Pardey is especially concerned about the level of funding in Africa. "We're not talking slowdown there," he says, "we're talking retreat from R&D. Some [of the loss] was picked up by donors, but now those donor funds are not even there." Annual research spending from all sources increased by only 0.8% in the 1980s, about a third of the average worldwide rate. An analysis of funding as a percentage of agricultural gross domestic product paints an even bleaker picture: The United States is about 2.5%, while Africa is only 0.5%. "And the gap between the intensity ratio of the developed and developing world is widening," says Pardey.

IFPRI economist Mark Rosegrant and three colleagues are refining an econometric model they unveiled in June to factor in the impact of different levels of R&D investments. But the overall picture seems clear. "We can, I think, feed everyone, even if we will continue to have problems distributing it equitably," Rosegrant says. "That's what the model indicates. But everything I'm saying could be destroyed if people stampede out of research funding." If that happens, he says, "the jig might be up."

—**Charles Mann**

How Much Food Will We Need in the 21st Century?

By William H. Bender

Seldom has the world faced an unfolding emergency whose dimensions are as clear as the growing imbalance between food and people.[1]

The world food situation has improved dramatically during the past 30 years and the prospects are very good that the 20-year period from 1990 to 2010 will see further gains.... If Malthus is ultimately to be correct in his warning that population will outstrip food production, then at least we can say: Malthus Must Wait.[2]

Ever since Malthus, society has worried periodically about whether it will be able to produce enough food to feed people in the future. Yet until recently, most of the debate surrounding the issue of food scarcity focused on the potential for increasing the food supply. The key questions were whether there would be enough land and water to produce the amount of food needed and whether technology could keep increasing the yields of food grains. Now, however, scientists are growing concerned that the intensive use of land, energy, fertilizer, and pesticides that modern agriculture seems to require jeopardizes the health of the environment. This anxiety has been integrated into the general debate about food scarcity, but interestingly enough, the question of the demand for food—including the specific physiological needs and dietary desires of different peoples—has not. In fact, relatively little attention has been paid to the issue of demand despite the fact that like energy and water, food can be conserved and the demand for it adjusted to meet human needs and lessen the burden that modern agriculture places on the environment.

Unlike with many other forms of consumption, there are limits to the physical quantity of food that people can consume. In a number of high-income countries, that limit seems to have been reached already. If global population does double by 2050, as many have predicted, providing everyone with a rich and varied diet (equivalent to that enjoyed by today's wealthiest countries) would only require a tripling of food production. Alternatively, with sufficient improvements in efficiency and adoption of a healthier diet in high-income countries, it would be possible to provide such a diet for the entire global population with just a doubling of food production. But even a doubling of current production could strain Earth's ecosystems, as critics of modern agriculture's intensive use of resources will attest. Clearly, then, increases in food demand will have to be slowed if we hope to achieve a sustainable agricultural system. Central to the issue of demand, however, is the question of how much food the world really needs.

From an analytical standpoint, the amount of food a given population (be it a country, a region, or the world) actually *needs* is the product of two factors: the number of people and the average (minimal) food requirement per person. The amount of food the population *consumes*, however, is determined not only by its basic needs but also by its income and dietary preferences. This difference is particularly important in high-income countries, where crops that could be consumed directly are instead fed to animals to produce eggs, meat, and milk. Finally, the amount of food a given population *requires* (i.e., has to produce or import) depends on how much is wasted in going from farm to mouth as well as on its level of consumption. In mathematical terms,

$$Req = Pop \cdot PFR \cdot Diet \cdot Eff,$$

where *Req* is the total number of food calories that has to be produced, *Pop* is population, *PFR* is the number of calories per person that is needed to sustain life and health, *Diet* is a factor reflecting the conversion of some plant calories to animal calories, and *Eff* is the ratio of calories available in the retail market to those consumed.

This article will address the neglected issue of food demand in terms of the four variables of this equation. In the process, it will question some of the assumptions previous analysts have made, particularly with regard to desirable diets and food system efficiency. Though not definitive, the analysis strongly suggests that the right policy choices can reduce the growth in the global demand for food. Indeed, the potential scope of such a reduction appears to be substantial: As Table 1 on the next page shows, vastly different numbers of people can be supported by a given amount of agricultural production depending on dietary habits and degrees of efficiency.

Population

Global population will play an important role in determining how much food we will require in the future. For this reason, attempts to calculate future food requirements depend upon projections of population growth. Although demographers generally agree that the current global population will double by the middle of the next century, considerable uncertainty accompanies these projections. The United Nations' estimates of the world's population in 2050, for example, vary from 7.9 billion to 11.9 billion. If global population reaches the higher value rather than the lower one, global food requirements will be 50 percent higher.

National and international policies that provide family planning services, maternal education, and social support systems can affect population growth, and these policies will undoubtedly have the single largest effect on food requirements in the 21st century. The availability of food will also play a role, however. Famine—the most dramatic example of lack of food—has fortunately been largely eliminated (except during wars) and no longer ranks as a major factor in global population growth. Even so, the relative abundance of food has a direct

34. How Much Food Will We Need?

Table 1. World population supportable under different conditions	
Conditions in	Population (billions)
United States	2.3
Europe	4.1
Japan	6.1
Balgladesh	10.9
Subsistence only	15.0
Addendum: Actual 1990 population	5.3

NOTE: This table shows the number of people that could be fed at the 1990 level of agricultural production if the dietary preferences and food system efficiencies in the countries (or area) shown prevailed throughout the world. Dietary preferences reflect both income and the extent to which cereal grains are fed to animals instead of being consumed directly. Food system efficiencies reflect the extent to which food is spoiled or wasted in going from farm to mouth.

SOURCE: Author's calculations.

effect on the other key factors that influence population growth, and combined with the subtle influences exerted by the food and agriculture sector, it can have a significant impact. For example, in rural agricultural societies, the demand for agricultural labor affects fertility rates, while reductions in child mortality (which are influenced by food availability) usually precede reduction in fertility rates.

Physiological Requirements

Physiological food requirements, represented by PFR in the equation, are determined by several factors, including the population's age and gender distribution, its average height and weight, and its activity level. One may compute such requirements in two different ways, using either actual circumstances or normative ones (such as desired heights and weights or activity levels).[3]

Around the world, actual per capita caloric consumption varies from a low of 1,758 calories per day in Bangladesh to a high of 2,346 calories per day in the Netherlands. Caloric consumption is higher in the Netherlands for several reasons. First, the population is generally older, and adults require more food than children. Second, people in the Netherlands are on average taller and heavier than those in Bangladesh and therefore need more food. (Lower activity levels in the Netherlands partially offset these factors, however.) If the actual consumption levels in these two countries were to change, either the weights of individuals or their activity levels would have to change accordingly. Caloric consumption levels vary by no more than one-third on a national basis—far less than the variation in caloric availability.

When making future projections, normative considerations can also be very important. A population's general health, for instance, affects the amount of food it needs. Parasites and disease can substantially increase an individual's energy requirements, with fever, for example, raising his or her basal metabolic rate (the number of calories he or she uses when at rest) approximately 10 percent for every one degree C increase in body temperature.[4] Disease can also impair the body's ability to absorb nutrients, while parasites siphon away food energy for their own use. Although not important globally, health factors are highly significant in certain low-income countries. In fact, in localized situations health interventions may be more effective than merely increasing the food supply in helping people to satisfy basic physiological food requirements.

Of course, to qualify as truly sustainable, the world's agricultural system has to produce enough calories to ensure food security around the globe. This is a normative concept, as is clear in the commonly accepted definition of food security: "access by all people at all times to enough food for an active, healthy life."[5] Thus, for future projections, we could consider a world with lower levels of undernutrition and stunting, leading to higher food requirements.

Table 2 shows estimates of physiological food requirements for the world as a whole, for high-income countries, and for low-income countries, all based on current circumstances. (The box on the next page discusses the way in which these estimates were prepared). High-income countries use much more than twice as much food per person. This variation is not due to differences in calories actually consumed but to differences in diet and the lower efficiency of food systems in those countries.

Dietary Patterns

Diets are largely determined by economic factors, particularly prices and incomes. In Africa, for example, people derive two-thirds of their calories from less expensive starchy staples (including cereals, roots, and tubers) and only 6 percent from animal products. In Europe, on the other hand, people derive 33 percent of their calories from animal products and less than one-third from starchy staples. The global diet falls somewhere in the mid-range between these two extremes.

As people's level of income increases, the share of starchy staples in their diet declines, and the shares of animal products, oils, sweeteners, fruits, and vegetables increase.[6] In fact, the absolute quantities of these products that people eat increase even faster than the shares because caloric availability overall also increases as incomes increase. This growing dietary diversity provides a substantial health benefit for people at the low to medium income level.

The overall increase in food availability over the last several decades, while a welcome development, has created problems of its own. As people consume more animal products, they tend to consume more animal fats than recent medical research has shown to be healthy. Currently, the World Health Organization (WHO) recommends that people limit their dietary intake of fat to no more than 30 percent of calorie consumption, and some foresee a revision to no more than 25 to 20 percent in the future.[7]

At present, 16.8 percent of the global population lives in high-income countries, where, on average, fat consumption exceeds the 30 percent level. But health concerns have clearly begun to affect consumption patterns in those countries: Despite rising incomes and relatively stable prices, beef consumption has declined in a number of countries since the mid-1970s. In the United States, for instance, per capita beef consumption has dropped 25 percent.[8] (Overall meat consumption in the United States has remained approximately constant, however, because people merely shifted to eating poultry.)

Clearly, public policy that encourages people to reduce their consumption of animal fat has two benefits. It improves the health of the population while reducing the pressure that increased food production places on the global agricultural system. Table 3 on page 197 shows the conversion rates of grain to animal products in terms of two common measures: kilograms and calories. For the past 30 years, approximately 40 percent of all cereal grains produced globally have been used for feed, with 50 percent being used for food. (The remaining 10 percent have gone to seed, been used in processing, or ended up as waste.) As Table 2 shows, however, the use of grain for feed is much higher in high-income countries.

Efficiency

The last factor affecting global food requirements is the efficiency with which food moves from farms to human

Table 2. Numerical estimates for key food variables, 1991

Variable	World	High-income countries	Low- and middle-income countries	Best practice/ medically preferred
	Calories per person per day			
Total food available[a]	3,939	6,964	3,007	n/a
Food available in retail markets	2,693	3,255	2,520	n/a
Physiological food requirements[b]	2,179	2,231	2,169	n/a
	Ratio			
Dietary conversion factor[c]	1.46	2.14	1.19	1.5
End-use efficiency factor[d]	1.24	1.46	1.16	1.3

n/a Not applicable
[a] Includes animal feed
[b] 1990 estimate
[c] Line 1 divided by line 2, except for last column, which is author's estimate
[d] Line 2 divided by line 3, except for last column, which is author's estimate

NOTE: Computational methods are described in the box on this page.

SOURCES: Line 1: Author's calculations; Line 2: United Nations Food and Agriculture Organization at http://www.foa.org; Line 3: Author's calculations.

The United Nations Food and Agriculture Organization (FAO) estimates that per capita caloric availability (i.e., the amount of food that appears in the retail market) ranges from a low of 1,667 calories in Ethiopia to a high of 3,902 calories in Belgium-Luxembourg. These two figures differ by 234 percent—much more than the 33 percent difference in physiological consumption. Because it is physiologically impossible for the population of an entire country to consume an average of 3,902 calories, we know that a substantial amount of food in high-income countries is never consumed. According to estimates, losses from end-use inefficiencies equal 30 to 70 percent of the amount of food actually consumed.10 With the exception of Belgium-Luxembourg, it is middle-income countries such as Greece, Ireland, Yugoslavia, Hungary, Bulgaria, Egypt, and Libya that have the highest levels of waste. But in every country where per capita income is more than $1,500 (U.S.), at least 20 percent more food is used than is consumed. The computed values for the end-use efficiency factor in Table 2 also reflect the discrepancy between high- and low-income countries.

It is unclear to what extent these losses are a necessary component of increased standards of living because little analysis has been done on the sources of this waste. Some intercountry comparisons provide useful insights, however. The Netherlands, Finland, Japan, and Sweden, which have comparable levels of income, waste only about 35 percent (on a per capita caloric basis),

mouths. Efficiency actually has two components, one pertaining to marketing and distribution and one pertaining to "end use." Losses in marketing and distribution, such as those due to rodents and mold, are important in low-income countries but decline steadily with increases in income.[9] Inefficiencies in end use, which include losses due to spoilage, processing and preparation waste, and plate waste, are most significant in high-income countries, however.

A NOTE ON COMPUTATIONS

Physiological food requirements. Estimating the number of calories that the average person in a given population actually consumes (as distinct from the number that is available in the retail market) entails a five-step procedure.[1] The first step is to determine the age-gender structure of the population, placing children in single-year age groups and adults in five-year age groups. The second step is to estimate the basal metabolic rates for each group based on group members' average heights and weights. The third step is to estimate the different groups' physical activity patterns and combine them with their basal metabolic rates to determine each group's energy requirements. The fourth step is to make allowances for such factors as pregnancy and infection rates and then multiply the average energy requirement for each group by the number of people in the group. The final step is to sum the energy requirements for the different groups and divide by the total population. Normative food requirements (i.e., the number of calories needed to maintain desired heights, weights, and activity levels) can then be determined by adjusting appropriately for heights, weights, and activity levels.

Dietary conversion factor. This factor reflects the number of calories "lost" in using grain to produce animal products. It is computed as the ratio of the total number of calories produced to the number available (in final form) in the retail market. The denominator is usually per capita caloric availability as estimated by the United Nations Food and Agriculture Organization (FAO).[2] The numerator is more difficult to determine because some animals graze rather than being fed grain. The procedure used in this article was to sum three factors: the number of plant calories available, excluding cereals, starchy roots, and tubers; the number of plant calories available from cereals, starchy roots, and tubers, whether used for feed or for human consumption; and the estimated number of animal calories derived from range feeding.

End-use efficiency factor. End-use efficiency—the proportion of calories produced that actually ends up in human mouths—is computed as the ratio of calories available in the retail market (from FAO) to calories consumed (as computed above).

---

1. See W. P. T. James and E. C. Schofield, *Human Energy Requirements* (Oxford, U.K.: Oxford University Press by arrangement with the United Nations Food and Agriculture Organization, 1990).
2. Available at http://www.fao.org.

while the United States, Belgium-Luxembourg, Switzerland, and Italy waste nearly 60 percent.[11] This suggests that there is scope for reducing food requirements without lowering standards of living, much as high-income countries have done with energy use since the 1970s.

Given the current distribution of food consumption and food system efficiency, if every middle- and high-income country were to reduce its level of waste to 30 percent, global food requirements would decline 7.4 percent. (If consumption of animal products were to decrease in proportion, requirements would decline 12.5 percent owing to the lower demand for feed.) Clearly, as global incomes increase and the number of people living in countries with low food system efficiency continues to grow, the level of end-use waste will become an increasingly important part of overall food requirements.[12]

Final Thoughts

By its very nature, agricultural production has significant impacts upon natural ecosystems and the environment. There is little question that agricultural production must increase to meet population growth, but the magnitude of the increase necessary to improve human welfare is very much a question of policy tradeoffs between demand management and supply promotion.

Food is the only sector of the economy that has reached satiation for a large portion of the world's population. Tripling world food production would provide sufficient food for a doubled global population to have a varied, nutritious, and healthy diet comparable to today's European diet. The same goal could be reached by slightly more than doubling agricultural production if an effort were made to improve food system efficiencies and if diets low in fat became commonplace. This change will only take place if public policy creates explicit incentives for healthier diets and more efficient food systems, however.

It is environmentally and medically prudent to prevent the levels of waste and fat consumption in the wealthier developing economies from rising to those seen in North America today. It is also fiscally prudent: Grain imports tend to rise rapidly in maturing developing economies, so that decreased food system efficiency and increased fat consumption can lead directly to the loss of vital foreign exchange. Therefore, self-interest can be used to dramatically improve the long-term sustainability of the global agricultural system.

William H. Bender is an economist with extensive international experience in food and nutrition. He has consulted with the United Nations Children's Fund, the United Nations Food and Agriculture Organization, the World Bank, the United States Agency for International Development, the European Community, and many other organizations. His address is P.O. Box 66036, Auburndale, MA 02166 (telephone: (617) 647-9210; e-mail: bender@tiac.net).

Notes

1. L. Brown et al., *State of the World, 1994* (New York: W. W. Norton & Company, 1994), 196.
2. D. O. Mitchell and M. D. Ingco, *The World Food Outlook* (Washington, D.C.: World Bank, 1993), 232.
3. Requirements are usually measured in terms of calories, both because food analysts tend to focus on producing sufficient calories and because even with a cereal-based diet, people can get adequate protein if they consume enough calories. (The exception to this rule lies in groups, such as infants, who have special nutritional needs.) Of course, consuming enough calories does not guarantee getting enough micronutrients such as iron, vitamin A, and iodine. To obtain those nutrients, people have to eat fruit, vegetables, and fat in addition to starchy staples. However, the inputs (e.g., land, water, and fertilizer) needed to provide a diverse diet are minor compared with those needed to provide enough calories.
4. R. E. Behrman and V. C. Vaughn, *Nelson Textbook of Pediatrics,* 13th ed. (Philadelphia: W. B. Saunders Company, 1987).
5. World Bank, *Poverty and Hunger: Issues and Options for Food Security in Developing Countries* (Washington, D.C., 1986).
6. Calculations of income elasticities reflect how consumption patterns change as income increases. Income elasticity is the percentage change in the demand for a particular good that results from a 1 percent increase in income. Demand is considered elastic when the percentage change is greater than 1, inelastic when it is less than 1. Animal products, for instance, are income-elastic goods because their share in diets tends to increase more rapidly than income. Researchers at organizations like the World Bank, the International Food Policy Research Institute, and the International Institute for Applied Systems Analysis use such elasticities in global models that attempt to simulate future developments in the world agricultural system. Models like these can help provide insight into questions such as the effect increased demand for animal products is likely to have on grain production.
7. World Health Organization Study Group on Diet, Nutrition and Prevention of Noncommunicable Diseases, *Diet, Nutrition and the Prevention of Chronic Diseases: Report of a WHO Study Group,* World Health Organization Technical Report Series 797 (Geneva: World Health Organization, 1992), 109.
8. L. A. Duewer, K. R. Krause, and K. E. Nelson, *U.S. Poultry and Red Meat Consumption, Prices, Spreads, and Margins,* Agriculture Information Bulletin Number 684 (Washington, D.C.: United States Department of Agriculture, Economic Research Service, 1993).
9. In low-income countries, such losses are as high as 15 percent for cereals and 25 percent for roots and tubers; in high-income countries they are less than 4 percent. See, for example, D. Norse, "A New Strategy for Feeding a Crowded Planet," *Environment,* June 1992, 6; and W. Bender, "An End Use Analysis of Global Food Requirements," *Food Policy* 19, no. 4 (1994): 381.
10. Bender, note 9 above, pages 388–90. In these countries, of course, data accuracy is also highest, giving us the most confidence in these estimates.
11. Ibid.
12. Ibid.

Table 3. Conversion rates of grain to animal products

Animal product	Kilograms of feed/ kilograms of output	Calories of feed/ calories of output
Beef	7.0	9.8
Pork	6.5	7.1
Poultry	2.7	5.7
Milk	1.0	4.9

NOTE: These conversions are very approximate, as the caloric density of both feeds and animal products can vary greatly. Furthermore, data units are often not specified or precisely comparable.

SOURCE: Column 1: Office of Technology Assessment, *A New Technological Era for American Agriculture,* OTA-F-474 (Washington, D.C., 1992); Column 2: Author's estimates.

The Changing Geography of U.S. Hispanics, 1850–1990

by Terrence Haverluk
DEPARTMENT OF ECONOMICS AND GEOGRAPHY, UNITED STATES AIR FORCE ACADEMY, USAFA, COLORADO

In 1930, the majority of Hispanics were of Mexican descent and lived in the five Southwestern states of Arizona, California, Colorado, New Mexico, and Texas. After World War II, the Latino migrant stream began to diversify and include large numbers of Caribbeans, and Central and South Americans who generally settled in the Eastern states and California. Hispanics of non-Mexican origin now account for 36 percent of the U.S. Hispanic population. Mexican immigration has continued, and large numbers of Mexican Americans live in regions far from the border states. The U.S. Hispanic population has increased from approximately one million in 1930, to approximately 32 million in 1997. County maps chronicle the changing distribution and numbers of Hispanics from 1850 to 1990. Key words: Hispanic, Latino, Hispano, Tejano, Spanish-language radio stations.

Knowing where Hispanics settle and why is a necessary first step toward understanding a number of contemporary social issues such as illegal immigration, bilingual education, English-only laws, multiculturalism, and assimilation. The U.S. Hispanic population has increased from four million in 1960, to 32 million in 1997, and the United States is now the fifth largest Spanish speaking country in the world after Mexico, Spain, Argentina, and Colombia (U.S. Department of Commerce 1993). This so-called "browning of America," is a popular topic among educators and academics, but little longitudinal mapping exists that chronicles the changing geography of Hispanics.

Definition of Terms and Methods

The United States is the only country in the world with a Hispanic minority. The term Hispanic comes from the Latin word for Iberia, *Hispania*. Widespread use of the term began in the late 1970s when the U.S. Census Bureau adopted it to describe all persons in the United States who are descendant from Spain or from a Spanish-speaking country of the New World (Garcia 1996, 197). Hispanic is based more on history and geography than ethnicity because Hispanics may be of any race—African, Asian, European, or Native American—as long as they can trace their ancestry to Spain or one of Spain's colonies. The most numerous Hispanics are Mexicans, followed by Puerto Ricans, and Cubans. Hispanics not part of these groups are labeled "Other" Hispanics by the Census Bureau. "Other" Hispanics include Dominicans, Spaniards, Central and South Americans, and native Hispanics such as Tejanos, Hispanos, and Chicanos. Recently, the term Hispanic has been losing ground to Latino. *Latino* originated from within the social group it

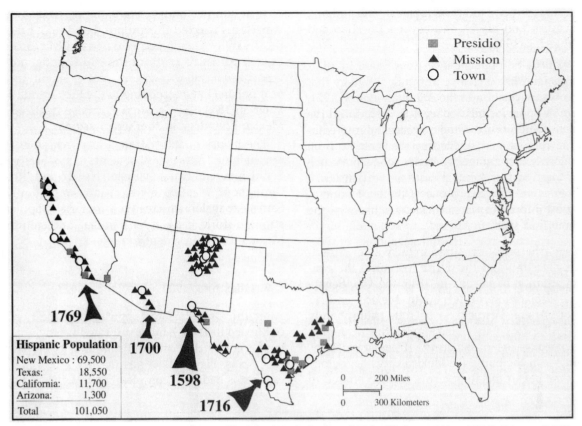

Figure 1. Hispanic population, 1850. *Source:* Martinez (1975), Carlson (1966), Simons and Hoyt (1992), Bannon (1979), and Bufkin (1975).

describes and is considered a more appropriate term. I use Latino and Hispanic interchangeably. The term Anglo-American, or simply Anglo, is commonly used to describe non-Hispanic whites and is used in this article.

This article presents a series of maps showing the distribution of U.S. Hispanics for 1850, 1900, 1930, 1960, and 1990. Accompanying the maps is an extensive historical geography that helps explain current and past Hispanic distributions. The 5 percent level is the lowest breakpoint used in the maps because it is an important political and social threshold. The 1975 amendment to the *Voting Rights Act* states that jurisdictions must provide bilingual ballots and bilingual election materials when 5 percent of its voting age population belongs to a single-language minority (Kusnet 1992, 15).

Hispanics in 1850

The first Europeans to settle permanently in what is now the United States were from Spain, not England. The Upper Rio Grande Valley, currently part of New Mexico, was settled by Juan de Oñate in 1598. In 1607, Santa Fe was founded, and in 1610 it became the capital of New Mexico, making it the oldest U.S. state capital (Carlson 1990).

Oñate's expedition was one of four broad *entradas* (entries) into the present boundaries of the United States. Figure 1 presents the dates, population, and settlement geography of the four entradas. The entradas were part of a Spanish strategy to provide a buffer from Russian and French advances and to Hispanicize Native Americans. The entradas were based on three components: 1) missions to christianize Native Americans, 2) towns for commerce and administration, and 3) *presidios* (forts) for protection.

The second entrada began a century after the Oñate expedition. Father Eusebio Kino's missionary work advanced into Pimería Alta, currently southern Arizona, in 1700. Father Kino helped establish the mission at San Xavier del Bac near present-day Tucson. The third entrada began in response to French settlements in Louisiana. Fearful of French advances, Spain established a presidio at Nacogdoches, in east Texas in 1716, followed by the presidio in San Antonio in 1718. The fourth and final entrada began in response to Russian settlements in the Pacific Northwest. Led by Father Junipero Serra, the Spanish established a string of missions along the California coast, beginning in San Diego, California, in 1769 (Nostrand 1970).

These four settlement areas constituted Spain's northern frontier; close to the hostile Apache and Comanche Indians, but far away from the civilization of Mexico City and the wealth of the silver mines on the Central Mexican Plateau. Because of this isolation, recruiting settlers was difficult. Although there were a few aristocratic Spaniards and Jesuits leading the entradas, the majority of the settlers were poor mestizos (Spanish and Indian), mulatos (Spanish and black), and coyotes (mestizo and Indian) (De la Teja 1995, 24).

Mexico gained its independence from Spain in 1821, a year that also marked a change in frontier strategy that included the secularization of the missions (Bolton 1921, 275). Lack of support for missionary activity reduced the population and importance of the missions, but more liberal Mexican trade policies opened up the Santa Fe Trail, leading to increased commercial activity between Mexico and the United States. Contact with an expansionist United States eventually led to one of the least remembered but most important wars in American history—the Mexican-American War.

Mexico's northern frontier stood in the way of American Manifest Destiny and the desire to control the continent from Atlantic to Pacific. In 1846, the U.S. Senate approved a declaration of war with Mexico and 17 months later General Winfield Scott "rode triumphantly through the [Mexico] City Square amid the deafening cheers of what was left of his army" (Eisenhower 1989, 342). Having won the battle for Mexico City, the victorious Americans began dictating terms. Mexico offered to cede California and Texas, but not New Mexico. The Americans rejected this proposal and under the terms of the Treaty of Guadalupe Hidalgo in 1848 forced Mexico to cede all lands north of the Rio Grande and north of a horizontal line drawn one marine league south of the port of San Diego extending to the Rio Grande at El Paso. Later, in 1853, the Gadsden Purchase finalized the U.S.-Mexico boundary in southern Arizona.

The invasion, annexation, and purchase of the northern half of Mexico left a legacy of Spanish speakers, Spanish architecture, Mexican food, and Roman law throughout the Southwest. More importantly, United States expansionism created a minority population with strong historic, geographic, familial, and cultural links to Mexico that are evident today.

U.S. Hispanics, 1900

After the United States appropriated the northern half of Mexico in 1848, Anglos migrated westward, especially to the gold fields of California and to the farmland of east Texas. Figure 2 reveals that by 1900 California and east Texas had been completely overwhelmed by Anglos.

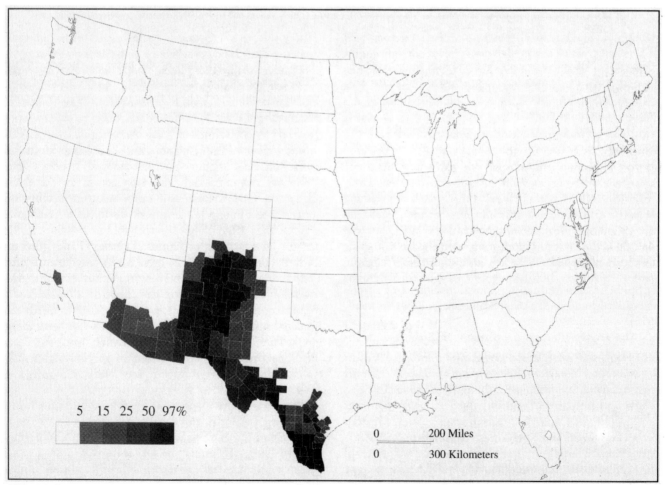

Figure 2. Percentage of Hispanics by county, 1900. *Source:* Nostrand (1980), De Leon and Jordan (1982), Hornbeck (1983), and Camarillo (1979).

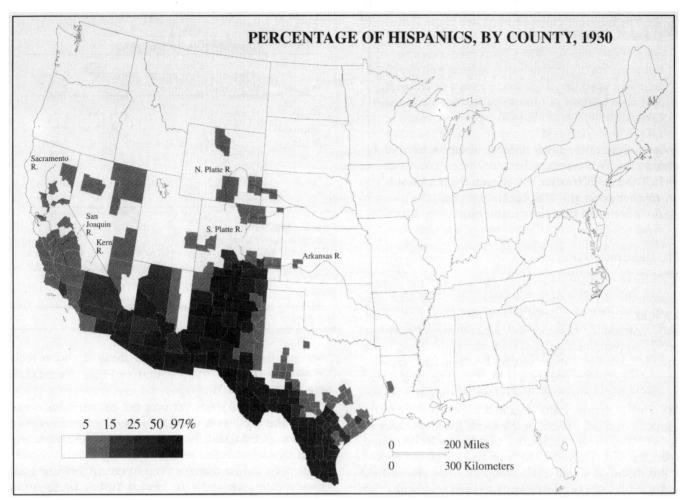

Figure 3. Percentage of Hispanics by county, 1930. *Source:* U.S. Bureau of the Census, 1930; Haverluk 1993.

In California, only Santa Barbara County remained at least 5 percent Hispanic. In southern Arizona, Hispanics had become the minority in every county. Mexicans were recruited to work the copper mines throughout southern Arizona, and as a result, several Arizona counties had sizable Hispanic proportions in 1900.

In contrast to California, the percentage of Hispanics in New Mexico and Colorado expanded during this period. The establishment of U.S. military power in New Mexico controlled Ute, Apache, and Comanche raids, thereby facilitating the expansion of New Mexican Hispanics, called Hispanos, throughout New Mexico and into Colorado. Hispanos moved into southern Colorado and established the state's first town, San Luis, in 1851 (Figure 2). The year 1900 marked the demographic and geographic apogee of what Richard Nostrand (1980) and Alvar Carlson (1990) call the Hispano Homeland. The Hispano Homeland is a distinctive cultural area, the core of which is in north-central New Mexico around Taos and Santa Fe. Some unique cultural attributes in the Hispano Homeland include archaic Spanish words, long lots along irrigation ditches, and Penitente *moradas* (meeting houses).

Arreola (1993) identified a second Hispanic homeland in South Texas. After the Anglo take-over, many Texas Hispanics, called Tejanos, fled to Mexico or to the relative safety of South Texas below the Nueces River. The 1900 core of the Tejano Homeland was along the United States side of the Rio Grande in the cluster of 13 counties greater than 50 percent Tejano. Like the Hispano Homeland, the Tejano Homeland has distinctive characteristics, including a high percentage of central town plazas, unique festival celebrations, and a distinctive music.

Although the Hispano and Tejano homelands were able to maintain their demographic and cultural dominance after 1850, contact with Anglo-Americans altered the economy and their role in society. With the Anglos came American capital and a religious determination to develop the land. Anglos immediately began to integrate the northern Mexican frontier into the expanding industrial capitalism of the American West, first in California and Texas, and later in Arizona, New Mexico, and Colorado. Development meant clearing land of mesquite and cactus, building dams, digging irrigation canals, constructing railroads, and expanding vegetable and cotton production. The primary source of labor to accomplish these tasks consisted of Mexicans and the recently conquered Mexican-Americans. One Anglo rancher of South Texas put it this way:

> ...if it were not for those Meskins [sic], this place wouldn't be on the map. It is very true about the Anglo know-how, but without those Meskin [sic] hands no one could have built up the prosperity we have in this part of the nation (Spillman 1979, 22).

Anglo immigration, the number of farms, and tillable acres in the West increased rapidly after 1900 (Worster 1985). The pre-1850 linkages between the northern Mexican frontier (now the U.S. Southwest) and the Mexican interior continued during the American period.

U.S. Hispanics, 1930

The historic linkages between Mexico and the United States were an essential component to American economic expansion as people, money, and ideas flowed virtually unimpeded between the two countries. From 1900 to 1925, approximately 700,000 Mexicans migrated to the United States. Since there was no border patrol, crossing was safe and easy (Daniels 1990, 326). Figure 3 shows the expanded geography of Mexican migration. Between 1900 and 1930, Mexicans migrated to several newly developed irrigated valleys of the West: the sugar beet regions of the North and South Platte Rivers in Wyoming, Nebraska and Colorado; the fruit and vegetable regions of the Arkansas River in Colorado and Kansas; and the cotton, fruit, and vegetable areas of the Sacramento, San Joaquin and Kern Rivers in Central California.

The establishment of Hispanic communities in counties that did not traditionally have Hispanic populations created new migrant streams that made it easier for subsequent generations of Mexican migrants to relocate—it also made it easier for subsequent generations of Anglo farmers and industrialists to hire Mexicans and Mexican-Americans.

Several thousand Mexicans were also recruited to work the factories and the fields of the Midwest. By 1930, more than 30,000 Mexicans were working in the factories in the Chicago-Gary area, but unlike the West, where Mexicans were the primary source of labor, in the Midwest they were only one of several sources, and as a result, no Midwestern county had Hispanic populations of at least 5 percent in 1930 (Taylor 1932).

The mass movement of Mexicans to the United States spurred the Census Bureau to create a new racial category, *Mexican,* in 1930. Census enumerators were instructed that "all persons born in Mexico, who are not definitely white, Negro, Indian, Chinese, or Japanese should be returned as Mexican" (U.S. Bureau of the Census 1930, 27). The 1930 census enumerated 1,422,533 Mexicans, 90 percent of whom lived in the West, primarily in Texas. Table 1 provides a regional and ethnic breakdown of the Hispanic population in 1930 and reveals that the U.S. Hispanic population was still overwhelmingly Mexican and Western.

After 1900, Mexicans became the primary source of labor in the West as a result of several amendments to U.S. immigration law. In California, Chinese were excluded by the *Chinese Exclusion Act of 1882,* which was followed by a Gentlemen's Agreement in 1907 to curb Japanese immigration to the United States. These Asian exclusion acts were the first in a series of amendments that restricted immigration, culminating in the *Immigration Act of 1924* and the creation of the U.S. Border Patrol in 1925 (Daniels 1990, 328).

The *National Origins Act of 1924* established a quota system based on the national origin of immigrants (Yang 1995, 23). The number of immigrants admitted was based on 2 percent of the number of foreign-born persons of a given nationality in 1890. The year 1890 was chosen because prior to that date immigrants were primarily from western and northern Europe. Between 1890 and 1920, southern and eastern Europeans dominated the migrant stream. By basing admission on 2 percent of the 1890 population, the amendment effectively restricted southern and eastern European immigration, as well as most Asian immigration—but it also created labor shortages (Daniels 1990, 283).

The state department worked in cooperation with industry and agriculture to keep immigration from Mexico open and, by the 1920s, Mexicans became the most important source of immigrant labor in the West. Mexican labor, it was argued, would be easier to send home during recessions, but Mexico was close enough and labor plentiful enough that labor streams could be re-established when necessary (Daniels 1990, 309). The idea that

Table 1

Hispanic Population, 1930

Region**	Hispanic Population	Percent by Region	Mexican* Origin	Puerto Rican
West	1,478,535	92	1,477,273	1,262
Midwest	59,227	4	59,227	NA
South	6,908	1	6,908	NA
East	58,882	4	7,370	51,512
Total	1,603,522	100	1,550,778	52,774

NA = Not Available
* Includes Hispanos
** West: WA, OR, CA, ID, NV, MT, WY, CO, NM, UT, ND, SD, NE, KS, OK, TX. Midwest: MN, WI, MI, OH, IN, IL, IA, MO. South: AK, FL, LA, MS, AL, GA, SC, NC, TN, KY, VA, WV. East: MD, DE, DC, PA, NJ, NY, CT, RI, MA, NH, VT, ME.

Source: U.S. Bureau of the Census, 1930.

35. Changing Geography of U.S. Hispanics

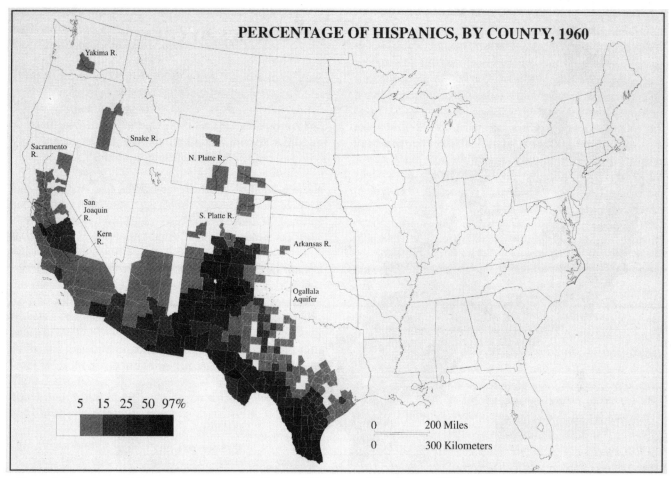

Figure 4. Percentage of Hispanics by county, 1960. Source: U.S. Bureau of the Census, 1960.

Mexican labor could be forcibly returned to Mexico during recessions is what I call the *repatriation strategy*.

The Great Depression led to the implementation of the repatriation strategy and thousands of Mexicans were forcibly returned to Mexico. The number of Mexicans in the United States declined throughout the 1930s and early 1940s, only to rebound again during and after World War II.

U.S. Hispanics, 1960

The end of the Depression and the onset of World War II led to labor shortages in agriculture and industry. Mexican workers were again seen as a partial solution to U.S. labor shortages. In 1942, the United States and Mexico established a guest worker system called the Bracero Program. The Bracero Program (from *brazo*, arm) guaranteed transportation, food, housing, and a minimum wage for braceros.

Figure 4 shows that the Bracero Program expanded Mexican settlement geography, this time to the cotton fields of the Texas south plains; the hops and orchard farms of the Yakima River Valley in Washington; and the vegetable growing areas of the Snake River Valley in Idaho. The Bracero Program was in effect between 1942 and 1964. During its peak in the mid-1950s, an average of 400,000 Mexicans entered the United States each year (Grebler et al. 1970, 176). Braceros were expensive, however. Many farmers bypassed the Bracero Program with its high cost and cumbersome bureaucracy and hired un-

Table 2

Hispanic Population, 1960

Region	Hispanic Population	Percent by Region	Mexican* Origin	Puerto Rican
West	3,561,668	78	3,519,318	42,350
Midwest	199,184	4	133,012	66,172
South	47,313	1	15,457	31,856
East	754,684	17	22,219	732,465
Total	4,562,849	100	3,690,006	872,843

* Includes Hispanos

Source: U.S. Bureau of the Census, 1960.

5 ❖ POPULATION, RESOURCES, AND SOCIOECONOMIC DEVELOPMENT

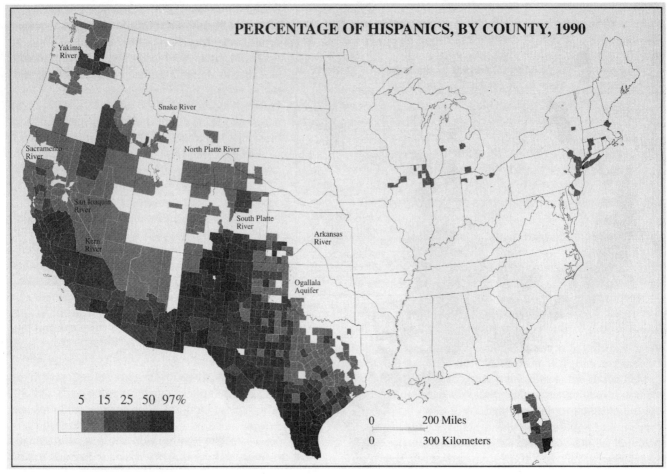

Figure 5. Percentage of Hispanics by county, 1990. Source: U.S. Bureau of the Census, 1993.

documented (illegal) Mexicans directly. Undocumented Mexicans were less expensive and easier to control than Braceros, and in 1953, over 800,000 undocumented Mexicans were apprehended along the border (Grebler et al. 1970, 68).

Figure 4 also reveals that the Hispano Homeland in New Mexico and Colorado contracted after 1900, while the Tejano Homeland was able to maintain its dominance. California's Hispanic proportions also increased, especially in the central valley around Fresno.

The period of high immigration in the 1950s was interrupted by a recession after the end of the Korean War, which again led to forcible repatriation. In 1954, the United States established a paramilitary organization called Operation Wetback, whose goal was to find and return undocumented workers. Operation Wetback was run by a retired Army general who successfully repatriated over 3.8 million undocumented Mexicans (many of whom were sent back more than once). Success at stemming the flow of undocumented Mexicans was only temporary because the western United States economy still relied on Mexicans in several economic sectors, especially agriculture. The end of the recession led to the termination of Operation Wetback, and in 1955, undocumented Mexican migration returned to pre-Operation Wetback levels (Grebler 1970, 176).

The Bracero Program was terminated in 1964 because of widespread abuse on both sides of the border, but immigration continued—albeit illegally. American farmers found it cheaper and easier to hire undocumented aliens. American businesses who hired undocumented Mexicans were breaking no law and after the termination of the Bracero Program in 1964, use of Mexican labor continued to be widespread.

Table 2 reveals that the Mexican population doubled from 1930 to 1960, but that the Puerto Rican population increased 16 times. Widespread Puerto Rican immigration to the United States began in the 1950s with the introduction of inexpensive flights from Puerto Rico. The overwhelming number of Puerto Ricans settled in New York City resulting in the increasing importance of Latinos in the East. In 1930, 92 percent of Hispanics lived in the West, in 1960 only 78 percent lived there. Changing settlement destinations among Puerto Ricans, a revolution in Cuba, and changes in U.S. immigration laws initiated new Latino migrant streams, continuing the trend of Latino diversification seen in Table 2.

U.S. Hispanics 1990

Puerto Ricans

Figure 5 reveals that by 1990 Puerto Ricans had become a highly visible minority group along the eastern seaboard. According to the 1990 census, there were 2.6 million Puerto Ricans in the United States, (triple the number from 1960) 70 percent of whom lived in megalopolis. The majority of Puerto Ricans still live in New York, but large numbers also live in New Jersey, Massachusetts, Connecticut, and Pennsylvania.

Puerto Rican migration to the United States is relatively easy because they are U.S. citizens. Puerto Rican citizenship stems from another U.S. colonial adventure. In 1898, the United States defeated Spain in the Spanish-American War and took control of the remnants of Spain's colonial empire, including the island of Puerto Rico. In 1910, there were only 1,513 Puerto Ricans in the United States. In 1917, Puerto Ricans became U.S. citizens. The number of Puerto Ricans on the mainland remained small until World War II when they were recruited by the War Man Power Commission to alleviate labor shortages in the East and Midwest. Puerto Ricans were given preference over other labor sources because they were U.S. citizens. At first, only skilled workers were admitted, but the need for more labor led to the recruitment of railroad workers for the Baltimore and Ohio line in 1944 (Moldanado-Denis 1976, 111).

After World War II, spurred by the expanding U.S. economy and the availability of inexpensive commercial flights to the mainland, Puerto Rican immigration increased markedly. During the 1960s, Puerto Ricans migrating to the United States overwhelmingly chose New York City, but agricultural workers were recruited to several eastern and midwestern states, especially Ohio and Pennsylvania.

In Megalopolis, the term Hispanic is associated with Puerto Ricans or "Other" Hispanics. Puerto Rican food, music, and immigration history are distinct from Mexican. Puerto Rican music is Bomba and Plena, not Mariachi and Tejano. Puerto Rican folk architectural influences can be seen on *casitas* (little houses) in many of New York's community gardens. The majority of my students in Maryland and Massachusetts, for example, had never heard of the word Chicano.

Cubans

Figure 5 reveals another important cluster of Hispanics in south Florida. This population is primarily Cuban and dates from the Revolution of 1959. In that year, Fidel Castro assumed control of Cuba and expropriated private land holdings, banks, and industrial concerns. Thousands of mostly upper- and middle-class Cubans who opposed the regime fled to the United States. Prohias and Casals (1973, 12) identified three stages of Cuban migration to the United States:

1. between 1959 and 1962 when commercial flights between Cuba and the U.S. were available,
2. between 1962 and 1965 when the Cuban missile crisis led to the suspension of flights, and
3. between 1965 and 1973 when daily air flights between Cuba and Miami resumed.

All Cubans who made it to this country were immediately granted refugee status and allowed to remain legally. Initially, the U.S. government attempted to relocate Cubans to other parts of the United States to lessen the impact of Cubans on south Florida, but most relocated Cubans eventually moved back to south Florida.

Two more stages must now be added to the list:

4. between 1973 and 1994, when, for the most part, Cubans used boats to sail the 90 miles between Cuba and south Florida, and

Table 3

Hispanic Population, 1990

Region	Hispanic Population	Percent by Region	Mexican Origin	Puerto Rican	Cuban	"Other" Hispanics*
West	14,325,314	66	12,186,656	220,777	108,114	1,809,772
Midwest	1,524,611	7	1,018,797	251,735	32,561	221,518
South	2,116,532	10	346,672	321,626	709,360	738,874
East	3,739,781	17	188,706	1,826,466	196,179	1,528,430
Total	21,706,243	100	13,740,831	2,620,604	1,046,214	4,298,594

* Includes Hispanos, Dominicans, Central Americans, South Americans, and Hispanics who identified themselves as Spanish, Chicano, Tejano, Californio, and so on.

Source: U.S. Bureau of the Census, 1990.

5 ❖ POPULATION, RESOURCES, AND SOCIOECONOMIC DEVELOPMENT

Table 4
"Other" Hispanic Populations, 1990

Region	"Other" Hispanics*	Percent by Region	Central Americans	South Americans	Dominicans
West	1,039,198	36	771,649	256,317	11,232
Midwest	104,646	4	45,444	54,330	4,873
South	493,846	17	229,350	224,650	39,846
East	1,227,775	43	276,095	496,560	455,120
Total	2,865,465	100	1,322,538	1,031,857	511,071

* Includes Dominicans, Central Americans, and South Americans.
Source: U.S. Bureau of the Census, 1990.

5. after August 1994, when President Bill Clinton revoked automatic refugee status for Cubans, thereby pulling up the 36-year old Cuban welcome mat in Florida.

Between 1959 and 1994, approximately 715,000 Cubans successfully relocated to the United States (*Washington Post* 1994), most to Florida. In Florida, the word Hispanic is associated primarily with Cubans, whose language and traditions are distinct from Puerto Ricans and Mexicans.

Table 3 presents the 1990 population of the most numerous Hispanic groups—Mexicans, Puerto Ricans, and Cubans, as well as "Other" Hispanics. The percentage by region reveals a continuation of the decreasing demographic dominance of the West—from 92 percent in 1930, to 66 percent in 1990, and the corresponding increase of the south and east. Until 1990, the South had fewer Hispanics than the Midwest. Table 3 also reveals the importance of the "Other" Hispanic population, which in 1990 was larger than the Puerto Rican and Cuban populations combined.

Other Hispanics

Between 1924 and 1965, most Latin Americans were restricted from legally immigrating to the United States. As already mentioned, the exceptions were Puerto Ricans, who are U.S. citizens, Cuban refugees, and Mexican braceros. Until 1965, U.S. Hispanics were overwhelmingly from Mexico and the Caribbean. After 1965, the Latin American migrant stream expanded to include Central and South Americans.

The social revolution in the 1960s and world-wide condemnation of the restrictive *National Origins Act of 1924* led to the emendation of U.S. immigration law. The *Immigration and Nationality Act of 1965* phased out the quota system and its restrictions and placed in its stead a hemispheric allocation system that admitted 170,000 persons from the eastern hemisphere and 120,000 from the western hemisphere for an annual ceiling of 290,000.

Available slots were meted-out based on family reunification, which favored Mexico with its large U.S. population. In 1976, the system was amended to allow only 20,000 from one country, which expanded the Latin migrant stream to Central and South America.

In 1980, Congress established the *Refugee Act*, which allowed 50,000 refugees to enter annually. The *Refugee Act* facilitated the migration of thousands of Central Americans caught in U.S.-sponsored wars against communism, especially in Nicaragua, El Salvador, and Guatemala. In 1980, thousands of anticommunist Nicaraguans, like Cubans 20 years earlier, fled their country and relocated to the United States, primarily to Florida (79,056). Salvadorans (338,769) and Guatemalans (159,177) preferred California, but several thousand also migrated to New York and south Florida (Daniels 1990, 341). Even with these changes, however, Mexicans are still the largest legal immigrant group not only among Latin Americans, but among all legal immigrant groups (Yang, 1995, 24).

Central Americans, South Americans, and Dominicans are lumped together by the Census Bureau under the title "Other" Hispanics. About one-half of all Central Americans live in California (637,656), especially Los Angeles County (453,048). About three-fourths of all Dominicans live in New York City (357,868). South Americans also prefer New York (279,101), but also live in California (182,384), Florida (170,531), and New Jersey (126,286). "Other" Hispanics are the fastest growing segment of the Latino population in the East and the South. "Other" Hispanic are the majority Latino population in Rhode Island (Guatemalans), Maryland (Salvadorans), and the District of Columbia (Salvadorans and Dominicans). In New York and New Jersey "Other" Hispanics are now almost as numerous as Puerto Ricans. Unlike the West and Midwest, where Mexicans are the overwhelming majority, the eastern and southern Hispanic populations are more heterogeneous. West of the Mississippi, only San Francisco, Los Angeles, and Houston have substantial "Other" Hispanic populations. Table 4

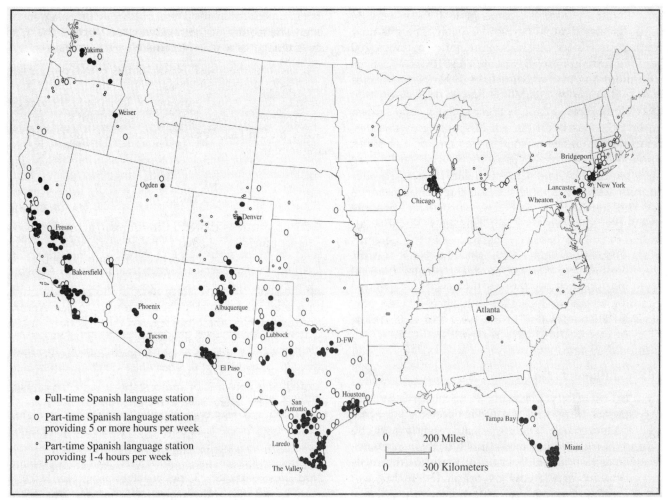

Figure 6. Spanish Language Radio Stations, 1990. Full-time formats in the West include Country and Western/Hispanic, bilingual Tejano, Hispanic/Anglo/Indian, Spanish music, Spanish hit radio, bilingual Hispanic, and Spanish. Full-time formats outside the West include Spanish news, Spanish urban Contemporary, Spanish, Spanish music, Spanish hit radio, Spanish/Caribbean, Spanish/Portuguese, Spanish talk, Hispanic, and Latin.
Source: *Broadcasting and Cable Yearbook, 1990.*

shows the distribution of the three principal "Other" Hispanic groups. The "Other" Hispanic population is essentially bicoastal, with about 90 percent of the population living along the northeastern seaboard, Florida, or California.

The Puerto Rican, Cuban, and "Other" Hispanic population is primarily eastern. In 1990, the East had 25 counties greater than 5 percent Hispanic, compared to zero in 1960. The same is true of Florida, which in 1960 was overwhelmingly Anglo, but in 1990 had 15 counties greater than 5 percent Hispanic, including Dade County, the only county east of the Mississippi greater than 50 percent Hispanic.

The Midwestern cluster of counties centered around Chicago is primarily Mexican (612,442 or about 70 percent of the Illinois Hispanic population). The large number of Mexicans in Chicago is a continuation of Mexican migration to Chicagoland factories that began during the 1920s. The Michigan Hispanic population is also primarily Mexican. Michigan was the second most important sugar beet producing state after Colorado in the 1920s, and thousands of Mexicans migrated to Michigan to work beets (Taylor 1932). Michigan and other parts of the Midwest still rely on Mexican agricultural labor.

Even more dramatic than Hispanic increases in the East and South is the continued growth and geographic expansion of the western Hispanic population shown in Figure 5. The Hispanic population west of the Mississippi is still overwhelmingly of Mexican descent. In 1960, the Northwest had one county, Yakima, greater than 5 percent Hispanic. In 1990, there were 34 counties with a Hispanic population greater than 5 percent. The migrant linkages established between Mexico and the Northwest during the Bracero Program still exist, and some Yakima Valley communities, such as Toppenish and Sunnyside, are now majority Hispanic. The Snake River Valley in Idaho has almost as many counties greater than 5 percent Hispanic (14) as the East (25). The number of Hispanics in the Northwest is not as large as in Megalopolis, but their proportions and their influence on the landscape and the culture are perhaps greater.

In Texas, the Tejano Homeland has expanded northward and westward—in 1960 Texas had 17 counties greater than 50 percent Hispanic, in 1990 there were 34 counties greater than 50 percent Hispanic. There is now a continuous cluster of Hispanic majority counties 700 miles long from El Paso in extreme West Texas to Corpus Christi along the Gulf of Mexico. In 1990, these counties had a population of 2.2 million and is one reason that Texas overtook New York to become the second largest state in 1994. Based on current Hispanic growth rates, several more counties in south Texas and on the South Plains will be majority Hispanic by the year 2000.

In New Mexico and Colorado the Hispano homeland has lost some of its demographic dominance while the southern part of New Mexico has become increasingly Mexican. In California and Nevada, almost every county has seen sharp rises in the percentage of Hispanics since 1960, yet only one county, Imperial, along the Mexican border is greater than 50 percent Hispanic. Even though the western Hispanic population is not as dominant as it once was, 66 percent of all Hispanics live in the West and it is America's most ethnically diverse region.

Hispanic Influence

Hispanics, especially Mexicans, have historically had a large impact on the cultural and economic geography of the United States. The American cowboy and the ranching (*rancho*) industry owe much of their existence to the Mexican *vaquero* (buckaroo). Many of the tools and vocabulary of the cowboy originated in Mexico: lariat (*la reata*), chaps (*chaparejos*), mustangs (*mesteños*) and rodeo (*rodear*, to round up), to name a few. Mexicans have also influenced music and contributed much of the "Western" to Country & Western music. In south Florida, the "Miami sound" is influenced by Caribbean rhythms and the Spanish language (Roberts 1979). Mexican food, which has always been popular out West, is now common throughout the United States and salsa now outsells ketchup (Minneapolis Star and Tribune, 13 November 1992). The increasing number and wider distribution of Hispanics in the United States means that the Spanish language is more widespread now than it has been historically, and many phrases such as nada, hasta la vista, mañana, mano a mano, and no más have entered the popular vernacular.

Hispanics, like the U.S. Asian population, are a diverse group from many different countries. Unlike the Asian population, however, Hispanics share a common language. Spanish was the first European language in North America, and the use of Spanish has increased with the Hispanic population. Spanish is reinforced through Spanish language media, including magazines, newspapers, TV, and radio. Hispanics, however, seem to prefer radio to print and TV (Greenberg 1983). Full- and part-time Spanish-language radio stations (SLRS) were once primarily a border phenomenon but are now common throughout the United States, especially in the West. SLRS can be an important diagnostic feature of the relative importance of the Spanish language in a place (Figure 6).

Texas has the most SLRS in the country—twice as many as California, which has double the Hispanic population of Texas. The dominance of Texas in SLRS suggests that Spanish may be more important there than in California, and in fact, Texas Hispanics have higher Spanish language usage rates than California Hispanics (U.S. Bureau of the Census 1993). One-half of Texas' full-time SLRS are in the Tejano Homeland, also known as "The Valley," where Tejano music originated.

Tejano music is rooted in traditional, accordion-based norteño music of Mexico but also incorporates synthesizers, salsa rhythms, and even hip-hop. Tejano music began in south Texas and has spread throughout Texas and the West. Many western radio stations shown in Figure 6 play predominately Tejano music. In San Antonio and Los Angeles, the highest rated radio stations in 1993 were Tejano stations (Dallas Morning News, 1993). When the well-known Tejano singer Selena was gunned down by a disgruntled fan in 1995, the news spread rapidly through the network of radio stations, and for days afterward there were Selena tributes as far north as the Yakima Valley. The popularity of Tejano may partially explain why there are more full-time SLRS in Lubbock, Texas, with 51,000 Hispanics than in New York City with over one million Hispanics. The state of Washington, with 206,018 Hispanics, has almost as many full-time SLRS as the East with 3 million Hispanics. The reason for the greater concentration of SLRS in the West is the result of several factors:

1. In the Texas and New Mexico homelands, Hispanics have always been the dominant population, thereby legitimizing Hispanic culture from a very early date;
2. Proximity to the border reinforces the use of Spanish among Hispanics and even many non-Hispanics;
3. Most western Hispanics are of Mexican descent and share more similar tastes and values than the more heterogeneous eastern Hispanic population;
4. The ability to procure airway space is easier in Lubbock, Texas, than in New York City, because the airways are more crowded in New York City; and
5. The West is where most U.S Hispanics live. My own research has shown that non-Hispanics living in areas with large numbers of Spanish speakers are significantly more likely to speak Spanish than other non-Hispanics (Haverluk 1993).

South Florida and Chicago, which are primarily Cuban and Mexican respectively, also have large numbers of full-time SLRS. Atlanta's full-time SLRS is perhaps emblematic of the "New South."

The Future

Large increases in non-European immigrants since 1965 have fueled anti-immigration sentiment and latent racism that culminated in the *Immigration Reform and Control Act of 1986* and Proposition 187 (Prop. 187) in California, in 1994. The Act of 1986 established four new immigration provisions:

1. Amnesty for illegal aliens in the U.S. since 1982. Over 3 million illegal immigrants, 70 percent of whom were Mexican, were accepted into the amnesty program and are now in various stages of the legalization process.
2. Requirements that employers verify the eligibility of all newly hired employees. This provision is the reason why all employees must now provide employers with proof of citizenship or proof of legal residence upon hiring. Instead of inhibiting immigration, this provision has led to the establishment of a sophisticated underground network of fraudulent document providers. Employers only have to ask for documentation, they do not have to verify its authenticity. Many western farmers still rely on illegal aliens and comply with the letter, but not the spirit, of the law.
3. Sanctions against employers that knowingly hire illegals. This provision was designed to mete out tough fines to persons or businesses that knowingly hire illegals. Many western farmers and businesses have relied on illegal Mexican labor for decades and are unwilling to blow the whistle on their neighbor for hiring illegals. Furthermore, in order to get the necessary votes to pass the bill, western growers insisted that the Immigration and Naturalization Service (INS) could not conduct raids during harvest, effectively taking the teeth out of the amendment.
4. Provisions to allow agricultural workers to be recruited during times of labor shortages. Clearly the most ironic, even hypocritical, component of a bill designed to reduce immigration is the Replenishment of Agricultural Workers (RAW) component of the Act. RAW established a mechanism to authorize an additional 250,000 agricultural workers each year who could eventually be returned to Mexico—again the repatriation strategy. Ten years after its implementation, the Act of 1986 has not been very effective at stemming Latino immigration (Daniels 1990, 342).

The Act's ineffectiveness led to Proposition 187 in California, which proposes that all state services be denied to illegal immigrants and their children, even if they were born in the United States. This latter component of the bill is probably unconstitutional, and a California court has blocked its implementation. At the national level, there is debate in Congress to amend current immigration and amnesty laws to reduce the number of immigrants from all areas.

Conclusion

Until the 1960s, the U.S. Latino population was overwhelmingly Mexican and western, and so was the Hispanic cultural imprint. After 1965, changes in immigration laws led to new migrant streams from Central and South America to the eastern and southern United States. These new migrant streams created a more geographically and socially heterogeneous Latino population whose influence is now felt beyond the West.

Increases in the U.S. Latino population since the 1960s have led to the reinstitution of the repatriation strategy in California and a general anti-immigrant sentiment. Attempts to restrict Latin American immigration, especially Mexican immigration to California and the West, will be unsuccessful in the long-term because many sectors of its economy—agriculture, landscaping, child care, and janitorial services—are predicated on the use of low-wage immigrant labor from Mexico (Bustamante et al. 1992; Jones 1995). Unless the U.S. economy changes in some fundamental way Mexican immigration, legal or illegal, will continue.

Continued Latino immigration combined with higher Latino birth rates means that the Hispanic population is growing seven times faster than the non-Hispanic population. Current projections suggest that there will be 81 million Hispanics, constituting 30 percent of the population, by 2050 (U.S. Department of Commerce 1993). In cities such as San Antonio, Miami, and El Paso, Hispanics are already the majority; in Los Angeles and New York they are approaching majority status. Culturally, these communities are quite different from other U.S. cities: the nightly news is given in Spanish and English; newspapers are bilingual; Latino holidays are celebrated as often as traditional "American" holidays; specialty crops are planted and/or imported to provide traditional Latino foods; and Latino history, sometimes taught in Spanish, is presented along with Anglo American history.

Increasing numbers of Latinos in the United States have led to new attempts to restrict immigration and immigrant services. English Only and/or Official English laws have been passed by many states and are being discussed at the national level. Bilingual education is being challenged, while at the same time many colleges and universities are requiring the study of foreign languages which, increasingly, means Spanish. In many respects, the entire socio-political framework of the United States is affected, especially in the shaded areas of Figure 5. The reason for these social debates is that unlike many previous immigrant groups, Latinos are more likely to maintain their language and culture even while they learn English and absorb American culture (Matovina 1995). Unlike other immigrant groups that have maintained their language and culture, such as the Greeks and the Chinese, Latinos are much more numerous and their source areas are closer to the United States.

The long-term maintenance of Mexican culture in the West, for example, has allowed many non-Hispanics to embrace aspects of Latino culture such as Santa Fe style clothing, Southwestern architecture, and New Mexican cuisine. Whereas some Anglos feel threatened by the existence of a large non-Anglo cultural group in society and wish to foster the Americanization of Hispanics, other Anglos feel enriched by Latino contributions to society. How the United States deals with these tensions will be a continuing theme in American politics and society well into the future. Let us hope we develop along the Swiss, rather than the Balkan, model.

References

Arreola, D. D. 1993. The Texas-Mexican homeland. *Journal of Cultural Geography* 13(2)6–74.

Bannon, C. 1979. *The Spanish borderlands frontier, 1513–1821.* Albuquerque, NM: University of New Mexico Press.

Bolton, H. E. 1921. The Spanish borderlands. *The Chronicle of America Series,* No. 23. New Haven, CT: Yale University Press.

———. 1964. *Bolton and the Spanish borderlands.* Norman, OK: University of Oklahoma Press.

Broadcasting and Cable Yearbook. 1990. *Broadcasting yearbook.* Washington, DC: Broadcast Publishing.

Bustamante, J. A., C. W. Reynolds, and R. A. Hinojosa. 1992. *US-Mexican relations: Labor market interdependence.* Stanford, CA: Stanford University Press.

Camarillo, A. 1979. *Chicanos in a changing society: From Mexican pueblo to American barrio in Santa Barbara and Southern California, 1848–1930.* Cambridge, MA: Harvard University Press.

Carlson, A. W. 1966. *The historical geography of the Spanish settlement in the middle Rio Grande Valley, 1598–1821.* Unpublished research paper. Minneapolis, MN: University of Minnesota.

———. 1990. *The Spanish-American homeland: Four centuries in New Mexico's Rio Arriba.* Baltimore, MD: Johns Hopkins University Press.

Dallas Morning News. 1993. Tejano Attracts the big labels. 10 October, 9C.

Daniels, R. 1990. *Coming to America.* New York: Harper Collins.

De Leon, A., and Jordan. 1982. *The Tejano community, 1836–1900.* Albuquerque, NM: University of New Mexico Press.

Eisenhower, J. S. D. 1989. *So far from God: The U.S. war with Mexico, 1846–1848.* New York: Doubleday.

García, I. M. 1996. Backwards from Aztlán: Politics in the age of Hispanics. In *Chicanos and Chicanas in contemporary society,* ed. Roberto M. de Anda, Needham Heights, NJ: Allyn and Bacon.

Grebler, L. et al. 1970. *The Mexican-American people: The nation's second largest minority.* New York: Free Press.

Greenberg, B. 1983. *Mexican Americans and the mass media.* Norwood, NJ.

Haverluck, T. W. 1993. *The Regional Hispanization of the United States.* Unpublished doctoral dissertation. University of Minnesota.

Hornbeck, D. 1983. *California patterns: A geographical and historical atlas.* Mountain View, CA: Mayfield Publishing.

Jones, R. C. 1995. *Ambivalent journey, U.S. migration and economic mobility in north central Mexico.* Tucson, Az: University of Arizona Press.

Kusnet, D. 1992. *Voting rights in America: Continuing the quest for full participation.* Edited by Karen McGil Arrington and William L. Taylor. Washington, DC: Leadership Conference Education Fund.

Maldonado-Denis, M. 1976. *En las entranas: Un analasis sociohistorico de la emigracion Puertorriquena.* Habana, Cuba: Casa de las Americas.

Matovina, T. M. 1995. *Tejano religion and ethnicity: San Antonio 1821–1860.* Austin, TX: University of Texas Press.

Minneapolis Star and Tribune. 1992. *Salsa savvy: Market is hot for spicy Mexican sauce.* 13 November 1992, E1.

Nostrand, R. L. 1970. The Hispanic-American horderland: Delimitation of an American cultural region. *Annals of the Association of American Geographers* 6O(4):638–661.

——— 1980. The Hispano homeland in 1900. *Annals of the Association of American Geographers* 70(3):382–396.

Roberts, J. S. 1979. *The Latin tinge: The impact of Latin American music on the United States.* Oxford University Press.

Simons, H., and C. Hoyt. 1992. *Hispanic Texas: A historical guide.* Austin, TX: University of Texas Press.

Spillman, R. C. 1979. *Hispanic population patterns in southern Texas, 1850–1970.* Research paper number 57. Eugene, OR: University of Oregon.

Taylor, P. S. 1932. *Mexican labor in the United States.* Berkeley, CA: University of California Press.

United States Bureau of the Census. 1990. *1990 social and economic characteristics.* Washington, DC: U.S. Government Printing Office.

———. 1960. *Persons of Spanish surname 1963.* Subject Reports PC(2)-1D. Washington, DC: U.S. Government Printing Office.

———. 1930. *Race and foreign origin 1933.* Washington, DC: U.S. Government Printing Office.

United States Department of Commerce. 1993. *We the Americans . . . Hispanics.* Washington, DC: Bureau of the Census.

The Washington Post. 1994. *Recurring waves of Cuban refugees.* 20 August.

Worster D. 1985 *Rivers of empire: Water, aridity and the growth of the American West.* New York: Pantheon.

Yang P. Q. 1995. *Post-1965 immigration to the United States: Structural determinants.* Westport: Praeger.

'Hispanics' don't exist

The fast-growing U.S. ethnic group isn't an ethnic group at all. It's a mishmash of many different groups. Herewith, a guide to the nation's 17 major Latino subcultures

BY LINDA ROBINSON

The growing proportion of Hispanics in the U.S. population constitutes one of the most dramatic demographic shifts in American history. The number of Hispanics is increasing almost four times as fast as the rest of the population, and they are expected to surpass African-Americans as the largest minority group by 2005. It's projected that nearly 1 of every 4 Americans will be Hispanic by the year 2050, up from 1 in 9 today. Yet other Americans often have no clear idea of just who these 29 million people are.

One reason is that the label *Hispanic* obscures the enormous diversity among people who come (or whose forebears came) from two dozen countries and whose ancestry ranges from pure Spanish to mixtures of Spanish blood with Native American, African, German, and Italian, to name a few hybrids. While most are bound by a common language, Spanish, many Hispanic-Americans speak only English. This diversity helps explain why Hispanics' political clout remains disproportionately slight. Hispanics even disagree on what they want to be called; most identify themselves by original nationality, while others prefer the term *Latino*.

A common Latino subculture doesn't really exist in the United States. True, there are some pockets of pan-Hispanic melding in major cities, and occasional alliances are struck on specific issues; with time, the differences may merge into a shared Latino identity. But for the present, it makes more sense to speak of Hispanics not as one ethnic group but as many. Mexicans are the largest, at 63 percent of the total Hispanic population, yet even they vary by region and experience.

How many Hispanic subcultures exist in the United States today? Ethnologists are bound to differ on this question, but *U.S. News* puts the number at 17. We have taken into account the largest communities as well as the smaller (yet, in our unscientific judgment, most culturally distinct) ones. what follows is an overview and taxonomy of the 17 major Latino subcultures in the United States, listed by geographic region.

CALIFORNIANS

Hispanics represent 30 percent of the population in California today and by 2020 are projected to outnumber non-Hispanic whites there. Many Latinos, of course, migrated to California back when it was still a part of Mexico. But more than 80 percent of Southern California's Hispanics came after 1970. In 1996, newly naturalized Latinos voted at higher rates than the general population. The galvanizing event was 1994's passage of Proposition 187, which sought to end school and health services for illegal immigrants. (A federal judge has blocked implementation of Prop. 187; the matter is expected to be appealed up to the Supreme Court.)

1. Immigrant Mexicans. Newcomers to Los Angeles traditionally settle in enclaves like East L.A., but in the past decade they've also poured into low-income black areas like South Central and Compton as well as Huntington Park, a formerly Anglo neighborhood that had become a ghost town. *"Ahora es México,"* says a man standing with his son at the corner of Florence and Pacific while his wife buys tamales and chicken in *mole* from a huge takeout store. "None of this was here when I came 15 years ago," he says, nodding at the Spanish-named car dealerships, shoe stores, bridal shops, and supermarkets stretching for blocks.

2. Middle-class Mexicans. Many Mexican-Americans in California have moved up the socioeconomic ladder, sometimes in a single generation. Overall, two thirds of Latinos in the United States live above the poverty line; half of Southern California's native Latino families, and one third of those from abroad, are middle class. New arrivals often hold two jobs, leveraging themselves or their children into such middle-income occupations as police officer, manager, and executive secretary. They have migrated from traditional ports of entry to more-prosperous neighborhoods and suburbs like San Gabriel and Montebello. There, Mexican-Americans buy three- and four-bedroom tract houses next door to Asians. Farther east, in Hacienda Heights, Mexican-American families' yards are bigger, the driveways parked with BMWs and Jeep Cherokees.

3. Barrio dwellers. Many Mexicans move up and out, but a growing number of second- and third-generation kids are getting trapped in ghettos. Boyle Heights' housing projects are the largest west of the Mississippi; 60 gangs with 10,000 members ("homeboys") run rampant over 16 square miles of urban wasteland.

4. Central Americans of Pico Union. As tough as life may be in the Mexican barrios, it's even grimmer in Pico Union, a gang-ridden section of L.A. just east of MacArthur Park that serves as the principal U.S. port of entry for Central Americans, the fastest-growing segment of L.A.'s population. Nearby Koreatown is also now predominantly Central American. Greater L.A. is home to half of all the Salvadorans and Guatemalans who live in the United States.

Even though 97 percent of U.S. Central Americans are working, incomes in Pico Union commonly range from $5,000 to $10,000. Everyone works, kids and parents. Most parents have less than a sixth-grade education; their children who work full time risk remaining at society's lower rungs. Still, two thirds of the families manage to stay above the poverty line, running little markets and shops along Eighth Street.

TEJANOS

Texas Mexicans argue with their California brethren over whose culture is more authentically Mexican-American. What's certain is the two groups couldn't be more different. In contrast to the majority of "Californios" who are recent arrivals, many Tejanos have been here for generations. They've brewed a cowboy culture that's equal parts Texas and Mexico. Tejano music, a widely popular blend of country and *ranchera,* epitomizes the hybrid. Tex-Mex conservatism on issues from abortion to immigration shocks California Mexicans.

5. South Texans. The most Mexican part of the United States is the lower Rio Grande Valley. In Laredo and Brownsville, Mexicans form 80 to 95 percent of the population. Their roots go back to the 1700s, giving them a strong sense of belonging. Hidalgo County, one of the nation's poorest, is also a cradle of Mexican culture and scholarship. Like California, Texas was the scene of bitter battles over job and school discrimination in the 1970s, but anti-immigrant sentiment is far less virulent here. Many Anglos speak Spanish, and intermarriage is common.

6. Houston Mexicans. In Houston, Latinos are still a minority. Anglos make up 41 percent of the population and hold most positions of political and economic power. But Hispanics—mostly Mexicans, but also a growing number of Central Americans—have grown from 18 to 28 percent since 1980. (The remaining 31 percent of Houston is mostly African-American and Asian.) Houston's Mexican-Americans are mostly working-class residents of ethnic enclaves even though 56 percent of them are U.S.-born. "South Texans who go to see their relatives in Houston feel sorry for the barrio dwellers' quality of life," says Joel Huerta of the University of Texas's Center for Mexican American Studies.

7. Texas Guatemalans. Houston's urban sprawl could not be more foreign to the Mayan Indians of Guatemala, who grew up in the rural highlands speaking their native Indian language. Because they have little chance of upward mobility in their own highly race- and class-conscious country, the Mayas have joined Houston's Central American working class. These short, full-blooded Indians tend to keep to themselves in their southwest Houston enclave—they have their own soccer leagues and Pentecostal churches—but they did join with African-American residents of one area they colonized, Stella Link, to form crime-watch groups and youth programs. In his new book, *Strangers Among Us: How Latino Immigration Is Transforming America,* journalist Roberto Suro recounts the trail of Guatemalans to Randall's, an upscale supermarket chain that ended up hiring 1,000 Mayas.

CHICAGO LATINOS

Latinos followed Irish, Polish, and other European immigrants to this city of ethnic neighborhoods. Only Los Angeles and New York have larger Hispanic populations than Chicago, which is projected to be 27 percent Hispanic in the year 2000. And Chicago's mix of Hispanic subgroups is more diverse than that of L.A. or New York. Among U.S. cities, Chicago ranks second in the number of Puerto Ricans, fourth in the number of Mexicans, and third in the number of Ecuadorans. Guatemalans and Cubans are also here in force.

8. Chicago Mexicans. The first of Chicago's nearly 600,000 Mexicans arrived to work on the railroad just after the turn of the century; more came to man steel mills during World War II. "Chicago's weather is so harsh that the only reason Latinos come here is jobs," says Rob Paral, research director of the Latino Institute. Chicago has absorbed the steady influx fairly well: Its manufacturing base remains strong and unemployment is low. Its Latinos mirror the national profile in that 60 percent are native-born and two thirds lack high school diplomas. But only one fourth are poor. (The national rate is 31 percent.) The commercial heart of Mexican Chicago, 26th Street, generates more tax revenue than any other retail strip except tony Michigan Avenue. It's lined with hundreds of stores like La Villita Dry Cleaner, a piñata shop, Nuevo León restaurant—but has just one Walgreen's.

9. Chicago Puerto Ricans. Two giant, steel Puerto Rican flags fly over Division Avenue by Roberto Clemente High School. They were erected to stake out the turf of Paseo Boricua, a strip of 80 mom and pop businesses, and the Puerto Rican-owned Banco Popular, the largest Hispanic-owned bank in the United States. Sitting in his sister's bakery across from the MDS education center

Where the growth is
Seventy-five percent of Latinos currently live in only five states (California, Texas, Florida, Illinois, and New York), but growth is fastest in other states, where many work in the meat-packing and poultry-processing industries.

1990-96 growth rate of Hispanics in U.S. population, by county
- More than doubled
- 50% to 100%
- 25% to 49%
- 0% to 24%
- Declined

Source: U.S. Census Bureau

he founded, community leader Jose Lopez says that urban renewal plans are pushing Puerto Ricans into suburban ghettos instead of helping them prosper. He launched the flag project as part of his drive to bolster Puerto Rican pride and identity. One of the great paradoxes of *puertorriqueños* is that while they have the benefit of being born U.S. citizens, they have fared worse economically than any other Hispanic group. They have the highest rates of poverty (38 percent), unemployment (11.2 percent), and households headed by single females (41 percent).

MIAMIANS

Miami is the one major city in the United States where Hispanics dominate numerically, politically, and economically. They make up about 60 percent of the population, a meteoric rise from only 5 percent in 1960. Miami is seen as a Cuban city, but other immigrants who have poured in since 1980 now make up 40 percent of Hispanics living here.

10. Cubans. Success stories are not hard to find among Miami's 1 million Cubans. Of the 80 Latinos in the United States worth $25 million or more (according to a recent survey in *Hispanic Business* magazine), 32 are of Cuban origin. Singer Gloria Estefan, the late exile leader Jorge Mas Canosa, and a handful of Miami builders made last year's list. Roberto Goizueta, the late head of Coca-Cola, topped it. U.S.-born Cubans have the highest incomes of any Hispanic subgroup, and over two thirds of them live in Florida.

For this influx of talented and successful immigrants, America has Fidel Castro to thank. The first wave of Cuban immigrants in the 1960s, following Castro's Communist takeover of Cuba, doubled their incomes in three years: Four thousand were doctors, and most had good educations. They started restaurants; clothing, furniture, and cigar businesses; and drive-up storefronts dispensing strong, sweet *café cubano*. They built subdivisions sprawling into the Everglades and provided jobs for tens of thousands of later, poorer Cuban immigrants. Alone among Hispanic subgroups, Cubans were warmly welcomed by the U.S. government throughout the cold war: They received financial assistance and, until 1995, automatic legal residency. As of 1990, 55 percent of Cubans had graduated from high school, and 20 percent held white-collar jobs. But one third do not speak English well or at all; many of them are older Cubans with little incentive to learn the language in a Spanish-speaking city.

11. Nicaraguans. During the 1980s, U.S.-backed rebel leaders plotted to overthrow Nicaragua's Communist government from offices near Miami's airport. As the war dragged on, young Nicaraguans came here to evade the military draft. After the Communists finally lost power in 1990, some 75,000 Nicaraguans remained in the United States. Congress recently granted them the right to stay, so many may eventually become U.S. citizens. Nicaraguan exiles were embraced by Cubans who sympathized with their flight from communism; they settled in Cuban areas like Hialeah and East Little Havana and found work in Cuban-owned businesses. Unlike Miami's Cubans, though, the Nicaraguan immigrants are mostly poor, rural folk, averaging 26 years of age and nine years of schooling. More than half don't speak English well or at all, and their median income of $9,000 in 1990 was the second lowest of all ethnic groups in Miami. (The lowest-ranked group was the 20,000 Hondurans who moved to Miami when the Nicaraguan war unsettled their country.)

12. South Americans. Miami's Hispanic upper crust is not just Cuban; it also includes Colombians, Peruvians, and other South Americans. These wealthy immigrants began coming to Miami when their countries' economies plunged into crisis in the 1980s. Business and professional people fled with their money, buying houses in Kendall, a Miami suburb, and condos in waterfront high-rises. They number well over 100,000.

NEOYORQUINOS

Puerto Ricans used to represent the vast majority of New York's Hispanics; now they are roughly half. Immigrants from the Dominican Republic, Colombia, and Cuba have swelled the metropolitan area's multiethnic mix to 3.6 million Latinos.

13. Puerto Ricans. During the 1950s, the decade when *West Side Story* came to Broadway, New York was home to 80 percent of all Puerto Ricans in the United States. Cheap, frequent flights ferried the islanders back and forth. One million immigrated to New York after World War II, forming the backbone of the city's manufacturing work force. By the 1960s, Puerto Ricans also owned some 4,000 businesses. Many were in Spanish Harlem, which was dotted with restaurants serving chicken *asopao* and *pasteles,* the Puerto Rican version of tamales made with green bananas. In the 1970s Puerto Ricans' American experience turned sour: Newer immigrants began displacing them, and then the industrial base of New York withered away. Unemployed Puerto Ricans headed back home, only to return to New York when they couldn't find jobs there either. In New York, they saw their median family income drop below that of African-Americans, which was rising. "Compared to the black community, our resources are so much weaker," says Angelo Falcon, director of the Institute for Puerto Rican Policy. "We don't have their church leaders or their colleges. We don't have a solid middle class."

14. Dominicans. Washington Heights is the expatriate capital of Dominicans, who now represent almost 10 percent of all Latinos in the New York area. They came to this rundown tip of upper Manhattan, named it Quisqueya—the Native American name for the Dominican Republic—and immediately went into business. They opened neighborhood stores called bodegas all over the city, and drove cabs that competed with yellow taxis. Some Dominicans also tapped their location by the George Washington Bridge to set up a huge drug distribution network serving the Atlantic Coast. Despite all this entrepreneurial activity and Dominicans' comparatively high median income ($10,000 to $15,000), their unemployment rate is 53 percent; 14 percent are on welfare; and 42 percent don't speak English well. New York's Dominicans have fared nearly as badly as Puerto Ricans, in part because they are overwhelmingly first-generation immigrants without high school degrees. They too suffer from a revolving-door syndrome that has kept them from putting down roots. Community leaders have yet to solve Quisqueya's many problems: discrimination against the mostly black and mulatto Dominicans, poor police relations (the 1992 killing of a Dominican immigrant sparked riots), drug-fueled crime, and high rents.

15. Colombians. Colombians have won the economic success that has eluded most Hispanics, but they're dogged by a stereotype that all Colombians are drug traffickers. Most are in fact legitimate businesspeople and successful professionals; yet to avoid stigma, some say they are from another country. New York is their principal U.S. destination, followed by Miami. Only 40 percent are U.S. citizens, although the number is increasing because Colombia now allows dual citizenship. Two thirds of Colombians have jobs, and their median income is close to that of non-Hispanic whites. One fifth of Colombian families earn $50,000 or more, in keeping with their reputation as South America's best entrepreneurs. But arrests of major Colombian traffickers and grisly murders in their Queens enclave of Jackson Heights have cemented a negative image in the public's mind.

ELSEWHERE IN THE U.S.
16. New Mexico's Hispanos. Northern New Mexico is home to the nation's most unusual and least-known group of Hispanics. They are descendants of the original Spanish conquistadors and as such belong to the oldest European culture within U.S. borders. In the valleys of Rio Arriba they farm ribbonlike plots bequeathed to their ancestors by the Spanish crown; live in ancient adobe homes; and cook pork in red *chile* sauce in outdoor ovens. A proud, poor people, they call themselves Hispanos to emphasize that they are not immigrants from Latin America. The Spanish they speak is a dialect from the time of Coronado, and the holidays they celebrate are Spanish ones commemorating events like the 1692 reconquest of New Mexico and the conquest of the Moors. A dwindling Catholic sect called the Penitentes practices self-flagellation in their ancestors *morados,* or temples. Another subgroup are descendants of *marranos,* Spanish Jews who fled the Inquisition and continued to observe Jewish rites secretly. Centuries of subdividing their farmland have forced young Hispanos to seek seasonal work elsewhere or to move away entirely. Unemployment hovers around 20 percent and welfare dependence is high.

17. Migrant workers. For decades, the demand for temporary farmhands has sent Hispanics all over the United States. The migrant farmhands still travel from crop to crop, living in camps straight out of a Steinbeck novel, but farm mechanization has reduced their numbers to about 70,000 for the Midwest harvest. Meanwhile, a second stream of Mexicans is being drawn to work in chicken- and beef-packing plants in places like Dodge City, Kan., where 4,000 Hispanics have arrived since 1990. In Maine, hundreds of Mexicans work on egg farms in Turner (pop. 5,000), which now has a bilingual school program. Siler City, N.C., had 200 Hispanics in 1990. Today, half its 6,000 residents are Hispanic, and the town has three churches offering services in Spanish and four Latin American grocery stores.

RUSSIA'S POPULATION SINK

In the former heart of the Soviet empire, deaths are far outpacing births.

Toni Nelson

Toni Nelson is a staff researcher at the Worldwatch Institute

In Nadvoitsy, a small Russian town near the Finnish border, an estimated 4,000 children have been poisoned by fluoride, which replaces calcium in the body, leaving its victims with blackened, rotting teeth and weakened bones. Although the town's aluminum plant no longer dumps fluoride into unlined landfills, the contamination persists because neither the authorities nor the company can afford a full-fledged clean-up. Today 5 to 10 percent of the town's kindergartners continue to exhibit signs of fluorosis.

Nadvoitsy's experience provides a glimpse into the myriad problems facing the countries of the Former Soviet Union (FSU). Years of environmental contamination have combined with economic instability to push the region into a public health crisis, and several FSU countries are now experiencing the most dramatic peacetime population decline in modern history. In Russia, which has more than half the FSU's population, the situation may be at its worst. As the country's birth rate falls and its death rate climbs, the population is expected to shrink by some 9 million between 1992 and 2005. More important, perhaps, is the rising incidence of birth defects and other health problems whose effects may linger for generations.

Russia's demographic decline began in the mid-1980s, well before the collapse of the Soviet Union in 1991 (see graph). Total live births in Russia dropped from a peak of 2.5 million in 1987 to 1.4 million in 1994, while total deaths climbed from 1.5 million to 2.3 million over the same period. The year 1994 brought the most precipitous decline on record, with deaths exceeding births by more than 880,000 and the population falling by 0.6 percent (excluding immigration, which compensated for two-thirds of the decline). Life expectancy, which provides the best general measure of a country's health conditions, also dropped sharply between 1987 and 1994, from 65 to 57 years for men, and from 75 to 71 years for women. This decline has no precedent in industrialized societies; Russian male life expectancy is now the lowest of all developed countries.

Russia's deteriorating social and ecological conditions have had serious consequences for the country's children as well. Infant mortality has climbed to at least 20 deaths per 1,000 live births, although some experts suggest the figure could be as high as 30 per 1,000—more than three times the U.S. rate and double that of Costa Rica, one of the most advanced developing countries. Birth defects occur in 11 percent of newborns, and 60 percent exhibit symptoms of allergies or the deficiency disease known as rickets, caused by a lack of vitamin D. Children's health tends to decline throughout childhood; scarcely one-fifth of Russia's children can be considered healthy by the end of their school years.

Maternal health, and the health of women of reproductive age in general, is also declining—a trend that will almost certainly intensify problems with infant health. Gynecological pathologies have been found in 40 to 60 percent of women in their child-bearing years, and even girls in their early teens are showing signs of reproductive abnormalities. Fully 75 percent of Russian women experience complications during pregnancy, and the death rate during childbirth is 50 per 1,000 births—more than six times the U.S. rate. Only 45 percent of Russian births qualify as normal by Western medical standards.

The factors underlying these trends are complex and numerous, but most can be traced to some combination of environmental contamination and economic instability. In part, the fertility decline is a matter of simple demographics: the number of marriages has decreased, and there are fewer women of childbearing age in the population due to a brief decline in births after World War II.

5 ❖ POPULATION, RESOURCES, AND SOCIOECONOMIC DEVELOPMENT

But life in the FSU is still haunted by the abrupt transition from communism, which provided work and housing for nearly everyone, to a capitalist system driven by competition and characterized by insecurity. The uncertain economic situation has prompted many couples to forgo childbearing, according to Carl Haub, Director of Information and Education at the Population Reference Bureau. In a survey of 3,000 Russian women in 1992, 75 percent cited insufficient income as a factor discouraging childbearing. As Haub observed in a 1994 report, "The recent birth dearth, not surprisingly, is a direct result of the collapse of the economy and a general lack of confidence in the future."

Births and Deaths in Russia
Source: Goskomstat, Russia (Moscow: 1995)

The rise in Russia's death rate illustrates even more dramatically the extent to which the country's social fabric is fraying. The incidence of stress-related conditions such as heart attacks and alcoholism has risen sharply. When the death rate jumped from 12.1 per 1,000 persons to 14.5 between 1992 and 1993, for example, three-quarters of the increase was due to cardiovascular disease, accidents, murder, suicide, and alcohol poisoning. According to Gennadi I. Gerasimov, a former spokesman for Mikhail Gorbachev, 100,000 Russians have killed themselves in the past two years—a suicide rate more than three times that of the United States. A sharp rise in alcoholism is causing increases in both alcohol poisoning and fatal accidents at work. And lack of adequate health care has exacerbated the problem—treatable conditions are increasingly fatal, and people often die after a relatively mild heart attack.

The death rate also has a clear environmental connection. According to Murray Feshbach, a professor of demography at Georgetown University in Washington, D.C., and an expert on Russia's environment, pollution plays a role in 20 to 30 percent of the country's deaths. In the most affected areas, that percentage is much higher. Feshbach cites the example of Nikel, a town of 21,000 on the Norweigan border in Russia's extreme north. For employees at the local smelter, life expectancy is just 34 years; for the town in general, it is 44 years.

Contamination, in one form or another, is undermining public health throughout the FSU. Water in as many as 75 percent of the region's rivers and lakes is unfit for drinking, according to Alexey Yablokov, chairman of a Russian Federation commission on the environment and a former environmental advisor to both Gorbachev and Boris Yeltsin. And at least 50 percent of tap water fails to meet sanitary norms for microbial, chemical, and other forms of pollution. Waterborne infectious diseases are epidemic (see "Environmental Intelligence," November/December 1995), and children are often the principal victims. Yablokov has classified about 15 percent of Russia's land area as ecological disaster zones.

Presumably, economic improvements will eventually stabilize Russia's demographic picture. But economic recovery alone will not erase the effects of decades of gross environmental abuse. Russia's public health crisis is a lesson in the human costs of unbridled development—a lesson that may have particular relevance for the most rapidly developing countries. In China, for example, where the economy has grown by 57 percent in the past four years, coal accounts for more than three-quarters of the commercial energy supply. Yet, nearly 1 million people a year may already be dying from lung diseases related to air pollution—and China plans to double its coal use over the next two decades.

Vanishing Languages

When the last speakers go,
they take with them their history and culture

DAVID CRYSTAL

THERE'S A WELSH PROVERB I'VE KNOWN FOR AS LONG as I can remember: "*Cenedl heb iaith, cenedl heb galon.*" It means, "A nation without a language [is] a nation without a heart," and it's become more poignant over the years as more and more families who live around me in North Wales speak in English instead of Welsh across the dinner table.

Welsh, the direct descendant of the Celtic language that was spoken throughout most of Britain when the Anglo-Saxons invaded, has long been under threat from English. England's economic and technological dominance has made English the language of choice, causing a decline in the number of Welsh speakers. And although the decline has steadied in the past 15 years, less than 20 percent of the population of Wales today can speak Welsh in addition to English.

The Welsh language is clearly in trouble. Someday, it may even join the rapidly growing list of extinct languages, which includes Gothic and Hittite, Manx and Cornish, Powhatan and Piscataway. If present trends continue, four of the world's languages will die between the publication of this issue of CIVILIZATION and the next. Eighteen more will be gone by the end of 1997. A century from now, one-half of the world's 6,000 or more languages may be extinct.

The decline is evident the world over. Consider the case of Sene: In 1978 there were fewer than 10 elderly speakers remaining in the Morobe province of Papua New Guinea. Or Ngarla: In 1981 there were just two speakers of the Aboriginal language still alive in northwest Western Australia. And in 1982 there were 10 surviving speakers of Achumawi out of a tribal population of 800 in northeastern California. Does it matter? When the last representatives of these peoples die, they take with them their oral history and culture, though their passing is rarely noticed. Sometimes, years later, we find hints of a culture's existence, in the form of inscriptions or fragments of text, but many of these—the Linear A inscriptions from ancient Crete, for example—remain undeciphered to this day.

There is some controversy over exactly how to count the number of languages in the world. A great deal depends on whether the speech patterns of different communities are viewed as dialects of a single language or as separate languages. The eight main varieties of spoken Chinese, for example, are as mutually unintelligible as, say, French and Spanish—which suggests that they are different languages. On the other hand, they share a writing system, and so perhaps are best described as dialects of the same language. If you opt for the first solution, you will add eight to your tally of the world's languages. If you opt for the second, you will add just one.

Taking a conservative estimate of 6,000 languages worldwide, one fact becomes immediately clear: Languages reveal enormous differences in populations. At one extreme, there is English, spoken by more people globally than any other language in history, probably by a third of the world's population as a first, second or foreign language. At the other extreme is Ngarla (and most of the other languages of the native peoples of Australia, Canada and the United States), whose total population of speakers may amount to just one or two. And then there are closely related groups of languages like the Maric family in Queensland, Australia, which consists of 12 languages. When it was surveyed in 1981, only one of these, Bidyara, had as many as 20 speakers. Most had fewer than five. Five of them had only one speaker each.

The loss of languages may have accelerated recently, but it is hardly a new problem. In the 19th century, there were more than 1,000 Indian languages in Brazil, many spoken in small, isolated villages in the rain forest; today there are a mere 200, most of which have never been written down or recorded. In

North America, the 300 or more indigenous languages spoken in the past have been halved.

People sometimes talk of "the beauty of Italian" or of "German's authority," as if such characteristics might make a language more or less influential. But there is no internal mechanism in a language that settles its fate. Languages are not, in themselves, more or less powerful. People don't adopt them because they are more precise. They gain ascendancy when their speakers gain power, and they die out when people die out or disperse. It's as simple as that.

A dramatic illustration of how a language disappears took place in Venezuela in the 1960s. As part of the drive to tap the vast resources of the Amazonian rain forests, a group of Western explorers passed through a small village on the banks of the Coluene River. Unfortunately, they brought with them the influenza virus, and the villagers, who lacked any immunity, were immediately susceptible to the disease. Fewer than 10 people survived. A human tragedy, it was a linguistic tragedy too, for this village contained the only speakers of the Trumai language. And with so few people left to pass it on, the language was doomed.

Other languages—such Welsh and Scottish Gaelic—have been threatened when indigenous populations have moved or been split up. Brighter economic prospects tempt young members of the community away from their villages. And even if they choose to stay, it doesn't take much exposure to a dominant culture to motivate ambitious young people to replace their mother tongue with a language that gives them better access to education, jobs and new technology.

A language's fortunes are tied to its culture's. Just as one language holds sway over others when its speakers gain power—politically, economically or technologically—it diminishes, and may even die, when they lose that prominence. Latin, now used almost exclusively in its written form, had its day as a world language because of the power of Rome. English, once promoted by the British Empire, is thriving today chiefly because of the prominence of the U.S.A., but it was once an endangered language, threatened by the Norman invaders of Britain in the 11th century, who brought with them a multitude of French words. In South America, Spanish and Portuguese, the languages of colonialists, have replaced many of the indigenous Indian tongues.

The death of languages is most noticeable in parts of the world where large numbers of languages are concentrated in a few small geographical regions. Travel to the tropical forests of the Morobe province in Papua New Guinea and you'll find five isolated villages in a mountain valley where fewer than 1,000 people speak the Kapin language. They support themselves by agriculture and have little contact with outsiders. Other tiny communities, speaking completely different languages, live in neighboring valleys. Linguists estimate that in the country as a whole there is approximately one language for every 200 people. Indeed, three countries, which together amount to less than 2 percent of the earth's land area, support 1,700—or a quarter—of the world's living languages: Papua New Guinea has 862; Indonesia, 701; and Malaysia, 140. These countries' isolation and physical geography account in large part for the existence of such concentrations, and it is hardly surprising to find that, as remote areas of the globe have opened up for trade or tourism, there has been a dramatic increase in the rate of language death. Valuable reserves of gold, silver and timber in Papua New Guinea, for example, are bringing speculators to the islands—and with them their languages.

There has been little research into exactly what happens when a language begins to die. The process depends on how long there has been contact between the users of the minority language and their more powerful neighbors. If the contact has been minimal, as in the case of Trumai in the Amazon, the minority language might remain almost unchanged until the last of its speakers dies. But if two languages have been in contact for generations, the dominant language will slowly erode the pronunciation, vocabulary and grammar of the minority language. Take the Celtic languages of northwest Europe. Following the death of the last mother-tongue speakers of Cornish (spoken in Cornwall until the 19th century) and Manx (spoken in the Isle of Man until the 1940s), the only remaining Celtic languages are Breton (in northwest France,) Irish and Scottish Gaelic, and Welsh. All have been in steady decline during the 20th century. Equally, all have been the focus of strenuous efforts to revive their fortunes (or, in the case of Cornish and Manx, to resurrect a new first-language base). But the effects of four centuries of domination by English are evident everywhere.

Walk into the stores in the strongly Welsh-speaking areas of North Wales, as I regularly do, and you will hear the Welsh language widely used—and apparently in good health. But there is also a great deal of recognizable English vocabulary scattered throughout the speech. Of course, all languages have what linguists refer to as "loan words"—words taken from other languages to supplement the vocabulary. English itself has tens of thousands of words borrowed from French, Spanish, Latin and other languages. But there is an important difference between traditional vocabulary borrowing and what takes place in an endangered language. When *arsenic, lettuce* and *attorney* came into English in the Middle Ages, it was because these items did not already exist in the English-speaking community: The nouns were introduced to describe new objects, and so to supplement the existing vocabulary. But in the case of an endangered language, the loan words tend to replace words that already exist. And as the decline continues, even quite basic words in the language are replaced.

I meet this phenomenon every day on the Welsh island of Anglesey, where I live. It's become quite unusual to hear locals referring to large sums of money in anything other than English. In a Holyhead butcher's shop recently, I overheard someone say "*Mae'n twelve fifty*" (It's twelve fifty), where the first part of the sentence is colloquial Welsh and the second part is colloquial English. As I waited for a train at the station the same day, I heard a porter calling out to disgruntled passengers "*Mae'n late*" (It's late). And I later overheard a group of people using the English word *injection* as they stood in a street describing in Welsh someone's visit to a doctor's clinic. In all these cases, perfectly good Welsh words already exist, but the speakers did not use them. Why they chose not to is not at all clear. Maybe they did not know the Welsh words, or maybe it is a sign of status or education to use the English equivalents. But when something as basic as its number system is affected, a language is clearly in danger.

Mixed languages are an inevitable result of language contact, and they exist all over the world, often given a dismissive label by more educated speakers: Wenglish, Franglais, Spanglish. Such mixed varieties often become complex systems of communication in their own right—and may even result in brand-new languages, or pidgins such as Tok Pisin, which is now spoken by more than 1 million people in Papua New Guinea. But when one of the languages in question has no independent existence elsewhere in the world, as in the case of Welsh, mixed languages are a symptom of linguistic decline.

In the West, when a population fears that its language is threatened, speakers often react defensively, establishing a committee or board to oversee and coordinate political policy and to plan dictionaries, grammars and local broadcasting. The best-known example is France, home of the Académie Française, where there is now a law banning the use of English words—such as *le week-end* and *le computeur*—in official publications if a native French term already exists (in these cases, *la fin de semaine* or *l'ordinateur*). Often two levels of language ability emerge as a consequence. There is an educated standard, used as a norm in education and the media. And there is a colloquial standard, used by the majority of the population (including many educated users, who thereby become bilingual—more technically, bidialectal—in their own language). It is the usage of the elite minority that is called by the majority the "proper" or "correct" language, even though it often represents a far more artificial style of speech than the language of the streets.

The plight of the indigenous languages of America was made vivid by James Fenimore Cooper as long ago as 1826, when the Indian chief Tamenund lamented that "before the night has come, have I lived to see the last warrior of the wise race of the Mohicans." There are 200 North American Indian languages, but only about 50 have more than 1,000 speakers, and only a handful have more than 50,000. Just over a year ago, Red Thunder Cloud, the last known fluent speaker of the Siouan language Catawba, died. The only surviving fluent speaker of Quileute is 80-year-old Lillian Pullen, of La Push, Washington. But at least the decline of American Indian languages has begun to attract widespread attention from politicians and the media—sources of support that are unlikely to help such equally threatened but less well known cases as Usku in Irian Jaya or Pipil in El Salvador.

In Europe, public attention is regularly focused on language rights by the European Bureau of Lesser Used Languages, headquartered in Brussels. A recent book, *A Week in Europe*, edited by the Welsh magazine editor Dylan Iorwerth, offers a glimpse of Western European life by journalists writing in minority languages. Some of these are minority uses of major languages, such as German in Denmark, Swedish in Finland, and Croatian in Italy; but in most cases the entire language-using community is found in a single region, such as Scottish Gaelic, Galician, Alsatian, Welsh, Catalan, Asturian, Breton, Fruilian, Basque, Sorbian, Occitan, Provençal, Frisian and Irish. Political concern over the status of minority languages is regularly voiced by the European Parliament, and occasionally words are backed up with financial commitment—to local newspapers and broadcasting, literary festivals and teaching programs.

When an endangered language (such as Gaelic) is spoken in a culture whose historical significance is widely appreciated—perhaps because it is associated with prowess in arts and crafts, or because it is known for its literary achievements—it may provoke widespread concern. And sometimes endangered languages that have suffered as a result of colonial expansion win support from speakers of the dominant language, who wish to distance themselves from the aggression of their ancestors. But in most cases, anxiety, like charity, begins at home. In the 1970s, Gwynfor Evans held a hunger strike as part of his (successful) campaign for a Welsh-language TV channel. And in 1952 in Madras, India, Potti Sriramulu died following a hunger strike in support of the Telugu language. Language, as that Welsh proverb reminds us, is truly at the heart of a culture. It is a matter of identity, of nationhood.

The rapid decline is most noticeable in parts of the world where large numbers of languages are concentrated into a few small geographical regions

With enough personal effort, time and money, and a sympathetic political climate, it is possible to reverse the fortunes of an endangered language. Catalan, spoken in northeast Spain, was allocated the status of an official regional language, and it now has more native speakers than it did 30 years ago. And the Hocak, or Winnebago, tribe in Wisconsin is hoping to develop a full Hocak-speaking school system. In an effort funded entirely by profits from the tribe's casinos, schoolchildren use interactive multimedia computer programs to gain familiarity with a language that was traditionally passed down orally from parent to child. Such advances generally depend upon collaboration between minority groups, such as those who united to form the European Bureau for Lesser Used Languages. Together they have a realistic chance of influencing international policies, without overlooking the vast differences between the political and cultural situations of minority languages: Welsh, Gaelic, Maori, Quechua and Navajo demand very different solutions.

Welsh, strongly supported by Welsh-language broadcasting and Welsh-medium schools, is alone among the Celtic languages in stopping its decline. The census figures for the last 20 years show a leveling out, and even some increase in usage among certain age groups, especially young children. A similar vigorous concern seems to be stimulating Navajo and several other American languages, as well as some minority languages in continental Europe. But it is quite clear that most of the endangered languages of the world are beyond practical help, in the face of economic colonialism, the growth of urbanization and the development of global communication systems. And, given the difficulty there has been in achieving language rights for such well-known communities as the Navajo or the Welsh, the likelihood of attracting world interest in the hundreds of languages of Papua New Guinea, each of which has only a few speakers left, is remote. Clearly, with some 3,000 languages at risk, the cost of supporting them on a worldwide scale would be immense. Can, or should, anything be done?

On an intellectual level, the implications are clear enough: To lose a language is to lose a unique insight into the human condition. Each language presents a view of the world that is shared by no other. Each has its own figures of speech, its own narrative style, its own proverbs, its own oral or written literatures. Preserving a language may also be instructive; we can learn from the way in which different languages structure reality, as has been demonstrated countless times in the study of comparative literature. And there is no reason to believe that the differing accounts of the human condition presented by the peoples of, say, Irian Jaya will be any less insightful than those presented by writers in French, English, Russian and Sanskrit. Moreover, the loss of a language means a loss of inherited knowledge that extends over hundreds or thousands of years. As human beings have spread around the globe, adapting to different environments, the distilled experiences of generations have been retained chiefly through the medium of language. At least when a dying language has been written down, as in the case of Latin or Classical Greek, we can usually still read its messages. But when a language without a writing system disappears, its speakers' experience is lost forever. The Bithynian, Cappadocian and Cataonian cultures are known today only from passing references in Greek literature. Language loss is knowledge loss, and it is irretrievable.

Such intellectual arguments may persuade the dispassionate observer, but most arguments in favor of language preservation are quite the opposite: They are particular, political and extremely passionate. Language is more than a shared code of symbols for communication. People do not fight and die, as they have done in India, to preserve a set of symbols. They do so because they feel that their identity is at stake—that language preservation is a question of human rights, community status and nationhood. This profoundly emotional reaction is often expressed in metaphors. Language nationalists see their language as a treasure house, as a repository of memories, as a gift to their children, as a birthright. And it is this conviction that has generated manifestoes and marches in Melbourne in support of Aboriginal languages; referendums, rioting and the defacing of public signs in Montreal on behalf of French; civil disobedience in India and Pakistan, in Belgium and in Spain.

Such demonstrations stand in stark contrast to places where cultural and linguistic pluralism works successfully, as in Switzerland and Sweden, where the dominant culture respects the identities and rights of its linguistic minorities, and provides educational opportunities for speakers. Successful multilingual communities such as Sweden's serve as examples for the United Nations, UNESCO, the Council of Europe, and the European Parliament as they act to preserve minority language use.

Conversely, several countries have actively repressed minority languages, such as Basque by the Spanish fascists, or Sorbian (a Slavic language spoken in southern Germany) by the Nazis. And laws forbidding the use of minority languages have been commonplace; children have been punished for using a minority language in school; street signs in a minority language have been outlawed; the publication of books in the language has been banned; people's names have been forcibly changed to their equivalents in the language of the dominant power. Whole communities, such as several in the Basque-speaking parts of northern Spain, have had their linguistic identity deliberately eliminated.

> There are 200 North American Indian languages, but only about 50 have more than 1,000 speakers, and only a handful have more than 50,000

Political arguments for and against preservation have been expressed with such vehemence that they tend to dominate any discussion of minority languages. Does the loss of linguistic diversity present civilization with a problem analogous to the loss of species in biology? Not entirely. A world containing only one species is impossible. But a world containing only one language is by no means impossible, and may not be so very far away. Indeed, some argue strongly in favor of it. The possibility of creating a unilingual world has motivated artificial-language movements (such as Esperanto) since the 16th century, and there are many who currently see the remarkable progress of English as a promising step toward global communication. They argue that mutual intelligibility is desirable and should be encouraged: Misunderstandings will decrease; individuals and countries will negotiate more easily; and the world will be more peaceful.

This kind of idealism wins little sympathy from language nationalists, who point out that the use of a single language by a community is no guarantee of civil peace—as is currently evident in the states of the former Yugoslavia or in Northern Ireland. But language nationalists are faced with major practical concerns: How can one possibly evaluate the competing claims of thousands of endangered languages? Is it sensible to try to preserve a language (or culture) when its recent history suggests that it is heading for extinction? In the next few years, international organizations may have to decide, on chiefly economic grounds, which languages should be kept alive and which allowed to die.

The publication in the early 1990s of major surveys of the world's languages has brought some of these issues before the public. UNESCO's Endangered Languages Project, the Foundation for Endangered Languages (established in the U.K. in 1995) and the Linguistic Society of America's Committee on Endangered Languages and their Preservation are fostering research into the status of minority languages. Information is gradually becoming available on the Internet—such as through the World Wide Web site of the Summer Institute of Linguistics. And a clearinghouse for the world's endangered languages was established in 1995, by request of UNESCO, at the University of Tokyo.

But after the fact-finding, the really hard work consists of tape-recording and transcribing the endangered languages before they die. The fieldwork procedures are well established among a small number of dedicated linguists, who assess the urgency of the need, document what is already known about the languages, extend that knowledge as much as possible, and thus help preserve languages, if only in archive form.

The concept of a language as a "national treasure" still takes many people by surprise—and even English has no international conservation archive. It is hard to imagine the long hours and energy needed to document something as complex as a language—and it's often a race against time. Thirty years ago, when anthropologist J. V. Powell began working with the Quileute Indians in Washington state, 70 members of the tribe were fluent speakers. Around that time the tribal elders decided to try to revitalize the language, writing dictionaries and grammars, and imagining a day when their children would sit around chatting in Quileute. "But," says Powell, "their prayers haven't been answered." Now they've scaled back to a more modest goal: basic familiarity rather than fluency. Powell recognized that they will not save Quileute, but it will be preserved in recordings for future scholars—and will serve as a symbol of the tribe's group identity. That may seem like a small success, but it's a far better fate than the one facing most endangered languages.

Article 39

RISKY BUSINESS:

WHO WILL PAY FOR THE GROWING COSTS OF GLOBAL CHANGE?

No other decade in history has hosted as many natural disasters as the 1990s. The list of catastrophic weather events—from Hurricane Andrew in Florida in 1992 to the recent flooding in Central Europe and Peru—reflects an alarming increase in the rate of events that, in the case of devastating floods, were thought to happen only once every few hundred years.

Behind many of these disasters lurks the specter of human-induced global change. Indeed, the Intergovernmental Panel on Climate Change (IPCC) concluded in its Second Assessment Report that the increased greenhouse gases due to human activity "are projected to change regional and global climate and climate-related parameters such as temperature, precipitation, soil moisture and sea level."

This increase of greenhouse gases, says the IPCC, will heat up the Earth's temperature, causing seas to rise and increasing the occurrence of, among other things, severe floods and droughts. Furthermore, the Second Assessment Report states that increased population density in sensitive areas leaves more humans vulnerable to natural catastrophes.

The global economic losses and human suffering from natural disasters are high and increasing. How best to address the costs and complexities surrounding the occurrence and aftermath of these disasters requires an interdisciplinary, international approach. Ideally, the approach should take into consideration the latest research advances on global change, as well as employ mathematical methods and computer modeling.

IIASA's Risk, Modeling and Policy (RMP) Project, with its history of risk and fairness analysis, as well as its long tradition of applying optimization techniques under uncertainties, is initiating such an approach. With an award-winning catastrophic risk model under its belt and an international network of researchers, public authorities, the insurance industry, and financial institutions at its disposal, the Project is poised to make IIASA a European center for catastrophic risk research.

Economic Losses are Mounting

Natural catastrophes during the past winter alone have resulted in millions of dollars in property damages. For example, in Peru, El Niño-related weather has devastated the fishing industry and has caused more than U.S.$800 million in damage. In Europe during the past two years, unprecedented amounts of precipitation have inundated areas in Poland, Germany, Austria, and the Czech Republic. Over the past 50 years, floods normally associated with melting snow in the spring have shifted their occurrence to summer because of heavier summertime precipitation.

The high costs associated with these destructive acts of nature are becoming more and more commonplace. According to figures published by Munich Re, an international re-insurance firm, in the last decade the number of major natural catastrophes (e.g., floods, hurricanes, earthquakes, wildfires, avalanches, sea surges, hail storms, and volcanic eruptions) is three times as great, and cost the world's economies eight times as much as in the decade of the 1960s. In 1997, the most frequent natural catastrophes were windstorms and floods, which accounted for 82 percent of the economic losses and no less than 97 percent of the insured losses.

The mounting costs affect a wide range of entities, from individuals to federal governments to the insurance industry. How best to control these costs and protect against catastrophic risks is a problem of increasing urgency, says Joanne Bayer, co-leader of the RMP Project. "Our past research provides a strong base for examining a range of issues associated with climate change and catastrophic risk management," says Bayer.

The Issues

Natural and technological disasters raise many important research and policy issues about how societies can protect themselves against catastrophic risks. What mitigation and other measures can and should be taken to reduce the losses from high-consequence events occurring very infrequently? Is there sufficient evidence to link an increase in storm severity to climate change? What role does the insurance industry play in reducing and mitigating the catastrophic

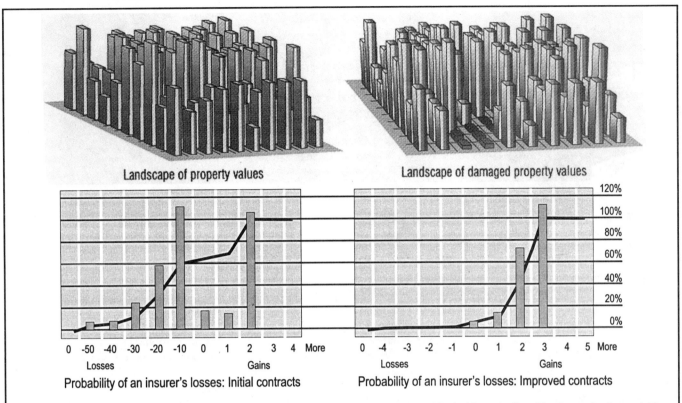

The RMP Project's model for catastrophic risk management can help improve industry-wide decisions on diversification and other variables and can generate policy strategies that decrease losses.

losses resulting from global change? Is there a risk of a mega catastrophe with widespread insurance insolvency, unpaid claims and a possible breakdown in the global insurance market?

These are just some of the issues the RMP Project has begun to address by examining the role of governments and the insurance industry in managing catastrophic events of all kinds.

The Model

The insurance industry is embracing computer models to aid their strategies for dealing with catastrophic risks by improving the insurability of rare events (with dependent claims) and reducing the vulnerability of the insurance industry to insolvencies.

Many of the models include "simulators" for various catastrophic events and their consequences. This approach is useful for analyzing possible alternative scenarios and the sensitivity of outcomes to frequencies of events, spatial patterns, and policy options. However, the large variety of "if-then" solutions generated by these models makes determining which strategy or policy to implement difficult.

The RMP Project has recently developed a model for catastrophic risk management that is concerned with choosing the best, most robust strategies without the user having to perform endless analyses of "if-then" strategies. The notion of "best, robust strategy" is explicitly specified in the model in terms of various indicators such as stability, insolvency, profits, losses, and available budgets. The model employs specific nonsmooth stochastic optimization techniques developed at IIASA to design the desirable strategies or policies.

The RMP Project's model can help improve industry-wide decisions on diversification, contracts and other decision variables. It can also generate policy strategies that decrease the risk of insurer insolvency and increase profitability, as well as decrease insuree losses. Because catastrophic risks are characterized by dependent losses, the model explicitly includes geographic diversification of insurance coverage. The model's simulations can be useful to insurers in decreasing their vulnerability, to the insured in increasing their security, and to regulators in gauging the amount of necessary intervention. This work received the Kjell Gunnarson Risk Management Prize from the Swedish Insurance Society at the annual meeting of the Society for Risk Analysis in June 1997.

Predictability of Natural Disasters

The RMP initiative includes work on a theory that may help to increase the predictability of natural catastrophes.

"The insurance industry assumes that big events are independent. This is incorrect for natural catastrophes," says IIASA Director Gordon MacDonald, who is also participating in the RMP initiative. MacDonald, a geophysicist, explains that natural catastrophes such as earthquakes and volcanoes happen in clusters due to memory of past events and so can be predicted to some extent. He is examining whether long-term prediction of catastrophic climate events is possible too.

MacDonald's historical analysis of catastrophic storms shows that there are time dependencies or inter-relationships for hurricanes and other destructive weather events. His research focuses on deep ocean currents as the driving force of time-dependent storms. "Atmospheric weather patterns allow us to predict

NASA's proposed activities would initially focus on the United States, Japan and Europe.

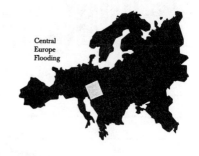

United States

Recent Natural Catastrophic Events:
Hurricane Andrew
Northridge Earthquake

Issues:
The social costs of disasters and the vulnerability of the insurance industry have raised difficult policy issues on the role of the private market and public authorities in reducing the damages from catastrophes and in equitably and efficiently spreading the losses.

Japan

Recent Natural Catastrophic Events:
Kobe Earthquake

Issues:
Limited private insurance, as well as limited mitigation measures, such as strict building codes, and the high concentration of certain types of industries, are factors contributing to Japan's vulnerability to catastrophic risks. An important policy issue is reducing this vulnerability through prudent risk management strategies, including siting decisions to avoid industry clustering, building codes and other mitigation measures, and financial diversification through insurance.

Europe

Recent Natural Catastrophic Events:
Windstorm Daria
Central Europe Flooding

Issues:
Catastrophic losses from natural and man-made disasters are worrying governments and insurers for several reasons. First, governments in Europe have traditionally taken primary responsibility for disaster aid and victim compensation, which has led to a reliance on the public authorities and little private responsibility for mitigating the risks.
Second, many of these governments do not have the financial resources for compensating the victims of large disasters; yet, for institutional and other reasons, these risks are often not spread across the industrial countries or even across Europe. The regulatory patchwork in Europe creates high costs for private insurers to enter the market on a large and geographically dispersed scale.

---

what the weather will be in the near future," says MacDonald. For distant-future predictions, he explains, "Ocean currents drive atmospheric conditions, but move much more slowly. By tracking these currents we could predict long-term weather patterns." He adds that last summer's severe flooding in Europe may be linked to Hurricane Hugo, which hit the U.S. nearly 10 years ago.

MacDonald says that long-term predictions could provide plenty of lead time for nations to prepare for catastrophic natural events. A better prepared populace may mitigate the high costs associated with these disasters.

International Dialogue

In addition to its risk modeling work and the research on natural catastrophe predictions, the RMP Project is initiating a collaborative research program and policy dialogue on global change and catastrophic risk management in the U.S., Europe and Japan. The dialogue will involve the international research community, private industry, public officials and non-profit organizations. Initial collaborators include the Joint Research Center of the European Union (JRC) and the Center for Risk Management and Decision Sciences of the Wharton School, University of Pennsylvania.

IIASA's goal to become a leading European center for catastrophic risk research may be ambitious, but as the costs associated with natural catastrophes mount, a range of groups from the insurance industry to policy makers to insurance policy holders will be looking for ways to mitigate these expenses. Responding to this urgent problem is one more example of IIASA applying its wealth of global change knowledge—as well as its long tradition in environmental risk management—to develop practical solutions.

Index

A

Abida Hussain, E. Syeda, 177
abortion services, U.S. aid and, 181
adult-onset diabetes, American Indians and, 173
Africa, southern, 71
aging, of Chinese population, 136
agriculture, Green Revolution and, 188–193
alcohol. *See* gasohol
algal blooms, 23–24
Aliyev, Heydar, 131, 132, 133
American Indians: casinos and, 161–171; 172, 173; poverty among Oglala Sioux and, 172–173
AM-1, 19–21
area studies (or regional) tradition, in geography, 10, 11–12
Argentina, 50, 51, 52
Arizona: Hispanics in, 201; Indian gaming casinos in, 165, 167–168
Arizona Indian Gaming Association, 168
Armenian Assembly of America, 132–133
Armenians, 132–133
art of mapping, 11
Article 907, 132–133
Asian clam, 26
Asian exclusion act, 202
Aster (Advanced Spaceborne Thermal Emission and Reflection Radiometer), 20
Aswan Dam, 68
automobile suburbs, 102–103, 143–144, 146
Azerbaijan, 130–134
Azerbaijan International Operating Co. (AIOC), 131–132

B

bachelorhood, in China, 137
Baku, Azerbaijan, 130
Bakun River Dam, 69, 70
Bangladesh, population growth in, 176–177, 178, 179–180, 182
Bayamon, Puerto Rico, 157
Bering Land Bridge, 148–153
biological diversity, loss of, 23
biomass energy, 40, 43–44. *See also* gasohol
bioregions, 94
bison, 151
BMW, 111–112
Boise, Idaho, 155
Bolivia, 50, 51, 52
Borlaug, Norman, 188–189
Bracero Program, 203–204, 207
Brazil, 50, 52
Broecker, Wallace, 56, 58
Bureau of Indian Affairs, 164, 172–173
Bureau of Land Management, 74
Bureau of Reclamation, 77

C

Cabazon Band, 162–163
CAFE (Corporate Average Fuel Economy) standards, 31
California Coastal Commission, 93
California, Hispanics in, 200, 201, 208, 211–212
California v. Cabazon Band of Mission Indians, 162–163
carbon dioxide, 24–25, 30–31, 64–65
Cardoso, Fernando Henrique, 52
"Cascadia," 104
casinos, American Indians and, 161–171; 172, 173
Cassman, Kenneth, 188–189, 191–192
Ceiba, Puerto Rico, 157
Central Americans, 206, 211–212
central business districts (CBDs), 142, 145, 146, 147
Ceres (Clounds and the Earth's Radiant Energy System), 20
Chicago Latinos, 212–213
Chile, 50–51, 52
China: aging of, 136; air and water pollution and, 83–84; arable land and, 84; deforestation and, 85–86; demographic problems in, and "One Child" policy, 135–137; economic assessment of environmental change in, 82–87; erosion in, 84–85; farmland losses in, 84; forest mismanagement in, 85; forests' contribution to climate control, 85; lost ecosystem services in, 85; marriage prospects in, 137; modernization and, 86; nitrate contamination in, 83; population problems in, 135–137; red tides in, 83; sex ratio in, 137; waterborne pathogens in, 84
Chinese Exclusion Act of 1882, 202
chorographic tradition, in geography. *See* area studies tradition
Church, Gardner, 104
Clean Air Act, 28, 29–30, 31
Clean Water Act, 28
climate change, 33–38; economics of environmental decisionmaking and, 30–31; El Niño and, 50–52, 58–59, 222; flip-flops in temperature and, 53–61; Mackenzie Basin Impact Study of, 62–66; Mission to Planet Earth of NASA and, 19–21; risk modeling of natural disasters and, 222–224
clouds, 35
coastal wetlands, 23
Colombia, 50, 51, 52
Colombians, 214
Colorado, Hispanics in, 201, 202, 204
Connecticut, Indian gaming casinos in, 164–165, 168
conservationism, 74
Consultative Group on International Agricultural Research (CGIAR), 193
contraceptives, population growth and, 176, 178, 179, 180, 181
core-and-boundary neighbourhood, 106
Corporate Average Fuel Economy (CAFE) standards, 31
countries, buying power by, 154–155
Cuba, 50, 205–206, 213
cultural areas, 94
Cuyahoga River, 28
cropland, 75, 76, 77

D

dams: cost-benefit analysis of, 67–68; resettlement of people and, 69; salt desposits and, 67; social penalties of, 68–69; waterlogging of soil and, 67
diabetes, American Indians and, 173
Digital Revolution, need for skyscrapers and, 159–160
Dominicans, 206, 213–214
Domenici, Pete, 163

E

Earth Observing System (EOS), 19–21
earth science tradition, in geography, 10, 12–13
economic cost, environmental, evaluation of, in China, 82–87
ecosystems, regions and, 92–93
Ecuador, 50, 51, 52
"Effective Buying Income" (EBI), 155
Egypt, 177
Ejeta, Gebissa, 190
El Niño, 50–52, 58–59, 222
Elchibey, Abullaz, 132
eminent domain, 76–77
endangered languages, 217–221
entertainment, land as form of, 15–16
environmental decision making, growing role of economics in, 28–32
Environmental Protection Agency (EPA), 28, 29, 30
Esmeralda County, Nevada, 155
ethanol. *See* gasohol
Ethiopia, 188, 189–191
Ethyl Tertiary Butyl Ether (ETBE) fuel, 42, 44
Europe, North Atlantic Current and, 54
externalities, 74
extinction: of languages, 217–221; as natural process, 26

F

Fairfax, Virginia, 155
family planning services, population growth and, 176, 177, 178–179
fisheries, 23
Food and Agriculture Organization (FAO), UN, 189
food requirements, 194–197; new Green Revolution and, 188–193
forest land, 75, 77–78
Former Soviet Union (FSU), 215–216
Foxwoods Casino/Hotel, 165–166
Freedom Support Act, 132–133
freshwater systems, 26
Fujimori, Alberto, 52

G

gambling, American Indians and, 161–171, 172–173
gasohol, 39–45; consumption of, 41; determinants of, use patterns, 41, 43; market share of, 41; sales of, 40–41
gasoline taxes, 46
geographical concentration, 114
geography and earth sciences, 12; social studies and, 12
global warming. *See* climate change
Goldin, Daniel S., 19, 20, 21
Goltz, Thomas, 133
Grameen Bank, 180
grasslands, 75, 78
Great Salinity Anomaly, 58
Greater-Toronto, 104
Green Revolution, 188–193
greenhouse gases, 30–31, 34, 35, 36, 64–65, 222
Greenville County, South Carolina, 108–113
Guangdong Province, 68–69, 98
Guanica, Puerto Rico, 157
Guatemalans, 206, 212
Guyana, 51

H

Hamilton, Indiana, 155
High Plains, 115–119; contraction of people and services in dryland farming areas and, 118; as global center of technology to improve efficiency in irrigation, 118; water conservation practices and, 118
Hispanics: changing geography of U.S., 198–210; subcultures of, 211–214
Hispano Homeland, 201, 204, 207–208
Hispanos, 201, 204, 207–208, 214
homelands, 94
Homestead Act, 16
Houston, Texas, 155
"human-dominated ecosystems," 22
human-dominated planet, 22–27

I

ice dams, 57–58
IIASA, Risk, Modeling and Policy Project of, 222–224
Immigration Act of 1924, 202
Immigration and Nationality Act of 1965, 206
Immigration and Naturalization Service (INS), 209
Immigration Reform and Control Act of 1986, 209
India, population growth programs in, 178, 179
Indian Gaming Regulatory Act (IGRA), 161, 163–164, 169
Intergovernmental Panel on Climate Change (IPCC), 222
International Center for Maize and Wheat Improvement (CIMMYT), 189, 193
International Crops Research Institute for the Semi-Arid Tropics (ICRISAT), 189, 190, 193
International Food Policy Research Institute (IFPRI), 192, 193
International Monetary Fund, 86
International Rice Research Institute (IRRI), 189, 191, 193
invasions, of floras and fuanas, 26
Iran, fertility rates in, 179
Isthmus of Panama, 56

J

Jacobs, Chuck, 173
Japan, 98–99, 100

K

Kalahari Desert, 71
Kamchatka Peninsula, 153
Kenya, 188
Kino, Eusebio, 199
knowledge, geographical, 14–18
Kocharian, Robert, 132
Kyoto protocol, 30

L

land bridges, 148–153
land cover, 73–81
land development, 75, 78; effects of, on arable land, 80; as offset by conversion of land from glassland and forest, 79–80
land reclamation, 74
land transformation, 22–27, 73–81
land use, 73–81
languages, decline in number of, 217–221
Las Vegas, Nevada, 155
Latinos. *See* Hispanics
Laurentide Ice Sheet, 153
Loess Plateau, 84
Los Angeles, California, 103, 155
Louisville, Kentucky, 155
Lummi Indian Nation Casino, 166–167

M

MacDonald, Gordon, 223–224
Mackenzie Basin Impact Study (MBIS), 62–66
mammoths, 151
mangrove ecosystems, 23
Manhattan, New York, 155
Manifest Destiny, 16
man-land tradition, in geography, 10, 12
Mashantucket Pequots, 165–166
Mayan Indians of Guatemala, in Texas, 212
metropolis: American, and growth of transportation, 140–147; sprawl and, 101–107
Metro-Toronto, 103, 104
Mexican-American War, 200
Mexican-Americans: barrio-dwelling, 211; in California, 200, 201, 208, 211–212; changing demographics of, in U.S., 198–210; in Chicago, 211; immigrant, 211; middle-class, 211; in New Mexico, 201, 204, 214; in Texas, 200, 201, 204, 207–208, 212
Mexico, 50, 52
Miami Hispanics, 213
Michigan, Indian gaming casinos in, 165, 168–169
middle class: in Mexico, 211; in Puerto Rico, 157
Middle East, conflict over water resources in, 120–129, 177
Midwestern United States, cropland use/cover of, 77
migrant workers, 202–204, 207, 214
Minsk Talks, 132, 133–134
Misr (Multi-angle Imaging Spectro radiometer), 20
Mission to Planet Earth, of NASA, 19–21
Modis (Moderate Resolution Imaging Spectroradiometer), 20
Mopitt (Measurements of Pollution in the Troposphere), 20
Mozambique, 191

N

Nagorno Karabakh, 132–133
NASA (National Aeronautics and Space Administration), Mission to Planet Earth of, 19–21
National Climatic Data Center, 35
National Forest System, 74
National Geographic Society, 14
National Indian Gaming Commission (NIGC), 163, 164
National Oceanic and Atmospheric Administration (NOAA), 35, 37
National Origins Act of 1924, 202
Native Americans. *See* American Indians
natural disasters, risk modeling of, 222–224. *See also* El Niño
negative externalities, 74
neighborhoods, 106
Neoyorquinos, 213–214
New Jersey, 105–106
New Mexico, Hispanics in, 201, 204, 207–208, 214
New York City, New York, Hispanics in, 213–214
Nicaraguans, 206, 213
nitrogen, 25
North American Free Trade Agreement (NAFTA), 98, 99
North Atlantic Current, 54
North Atlantic Oscillation, 57, 58–59
Northeastern United States, forest cover of, 76

O

ocean-atmosphere climate models, 34
Ogallala aquifer, 77, 115–119; contraction of people and services in

dryland farming areas and, 118; depletion of, 117; dry Arkansas River and, 117; high quality of, 117; irrigation and, 118; irrigation in 1930s and, 117; water conservation practices and, 118; water rights and, 118–119
Oglala Sioux, poverty among, 172
oil: in Azerbaijan, 130–132, 134; decline in global production of conventional, 183–187
Oñate, Juan de, 199
"One Child" policy, of China, and future demographic problems, 135–137
Oneida Tribe, 165
Operation Wetback, 204
Organization of Petroleum Exporting Countries (OPEC), 183–184, 185, 187

Paddock, William C., 188
Pakistan, population growth in, 176–177, 178–179
Panama, 51
Paraguay, 50, 52
Peru, 50, 51, 52, 222
Pico Union, Central Americans of, 211
Pine Ridge Reservation, of Oglala Sioux, 172–173
Ponce, Puerto Rico, 157
Population and Community Development Association, 179
population growth, 176–182; demographic problems of "One Child" policy of China and, 135–137; food requirements and, 194–195
Porter, John, 133
Portland, Oregon, 104, 105
positive externalities, 74
preservationism, 74
promotional campaigns, population growth and, 179–180
Proposition 187, 209
public housing, suburbs and, 46–47
Puerto Ricans, 204, 205, 212–213
Puerto Rico, 156–157

railroad suburbs, 102
Reagan, Ronald, 181
recovery factors, 186
reduced environmental services: impossibility of determining cost or price of, 82; biodiversity and, 86; greenhouse gas emissions, 86
Refugee Act, 206
region states: amenities of, 98; average population of, 98; boundaries of, 98; economies of scale of, 98; marketing and, 98; nation-states and, 99; as natural economic zones, 98; new, 98; as overlapping existing natural boundaries, 92; primary linkages of, 98; trade protection and, 98, 99; United States as collection of, 98
regional authority, 93
regional identity, 92
regionalism, 91
regions, 90–96; ethnic, 94; vernacular, 93–94
Relief International, 133
repatriation strategy, for Mexican Americans, 202–203
Replenishment of Agricultural Workers (RAW), 209
Risk, Modeling and Policy (RMP) project, costs of global climate change and, 222–224
Roaring Eagle Casino, 169
Rocky Mountains, 17
Russia, 215–216

Saginaw Chippewa Indian Tribe, 168
Salt Lake, Utah, 155
San Miguel County, Colorado, 155
Sasakawa Global 2000 (SG2000), 189–191
satellite imagery, 22
satellites, Mission to Planet Earth of NASA and, 19–21
Save the Children, 133
Scott, Winfield, 200
Seattle, Washington, 104
Seltzer, Ethan, 105
Seminole Tribe of Florida v. Florida, 162, 164
sense of community, 17
sense of place, 17, 18
sex ratios, "One Child" policy of China and, 137
"shock cities," 103
Sierra Nevada Ecosystem Project (SNEP), 92–93
skyscrapers, digital revolution and need for, 158–160
Soaring Eagle Casino, 168
sorghum, 190
South Americans, in Miami, 213
southern United States, balanced mix of land types, 76
Spanish-language radio stations (SLRS), 208
spatial, or geometric, tradition, in geography, 10, 11
sprawl, 101–107, government policy and, 46–47
St. Louis, Missouri, 47
Steele, John Yellow Bird, 172, 173
Striga hermonthica, 190

Stommel, Henry, 56, 57
street-neighborhoods, 106
suburbs, 102–103, 143–144, 146; automobile, 102–103; government policy and, 46–47; railroad, 102
Suriname, 51

T

Tejano Homeland, 201, 204, 207–208
Tejano music, 208
Tejanos, 201, 204, 207–208, 212
Ter-Petrossian, Levon, 132, 133–134
Texas, Hispanics in, 200, 201, 204, 207–208, 212
textile industry, in Greenville County, South Carolina, 109–110
Thailand, 179
theme parks, 16
Toronto, Canada, 103–104
Treaty of Guadalupe Hidalgo, 200

U

Uruguay, 50
U.S. Border Patrol, 202
U.S. National Resources Inventory, 75–76

V

Vancouver, Canada, 104
Venezuela, 50
vernacular regions, 93–94
Viravidaiya, Mechai, 179
Voting Rights Act, 199

W

Washington, Indian gaming casinos in, 163, 164, 166–167
water resources, conflict over, in Middle East, 120–129, 177
watersheds, 94
Weinstein, Richard, 103
West Coast, of U.S. 14
western United States, rangeland of, 76
wetlands, 75, 78
Wilmington, Delaware, 155
witchweed, 190
women, population growth and, 178, 180
wooly mammoths, 151
World Bank, 67, 68, 69, 86

Yellowstone, 76
Younger Dryas, 54–55

Z

Zimbabwe, 191

AE Article Review Form

We encourage you to photocopy and use this page as a tool to assess how the articles in **Annual Editions** expand on the information in your textbook. By reflecting on the articles you will gain enhanced text information. You can also access this useful form on a product's book support Web site at **http://www.dushkin.com/online/**.

NAME: DATE:

TITLE AND NUMBER OF ARTICLE:

BRIEFLY STATE THE MAIN IDEA OF THIS ARTICLE:

LIST THREE IMPORTANT FACTS THAT THE AUTHOR USES TO SUPPORT THE MAIN IDEA:

WHAT INFORMATION OR IDEAS DISCUSSED IN THIS ARTICLE ARE ALSO DISCUSSED IN YOUR TEXTBOOK OR OTHER READINGS THAT YOU HAVE DONE? LIST THE TEXTBOOK CHAPTERS AND PAGE NUMBERS:

LIST ANY EXAMPLES OF BIAS OR FAULTY REASONING THAT YOU FOUND IN THE ARTICLE:

LIST ANY NEW TERMS/CONCEPTS THAT WERE DISCUSSED IN THE ARTICLE, AND WRITE A SHORT DEFINITION:

ANNUAL EDITIONS revisions depend on two major opinion sources: one is our Advisory Board, listed in the front of this volume, which works with us in scanning the thousands of articles published in the public press each year; the other is you—the person actually using the book. Please help us and the users of the next edition by completing the prepaid article rating form on this page and returning it to us. Thank you for your help!

ANNUAL EDITIONS: Geography 99/00

ARTICLE RATING FORM

Here is an opportunity for you to have direct input into the next revision of this volume. We would like you to rate each of the 39 articles listed below, using the following scale:

1. **Excellent: should definitely be retained**
2. **Above average: should probably be retained**
3. **Below average: should probably be deleted**
4. **Poor: should definitely be deleted**

Your ratings will play a vital part in the next revision. So please mail this prepaid form to us just as soon as you complete it. Thanks for your help!

We Want Your Advice

RATING

ARTICLE

1. The Four Traditions of Geography
2. The American Geographies
3. NASA Readies a 'Mission to Planet Earth'
4. Human Domination of Earth's Ecosystems
5. Counting the Cost: The Growing Role of Economics in Environmental Decisionmaking
6. The Coming Climate
7. The Emergence of Gasohol: A Renewable Fuel for American Roadways
8. America's Rush to Suburbia
9. The Season of El Niño
10. The Great Climate Flip-Flop
11. Temperature Rising
12. "Dammed If You Do . . . "
13. Past and Present Land Use and Land Cover in the USA
14. China Shoulders the Cost of Environmental Change
15. The Importance of Places, or, a Sense of Where You Are
16. The Rise of the Region State
17. Metropolis Unbound: The Sprawling American City and the Search for Alternatives
18. Greenville: From Back Country to Forefront
19. Does It Matter Where You Are?

RATING

ARTICLE

20. Low Water in the American High Plains
21. Water Resource Conflicts in the Middle East
22. Boomtown Baku
23. Demographic Clouds on China's Horizon
24. Transportation and Urban Growth: The Shaping of the American Metropolis
25. Bridge to the Past
26. County Buying Power, 1987–97
27. Puerto Rico, U.S.A
28. Do We Still Need Skyscrapers?
29. Indian Gaming in the U.S.: Distribution, Significance and Trends
30. For Poorest Indians, Casinos Aren't Enough
31. Before the Next Doubling
32. The End of Cheap Oil
33. Reseeding the Green Revolution
34. How Much Food Will We Need in the 21st Century?
35. The Changing Geography of U.S. Hispanics, 1850–1990
36. 'Hispanics' Don't Exist
37. Russia's Population Sink
38. Vanishing Languages
39. Risky Business: Who Will Pay for the Growing Costs of Global Change?

(Continued on next page)

ANNUAL EDITIONS: GEOGRAPHY 99/00

NO POSTAGE
NECESSARY
IF MAILED
IN THE
UNITED STATES

BUSINESS REPLY MAIL
FIRST-CLASS MAIL PERMIT NO. 84 GUILFORD CT

POSTAGE WILL BE PAID BY ADDRESSEE

Dushkin/McGraw-Hill
Sluice Dock
Guilford, CT 06437-9989

ABOUT YOU

Name _____ Date _____

Are you a teacher? ☐ A student? ☐
Your school's name _____

Department _____

Address _____ City _____ State _____ Zip _____

School telephone # _____

YOUR COMMENTS ARE IMPORTANT TO US!

Please fill in the following information:
For which course did you use this book?
_____

Did you use a text with this *ANNUAL EDITION*? ☐ yes ☐ no
What was the title of the text?
_____

What are your general reactions to the *Annual Editions* concept?
_____

Have you read any particular articles recently that you think should be included in the next edition?
_____

Are there any articles you feel should be replaced in the next edition? Why?
_____

Are there any World Wide Web sites you feel should be included in the next edition? Please annotate.
_____

May we contact you for editorial input? ☐ yes ☐ no
May we quote your comments? ☐ yes ☐ no